T0401223

NEW DEVELOPMENTS IN MEDICAL RESEARCH

ADVANCES IN MEDICINAL CHEMISTRY RESEARCH

NEW DEVELOPMENTS IN MEDICAL RESEARCH

Additional books and e-books in this series can be found on Nova's website under the Series tab.

NEW DEVELOPMENTS IN MEDICAL RESEARCH

ADVANCES IN MEDICINAL CHEMISTRY RESEARCH

EDEILDO FERREIRA DA SILVA-JÚNIOR
EDITOR

NOTICE TO THE READER

The Publisher has taken reasonable care in the preparation of this book, but makes no expressed or implied warranty of any kind and assumes no responsibility for any errors or omissions. No liability is assumed for incidental or consequential damages in connection with or arising out of information contained in this book. The Publisher shall not be liable for any special, consequential, or exemplary damages resulting, in whole or in part, from the readers' use of, or reliance upon, this material. Any parts of this book based on government reports are so indicated and copyright is claimed for those parts to the extent applicable to compilations of such works.

Independent verification should be sought for any data, advice or recommendations contained in this book. In addition, no responsibility is assumed by the Publisher for any injury and/or damage to persons or property arising from any methods, products, instructions, ideas or otherwise contained in this publication.

This publication is designed to provide accurate and authoritative information with regard to the subject matter covered herein. It is sold with the clear understanding that the Publisher is not engaged in rendering legal or any other professional services. If legal or any other expert assistance is required, the services of a competent person should be sought. FROM A DECLARATION OF PARTICIPANTS JOINTLY ADOPTED BY A COMMITTEE OF THE AMERICAN BAR ASSOCIATION AND A COMMITTEE OF PUBLISHERS.

Additional color graphics may be available in the e-book version of this book.

Library of Congress Cataloging-in-Publication Data

ISBN: 978-1-53616-368-1
Library of Congress Control Number:2019950427

Published by Nova Science Publishers, Inc. † New York

Contents

PREFACE

According to the International Union for Pure and Applied Chemistry (IUPAC), medicinal chemistry is defined as a chemistry-based discipline that involves biological, medicinal and pharmaceutical aspects. Also, it concerns to the invention, discovery, design, identification, and synthesis of potentially active agents, where their metabolism, binding mode, and structure-activity relationship (SAR) should be considered, as well. In this context, medicinal chemistry represents a field that overlaps the disciplines of biochemistry, chemistry, physiology, and pharmacology, which plays a crucial role in the construction of a specific knowledge that could be employed on the design and development of new agents.

It was a great pleasure to organize this book entitled *Advances in Medicinal Chemistry Research* that aimed at undergraduates and postgraduates who have a background in chemistry with a focus on the pharmaceutical and medicinal chemistry. Additionally, it is of particular interest to students who might be considering a future career as a researcher at the University or as an employer in the pharmaceutical industry.

This book comprises a variety of scientific material spanning over seven chapters, which were written by active researchers, yielding an excellent scientific production in the field of medicinal chemistry of the most deadly and limiting diseases at the moment, such as obesity, multidrug-resistant infections, neglected tropical diseases, Alzheimer, and cancer. Also, it brings together an overview of current discoveries and trends in these fields while urging for dedicated and extended collaborative research. The facts presented and their discussions in every chapter are exhaustive, precise, and deeply informative and didactic; hence, the book would serve as a critical reference for new developments in the frontier research on medicinal chemistry of obesity, multidrug-resistant bacteria, neglected tropical diseases, Alzheimer, and cancer, and would also be of much use for scientists working in these promising fields.

Advances in Medicinal Chemistry Research is a book organized into seven chapters:

Initially, the book offers an excellent overview of the obesity topic (Chapter 1), showing that the obesity is a significant health problem in developed and developing countries worldwide, which accounts to more than 1.9 billion people overweight; and among these, 650 million obese. In sense, the human pancreatic lipase represents a valuable molecular target, which has a domain with a catalytic triad, composed by Ser^{153}, Asp^{177}, and His^{263}. In Chapter 2, the medicinal chemistry of the bacterial resistance is explored, focusing on five main families of efflux pumps (MFS, SMR, RND, ABC, and MATE) and their potential inhibitors from natural and synthetic sources. Considering the global impact of neglected tropical diseases, Chapter 3 provides a full and deeply overview of the thiophene-containing compounds as potential antituberculosis, antileishmanial, and anti-hepatitis agents. The SAR studies are discussed, and virtual approaches, such as molecular docking and virtual screening are explored and discussed, as well. Chapter 4 brings one of the most limiting diseases today – Alzheimer's disease – which approximately affects 10 million people every year. Then, the discovery of some inhibitors upon the human glutaminyl cyclase as promising anti-Alzheimer's agents are evaluated and discussed, considering SAR aspects. Currently, cancer has been recognized as a significant and global problem, which affected 18.1 million people in 2018, leading to the number of 9.6 million of deaths. Thus, Chapters 5–7 are devoted to discuss all aspects involving the computational techniques, such as ligand-based drug design (LBDD), structure-based drug design

(SBDD), docking, dynamic molecular (DM), quantitative structure-activity relationship (QSAR), and quantum mechanics/molecular mechanics (QM/MM); and their application as tools to design new anticancer agents. Also, indoleamine 2,3-dioxygenase 1 inhibitors are explored as a promising immunotherapy for treating cancer, *via* inhibition of Phe^{226} and Arg^{231} amino acid residues. Finally, an excellent update on Eg5 kinesin inhibitors is provided, regarding kinesin spindle protein. It shows that novel heterocyclic compounds able to target this enzyme could affect mitosis without disrupting microtubule dynamics.

I believe that this exciting book will serve not only as a valuable resource for researchers in their fields to predict promising lead compounds for the development of pharmaceutical products to treat these wide-reaching diseases, but also to encourage and motivate upcoming young scientists in the dynamic field of the medicinal chemistry research.

I want to express my sincere thanks once again to all the authors for the excellent reviews they have produced. Still, their participation and collaboration made my efforts to organize this book possible. I am sure that this book entirely contributes to the advances in the medicinal chemistry field.

Finally, I would like to express my gratitude sense to the editorial and publishing staff members associated with Nova Science Publishers for the invitation and their support all the time.

Edeildo F. da Silva-Júnior
Pharmaceutical Sciences Institute, Federal University of Alagoas
Maceió, Brazil

ACRONYMS

ABC	ATP-binding cassette family
ACR	Acriflavine
AD	Alzheimer's disease
ADME(T)	Absorption, distribution, metabolism, excretion (toxicity)
AhR	Aryl hydrocarbon receptor
AhR-ARNT	Aryl hydrocarbon receptor-nuclear translocator
ALK-5	Activin-like kinase 5
AMK	Amikacin
AMP	Ampicillin
APOE	Apolipoprotein E
APP	Amyloid precursor protein
ATP	Adenosine 5'-(tetrahydrogen triphosphate)
A*β*	Amyloid-*β*
BCL	Benzalkonium chloride
BMI	Body mass index
CASTp	Surface protein surface atlas
CCCP	Carbonyl cyanide *m*-chlorophenyl hydrazone
CFT	Ceftriaxone
CIP	Ciprofloxacin
CLA	Clarithromycin

CLM	Chloramphenicol
CoMFA	Comparative molecular field analysis
CoMSIA	Comparative molecular similarity index analysis
CPM	Capreomycin
CSF	Cerebrospinal fluid
CTR	Cetrimide
CYP450	Cytochrome P450
DAMP	Damage-associated molecular patterns
DCs	Dendritic cells
DFT	Density Functional Theory
DNA	Deoxyribonucleic acid
EDG	Electron-donating
eIF2α	Eukaryotic initiation factor 2α kinase
EmrD-3	Multidrug resistance protein D
ENO	Enoxacin
ERY	Erythromycin
ETB	Ethambutol
ETH	Ethionamide
EWG	Electron-withdrawing
FAS II	Type-II fatty acid synthase
FDA	Food and Drug Administration
FW	Fisher's weights
GAT	Gatifloxacin
GCN2	General control nonderepressible 2
GEN	Gentamicin
GI%	Percentage of growth inhibition
GI_{50}	Antiproliferative activity for 50%
GnRH	Gonadotropin-releasing hormone
HAV	Hepatitis A virus
HBV	Hepatitis B virus
HCS	High-content screening
HCV	Hepatitis C virus
HDV	Hepatitis D virus

HEK-293T	Human Embryonic Kidney
HEV	Hepatitis E virus
HHPQs	Hexahydropyranoquinoline
HTS	High-throughput screening
IC_{50}	Inhibitory concentration for 50%
IDO	Indoleamine 2,3-dioxygenase
IFN-γ	Interferon gamma
IL-12	Interleukin 12
INH	Isoniazid
iNOS	Inducible nitric oxide synthase
KAN	Kanamycin
K_i^{app}	Basal ATPase inhibitory activity
KM	Kanamycin
KP	Kynurenine pathway
KSP, Eg5	Kinesin spindle protein
KYN	Kynurenine
LBDD	Ligand-based drug design
LBVS	Ligand-based virtual screening
LE	Ligand efficiency
LEV	Levofloxacin
LIN	Linezolid
LNC	Lincomycin
LOM	Lomefloxacin
LPS	Lipopolysaccharides
MATE	Multidrug and toxic compound extrusion family
MD	Molecular dynamics
MDR-TB	Multidrug-resistant tuberculosis
MF	*(+)*-Morelloflavone
MFS	Major facilitator superfamily
MI_{20}	Mitotic accumulation against 20% of the cells tested
MIC	Minimal inhibitory concentration
MIL	Millepachine
MOX	Moxifloxacin

MRSA	Methicillin-Resistant *Staphylococcus aureus*
mTORC1	Target of rapamycin complex 1
NAD^+	Nicotinamide adenine dinucleotide
NB	Naïve Bayesian
NI	Nucleoside inhibitor
NK	Natural killer cells
NKT	Natural killer T cells
NMDA	*N*-methyl-*D*-aspartic acid
NMP	1-(1-Naphthylmethyl)-piperazine
NMR	Nuclear Magnetic Resonance
NNI	Non-nucleoside inhibitor
NOR	Norfloxacin
NR-2B	Glutamate receptor subunit epsilon-2
NS	Non-structural proteins
NTD	Neglected tropical diseases
OFL	Ofloxacin
OPS	Ordered predictor selection
PAβN	Phenylalanine-Arginine β-Naphthylamide
PCA	Principal component analysis
PER	Perfloxacin
pGlu	Pyroglutamic acid
*p*Glu-Aβ	*N*-Terminal glutamate of β-amyloid peptides
PKCθ	Protein kinase C theta
PL	Pancreatic lipase
PLS	Partial least square
PZA	Pyrazinamide
QC	Glutaminyl cyclase
QM/MM	Quantum mechanics and molecular mechanics
QSAR	Quantitative structure-activity relationship
QSPR	Quantitative structure-property relationship
QSTR	Quantitative toxicology-property relationship
RIF	Rifampicin
RNA	Ribonucleic acid

RND	Resistance-nodulation-cell-division family
ROC	Receiver operating characteristics
SAR	Structure-activity relationship
SBDD	Structure-based drug design
SBVS	Structure-based virtual screening
SI	Selectivity index
SMR	Small multidrug resistance family
SMT	Sulfamethoxazole-trimethoprim
SPR	Sparfloxacin
SSRI	Selective serotonin reuptake inhibitors
STLC	*S*-trityl-*L*-cysteine
STR	Streptomycin
TA	Tumor antigens
TB	Tuberculosis
TDO	Tryptophan 2,3- dioxygenase
TerE	Terpendole-E
TET	Tetracycline
TetK	Tetracycline efflux pump
TLM	Thiolactomycin
TRH	Thyrotropin-releasing hormone
tRNA	Transfer ribonucleic acid
UV	Ultraviolet
VAN	Vancomycin
VS	Virtual screening
WHO	World Health Organization
XDR-TB	Extensively drug-resistant tuberculosis

In: Advances in Medicinal Chemistry … ISBN: 978-1-53616-368-1
Editor: E. Ferreira da Silva-Júnior

Chapter 1

ANTI-OBESITY DRUGS AGAINST FAT ABSORPTION: CURRENT STATUS AND FUTURE PROSPECTS OF LIPASE INHIBITORS

Giovanni Ortiz Leoncini[1,2,*], PhD,
Nathan Araújo de Melo[1]
and João Xavier de Araújo-Júnior[1,2], PhD
[1]Pharmaceutical Sciences Institute,
Federal University of Alagoas, Maceió, Alagoas, Brazil
[2]Institute of Chemistry and Biotechnology,
Federal University of Alagoas, Maceió, Alagoas, Brazil

ABSTRACT

Obesity is a significant health problem in developed and developing countries, where it is associated with many health problems, including type 2 diabetes mellitus, hypertension, coronary heart disease, stroke, gallbladder disease, osteoarthritis, sleep apnea, and various cancers. The development of new anti-obesity drugs has become a priority of

[*] Corresponding Author's E-mail: giovanni.leoncini@iqb.ufal.br.

pharmaceutical research, not least because of the tremendous demand for safe and effective drugs needed for treatment of its complications. Human pancreatic lipase (EC 3.1.1.3) is a digestive enzyme that is targeted in the treatment of obesity. Typically, lipases hydrolyze triacylglycerides, a fundamental step for fat absorption by intestinal enterocytes that is essential for the efficient digestion of fat and transport of fat-soluble vitamins. Its typical interactions with emulsions or aggregated substrates that have water-lipid interface allow the opening of the domain responsible for the access to the catalytic site, composed of Ser^{153}, Asp^{177}, and His^{263} amino acid residues. Lipstatin from *Streptomyces toxytricini* is the active ingredient that contains a β-lactone capable of irreversibly inhibiting fat absorption. This chapter focuses on the causes of obesity, anti-obesity drugs, the mechanism of pancreatic lipase inhibition, and new perspectives on synthetic and natural lipase inhibitors.

Keywords: obesity, anti-obesity drugs, pancreatic lipase, lipase inhibition

INTRODUCTION

Obesity is a public health problem that affects developed and developing countries alike. It is characterized by abnormal increases in the size of adipocytes; obesity is also associated with various pathological disorders [1]. It is a complex condition that has many contributing factors, including genetic, environmental, social, psychological, and others, some of which are beyond the control of the individual [2]. Recently, the World Health Organization (WHO) reported that global obesity rate has nearly tripled since 1975. In 2016, more than 1.9 billion adults were overweight; among these, 650 million were obese. In the same year, 41 million children under five years old were overweight or obese, and more than 340 million children and adolescents between 5 and 19 years of age were overweight or obese [3].

Currently, obesity is defined using body mass index (BMI). However, this should not be considered as the main factor [2]. BMI is calculated using height and weight, and patients are classified as underweight, normal weight, overweight or obese; the last is defined as BMI greater than or equal to 30 kg/m^2 (Table 1) [4]. Rare healthy exceptions include trained athletes who may achieve BMIs above 30 [5]. According to the American College of Physicians (United States of America), pharmacological therapy is an alternative for obese patients who have failed to achieve adequate weight loss only by non-pharmacological means, including diet and exercise [1].

Table 1. Classifications for BMI (kg/m^2)

Classification	**BMI (kg/m^2)**
Underweight	< 18.5
Normal weight	18.5 – 24.9
Overweight	25 – 29.9
Obesity (Class 1)	30 – 34.9
Obesity (Class 2)	35 – 39.9
Extreme Obesity (Class 3)	≥ 40

Studies have consistently found that physical activity reduces or eliminates the risks associated with obesity. Sedentary lifestyle accompanying obesity promotes the development of chronic diseases, morbidities, and mortality [2, 6]. Obese people are predisposed to various diseases, primarily mediated by fat accumulation in visceral adipose tissue [7].

Diseases and Obesity

Obesity is associated with type 2 diabetes' risk, hypertension, coronary and cerebrovascular diseases, osteoarthritis, sleep apnea, and certain types of cancer [7, 8]. Obesity also tends to be associated with depression, infertility, problems related to pregnancy, and anomalies related to reproductive hormones [8].

Over one decade, a study involving obese patients of both genders revealed an increased risk of diabetes, gallstones, and hypertension in women, while in men the results showed an increased risk of heart disease and stroke, as well as diabetes, gallstones, and hypertension [9]. Currently, most of the medical institutes worldwide recognize weight loss and physical activities as the best methods of preventing and treating against obesity [1, 9].

Modest weight loss, approximately 5% of initial body weight, can lead to improvements in blood pressure, serum lipid concentrations, increased insulin sensitivity, and improved glucose levels. Weight loss is also associated with a significant reduction in the risk of diabetes. In a randomized study involving patients with type 2 diabetes, patients lost an average of 8.6% of body weight in 12 months, resulting in clinically significant weight loss associated with better control of diabetes and reduction of risk factors for cardiovascular diseases [9, 10].

WAYS TO ABSORB FAT

Currently, the urbanization in most of the world has led to changes in societal dietary habits [11, 12]. The increase in fat and sugar consumption, along with a decreased consumption of cereal and fiber, has contributed to growing obesity rates, in some cases associated with sedentary lifestyles [13]. In recent decades, more high-fat food products have become more available with low prices and attractive flavors. Fat is the densest macronutrient abundant in terms of energy, compared to carbohydrates and proteins [14]. Environmental influences including sedentary lifestyles and consumption of high-calorie diets cause an imbalance between energy expenditure and consumption [13].

The genetic predisposition of obesity involves both energy intake and energy expenditure reduction, both of which can occur via reduced rates of basal metabolism, macronutrient oxidation, altered adipogenesis, food intake deviations, hormonal profile, and thermogenesis [15]. The determination of the contributions of external influences and the genetic profile represents challenges for proposing real-world parameters regarding weight gain. Epidemiological aspects are not explained by genetic mutations [15, 16]. Despite the fact that genetic factors play an essential role, psychological, social, and environmental factors are more common parameters responsible for the current obesity epidemic [16].

In general, hyperlipidic diets play a crucial role in the prevalence of obesity, disrupting cumulative fat energy balance; this has implications for public health in the form of reduced life expectancy [12, 14]. Therefore, nutritional approaches involving control of lipid, protein, and carbohydrate intake are strategies frequently used to manage obesity [13].

RISKS OF ANTI-OBESITY TREATMENT

Bariatric surgery has been useful in the treatment of morbid obesity. Chakhtoura and coworkers [17] investigated mean weight loss in 22,904 patients who underwent bariatric surgery; the authors reported a 61% improvement in diabetes, hyperlipidemia, hypertension, and obstructive sleep apnea, and a decrease in the risk of premature death. Nevertheless, bariatric surgery causes several complications. Bikram and collaborators [18] reported vitamin deficiencies in a study of 318 patients followed for 12 months after surgery: vitamin A deficiency (11%), vitamin C (34.6%), vitamin D (7%), thiamine (18.3%), riboflavin (13.6%), vitamin B6 (17.6%), and vitamin B12 (3.6%).

Some pharmacological therapies are recommended before and after the surgery, depending on the side-effects of the medications and the tolerance profile of the patient. These therapies can be developed in various ways, such as inhibition of lipase, suppression of food intake, stimulation of energy expenditure, inhibition of adipocyte differentiation, and regulation of lipid metabolism [1, 19].

Anti-obesity drugs are thought of as additions to lifestyle interventions, facilitating weight loss, and promoting long-term weight maintenance [13, 20]. For patients with severe obesity (class III), several medications, as well as surgery can be considered, increasing risks generated from the adverse effects of medications, added to post-surgical weakness [20]. Medications are typically characterized by side-effects in the gastrointestinal system, kidneys, central nervous and cardiovascular systems. As a result, several anti-obesity drugs have been withdrawn from the market (*e.g.*, sibutramine, rimonabant, contrive, and phentermine) or suspended because of their dubious safety profile. Only Orlistat, which inhibits pancreatic lipase has remained for long-term use despite moderate efficacy [21].

Characteristics of Pancreatic Lipase as a Therapeutic Target

Pancreatic lipase (PL) is an enzyme of fundamental importance for the efficient digestion of triacylglycerides. After food ingestion, gastric lipase acts in the stomach and is responsible for the hydrolysis of 15 to 20% of dietary lipids [22]. The digestion of lipids in the intestine is concluded by mixing with pancreatic juice that contains several lipases, where PL has higher performance primarily with respect to hydrolysis of long-chain triacylglycerides and is responsible for the hydrolysis of 50 to 70% of the fat present in food [22, 23]. This efficient fat digestion is the result of the formation of micelles with bile salts, where the properties of the emulsion particles create a water-lipid interface that generates the most favorable environment for the action of PL [23, 24]. Studies with triacetin have demonstrated this behavior in lipases, in which there is an abrupt increase in enzymatic activity of PL after formation of micelles [24].

The structure and function of human PL is determined by expression of the *PNLIP* gene, which contains structural features similar to those of other members of the gene family, including lipoprotein lipase [22, 23, 25]. The folding pattern of these enzymes are similar, both of which have an *N*-terminal domain containing the pocket of the active site, and the *C*-terminal domain for binding of colipase, a physiological cofactor that controls access to the catalytic triad (Ser^{153}, Asp^{177}, and His^{263}) at the active site [26, 27]. Using 3D-dimensional structures, Winkler and coworkers [28] demonstrated interfacial activation of the *C*-terminal region in the presence of an amphiphilic loop covering the active site of the enzyme in solution, now called the “lid.” Brzozowski and coauthors [29] suggested that this loop undergoes a conformational rearrangement at the lipid-water interface that makes the active site accessible to the substrate.

Access to the active site by substrates or inhibitors is facilitated by steric arrays of the external amino acid residues using a force field generated by the molecular recognition extending from 10 to 15 Å. The availability of the active site in the *N*-terminal allows access to the catalytic triad (Ser^{153}, Asp^{177}, and His^{263}) responsible for the hydrolysis mechanism of triacylglycerides [26]. The formation of the tetrahedric hemiacetal intermediate with triacylglycerides allows the hydrolysis of the ester bond and the release of diacylglyceride; the presence of water in the site reacts with the acylated serine, expelling the fatty acid from the active site cleft [30].

The search for lipase inhibitors focuses on structural features involving hydrophobic moieties that are essential groups for reaching the *C*-terminal region, promoting the opening of the "lid" [31]. The ester groups in the molecules have been exploited in a widely used strategy. The rational drug design of these inhibitors is related preferentially to interfacial activation loop and nucleophilic attack promoted by the serine residue from the active site, maintaining the similarity of physiological substrates that are commonly amphipathic molecules [32, 33].

SYNTHETIC LIPASE INHIBITORS

Substitution by various functional groups bound to heterocycles leads to an extensive series of bioactive products. Indole rings present a wide range of biological activities easily obtained from several natural products. Buduma and colleagues [34] performed an investigation of bisindoles anti-obesity activity by inhibition of lipase from porcine pancreatic type II (Table 2). Compounds **(7)** and **(9)** were found to be the most active; they contain dichlorobenzyl and nitrobenzyl substituents, respectively. These compounds were found to be more potent than Orlistat in this experimental context.

Table 2. Indole derivatives and inhibition of lipase activity

(1) - (11)

Compound	R	IC_{50} (μM)
Orlistat	-	62.25 ± 0.12
(1)	Methyl	N.A
(2)	Benzyl	N.A
(3)	2-Fluorobenzyl	56.41 ± 0.21
(4)	4-Fluorobenzyl	N.A
(5)	2-Chlorobenzyl	N.A
(6)	3-Chlorobenzyl	N.A
(7)	2, 4-Dichlorobenzyl	14.89 ± 0.27
(8)	4-Bromobenzyl	N.A
(9)	4-Nitrobenzyl	30.88 ± 0.19
(10)	2-Cyanobenzyl	N.A
(11)	1,1'-Phenyl-2-cianobenzyl	N.A

N.A: not active.

Other five-membered *N*-heterocyclic compounds, including carbazoles, represent an essential class of indole alkaloids that have broad pharmacological activity. Sridar and coworkers [35] identified carbazole-fused oxoacetoamides with potential pancreatic lipase inhibitory activity (porcine pancreas type II). The study aimed to use acetamides to mimic ester groups by having reactive carbonyl groups acting as electrophiles toward Ser^{153} at the active site. It was also determined that the carbazole's hydrophobic interactions are essential for opening the lid. In this study, the introduction of the methoxyphenyl **(16)**, 3,4-dimethoxyphenyl **(17)**, and 3,4,5-trimethoxyphenyl **(27)** groups at amine position of carbazol ring gave promising IC_{50} values of 6.31, 8.72, and 9.58 μM, respectively (Table 3). Recently, Sridhar and coauthors [36] reported antilipase activity when an indole nucleus replaced the carbazole moiety. The compounds **(49)** and **(50)** presented IC_{50} values of 5.83 and 4.92 μM, respectively (Table 3).

Table 3. Oxoacetamides derivatives and inhibition of lipase activity

Série 1 (12) - (35)   Série 2 (36) - (53)

Compound	R	Ar	IC$_{50}$ (μM)
Orlistat	-	-	0.99 ± 0.11
(12)	4-Chlorobenzyl	Phenyl	13.91 ± 2.10
(13)		4-Methylphenyl	13.53 ± 1.67
(14)		3-Methoxyphenyl	26.61 ± 0.97
(15)		4-Methoxyphenyl	14.83 ± 2.17
(16)		3,4-Dimethoxyphenyl	6.31 ± 0.56
(17)		3,4,5-Trimethoxyphenyl	8.72 ± 0.47
(18)		4-Fluorophenyl	27.98 ± 2.25
(19)		4-Pyridyl	19.87 ± 1.08
(20)		*N, N'*-dimethylaminophenyl	14.78 ± 1.50
(21)		6-Quinolyl	23.05 ± 2.49
(22)		2-(5-Methyl)thiazolyl	32.76 ± 6.05
(23)	H	Phenyl	29.73 ± 5.62
(24)		3-Methoxyphenyl	33.02 ± 2.30
(25)		4-Methoxyphenyl	19.31 ± 0.80
(26)		3,4-Dimethoxyphenyl	44.12 ± 5.18
(27)		3,4,5-Trimethoxyphenyl	9.58 ± 1.24
(28)		4-Pyridyl	29.35 ± 2.17
(29)	Methyl	Phenyl	36.18 ± 1.88
(30)		3,4,5-Trimethoxyphenyl	23.12 ± 2.06
(31)	Ethyl	Phenyl	24.22 ± 0.30
(32)		3,4-Dimethoxyphenyl	16.37 ± 2.09
(33)		3,4,5-Trimethoxyphenyl	11.37 ± 1.12
(34)		*N, N'*-dimethylaminophenyl	56.18 ± 4.34
(35)		6-Quinolyl	15.48 ± 1.87
(36)	H	Phenyl	27.49 ± 1.68
(37)		4-Methylphenyl	26.13 ± 0.92
(38)		3,4-Dimethylphenyl	26.07 ± 1.14
(39)		4-Methoxyphenyl	23.72 ± 0.67
(40)		4-Methoxybenzyl	17.39 ± 0.53
(41)		3,4,5-Trimethoxyphenyl	17.28 ± 0.49
(42)		4-Bromophenyl	43.36 ± 1.67
(43)		3,4-Dichlorophenyl	44.27 ± 1.21

Table 3. (Continued)

Série 1 (12) - (35)   Série 2 (36) - (53)

Compound	R	Ar	IC_{50} (μM)
(44)		4-Fluorophenyl	47.62 ± 1.48
(45)	Benzyl	Phenyl	18.24 ± 0.74
(46)		4-Methylphenyl	12.9 ± 0.58
(47)		3,4-Dimethylphenyl	10.62 ± 0.66
(48)		4-Methoxyphenyl	10.86 ± 0.71
(49)		4-Methoxybenzyl	5.83 ± 0.64
(50)		3,4,5-Trimethoxyphenyl	4.92 ± 0.29
(51)		4-Bromophenyl	25.76 ± 0.97
(52)		3,4-Dichlorophenyl	26.72 ± 0.79
(53)		4-Fluorophenyl	24.18 ± 0.86

Using virtual screening, Vu and collaborators [33] analyzed 102 chalcone derivatives, and found molecules with the highest potential antilipase against porcine pancreas lipase. The results suggested that the benzylamino chalcone derivatives had moderate inhibitory activity (Table 4). Nevertheless, access to the catalytic triad at the active site allowed optimization of inhibitory activity through other substituents in strategic positions.

Sankar and colleagues [37] reported a series of potent inhibitors of phenacyl esters of pancreatic lipase. The compounds **(63)**, **(64)**, **(65)**, **(66)**, **(67)**, and **(68)** were found to be suitable pancreatic lipase inhibitors, with IC_{50} values ranging from 1 to 3 μM (Table 5). A bromated-compound **(62)** exhibited the maximum inhibition from this series, with an IC_{50} value of 0.9 μM. However, the %-inhibition was lower than that of Orlistat. This study suggested that the phenacyl ester structures could be used as promising compounds for conducting *in vivo* and preclinical studies.

Table 4. Benzyl amino chalcone derivatives and inhibition of lipase activity

(54) - (61)

Compound	R	IC_{50} (µM)
(54)	*m, p*-Cl_2	113.34
(55)	*m*-Br	57.89
(56)	*p*-OC_2H_5	100.64
(57)	*o*-Cl	87.70
(58)	*o, o'*-Cl, F	50.51
(59)	*m*-NO_2	76.18
(60)	*p*-NO_2	109.98
(61)	*p*-OCH_3	> 120

Table 5. Phenacyl esters derivatives and inhibition of lipase activity

(62) - (68)

Compound	n	R	IC_{50} (µM)
Orlistat	-	-	0.006
(62)	2	*p*-Br	0.9
(63)	3	*p*-OCH_3	1.46
(64)	3	*p*-NO_2	1.63
(65)	1	*p*-NO_2	1.72
(66)	1	*p*-Br	1.77
(67)	3	*m*-NO_2	1.79
(68)	3	H	2.61

Sridar and coworkers [38] synthesized thiazolidinone derivatives and evaluated pancreatic lipase activity. They found that, among the two series, compounds **(76)**, **(77)**, **(78)**, **(79)**, and **(80)** from the *p*-nitrobenzyl series showed significant inhibition with IC_{50} < 10 µM (Table 6). Of these, compound **(79)** showed an IC_{50} value of 4.81 μM, the most potent inhibitor. Finally, molecular docking studies revealed superposition with Orlistat; this compound proximity interacts to Ser^{152} at the active site.

Table 6. Thiazolidinones derivatives and inhibition of lipase activity

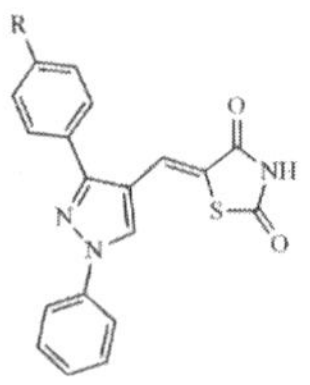

Série 1 (12 - 17) Série 2 (18 - 23)

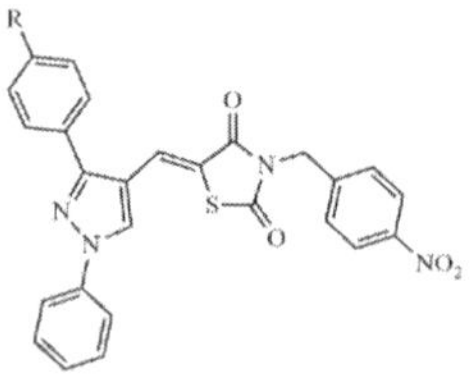

Compound	R	IC_{50} (μM)
Orlistat	-	0.99 ± 0.11
(69)	H	18.81 ± 0.21
(70)	CH_3	15.12 ± 0.83
(71)	OCH_3	13.05 ± 2.81
(72)	F	12.90 ± 0.89
(73)	Cl	10.30 ± 1.27
(74)	NO_2	17.32 ± 0.48
(75)	H	12.46 ± 2.92
(76)	CH_3	9.87 ± 1.03
(77)	OCH_3	9.40 ± 0.88
(78)	F	5.42 ± 0.43
(79)	Cl	4.81 ± 0.82
(80)	NO_2	8.44 ± 0.32

Previous results obtained by Mentese and coauthors [39] evaluated that the 4,5-dichlorobenzimidazole derivatives, an inhibitor of viral RNA synthesis, showed potent activity against porcine pancreatic lipase. Compounds **(81)**, **(82)**, **(83)**, **(84)**, **(85)**, **(86)**, and **(87)** showed IC_{50} values ranging from 1 to 6 μM, while compound **(88),** containing fluorine at the C6 position and piperazine at the C7 position, gave the most active result, with an IC_{50} value of 0.98 μM. *In silico* studies indicated affinity for the same site of the substrate (Table 7), although it was not more efficient than

Orlistat. Finally, ADME prediction gave pharmaceutical properties in the range of 95% of drugs for the leading compounds.

Table 7. Benzimidazole derivatives and inhibition of lipase activity

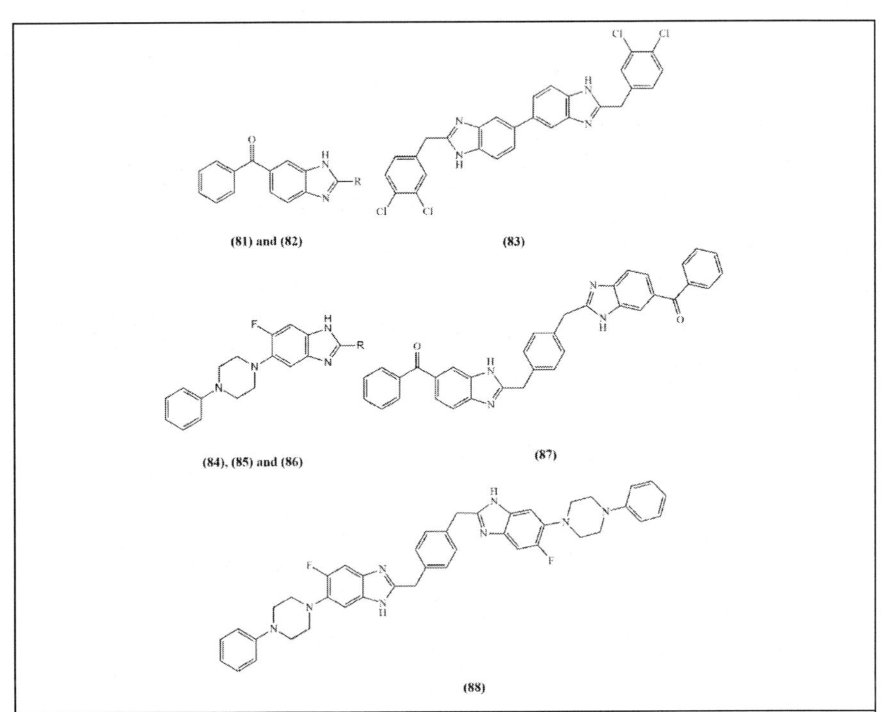

Compound	R	IC_{50} (μM)
Orlistat	-	0.0007 ± 0.00003
(81)	3, 4-Dichlorobenzyl	5.64 ± 1.47
(82)	4-Nitrobenzyl	5.38 ± 0.13
(83)	-	3.25 ± 0.06
(84)	Thiophen-2-ylmethyl	1.72 ± 0.12
(85)	4-Methoxybenzyl	3.31 ± 0.19
(86)	4-Nitrobenzyl	3.34 ± 0.01
(87)	-	1.92 ± 0.28
(88)	-	0.98 ± 0.09

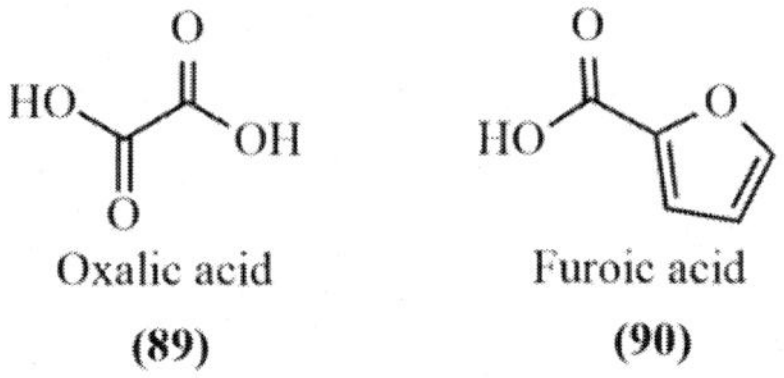

Oxalic acid
(89)

Furoic acid
(90)

Figure 1. Chemical structures of simple acids with anti-lipase potential.

Chlorogenic acid
(91)

Cryptochlorogenic acid
(92)

Neochlorogenic acid
(93)

Isochlorogenic acid A
(94)

Isochlorogenic acid B
(95)

Isochlorogenic acid C
(96)

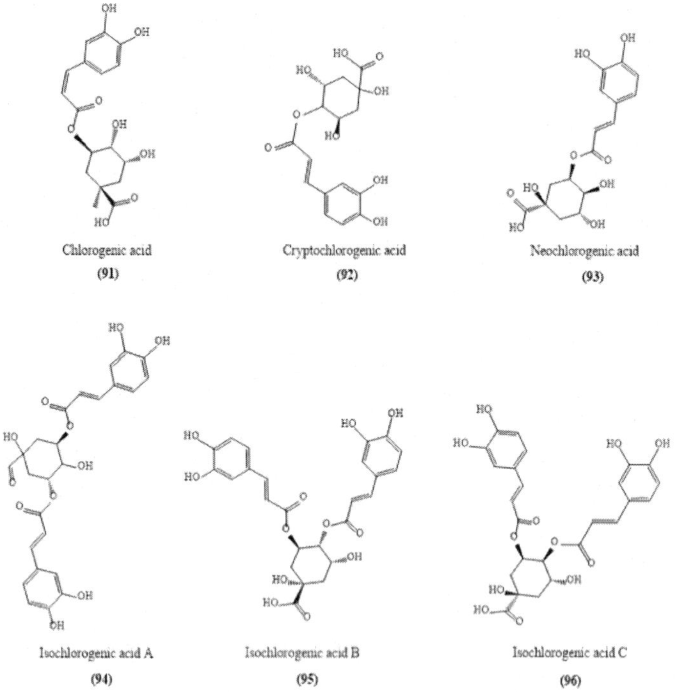

Figure 2. Chemical structures of caffeoylquinic acids with anti-lipase potential.

Studies with small molecules related to ascorbic acid oxidation for reducing lipid content were performed by Liu and coworkers [40], who found possible inhibitory action of oxalic and furoic acids. Oxalic acid **(89)** showed an IC_{50} value of 15.05 mM, while the furoic acid **(90)** had a value of 2.12 mM, both against *Mucor miehei* lipase (Figure 1). Although the study did not adequately evaluate treatment for obesity, and the results are valuable for the development of new synthetic or natural products that possess characteristics of both acids.

Hu and collaborators [41] reported inhibitory activity against porcine pancreatic lipase by caffeoylquinic acids, the main polyphenolic compounds present in coffee, which are the esters of caffeic and quinic acids. The bioactive characteristics of this class of compounds are attributed to the inhibition of lipase activity by the caffeoylquinic acid and its isomers. Compounds **(91)**, **(92)**, and **(93)** showed IC_{50} values of 1.10, 1.24, and 1.23 mM, respectively. Isomers **(94)**, **(95),** and **(96)** were more potent, exhibiting IC_{50} values of 0.591, 0.252, and 0.502 mM, respectively (Figure 2). *In silico* analysis showed the binding at the active site occurs near the Ser^{153} amino acid residue, characterizing a competitive inhibition mode.

LIPASE INHIBITORS FROM NATURAL SOURCES

Several classes of compounds are found in natural products from vegetal sources, including polyphenols, saponins, terpenes, and stilbenes; they are also derived from microbial sources such as valylactone, pancycline, ebactolactone, esterastine, vibralactone, percykinin, and lipstatin. All of these are promising chemical classes with activity against lipase [21].

Vegetal Sources

Polyphenols

Phenolic substances have aromatic nuclei containing hydroxylated substituents, including flavonoid compounds, which are known as catechins and are the best-known representatives from the phenolic group [42].

Camellia sinensis

Three teas are obtained from the leaves of *C. sinensis* (green, white, and black teas), where the differentiation between them occurs during the processing [43]. Before drying, the black tea goes through a stage of fermentation; for the green tea, after collection, the leaf is heated [44, 45]. White tea has its leaves naturally dried, where the leaves are slightly fermented and cooked in order to maintain the most of polyphenols [46]. Black tea has some benefits as an anticarcinogenic and antimutagenic agent and action against free radicals [47]. Furthermore, it can decrease cardiovascular diseases [48].

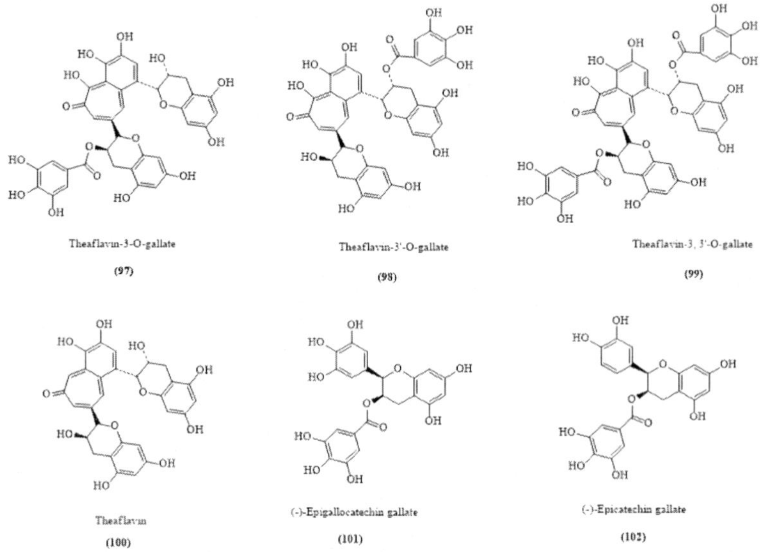

Theaflavin-3-O-gallate (97)
Theaflavin-3'-O-gallate (98)
Theaflavin-3, 3'-O-gallate (99)
Theaflavin (100)
(-)-Epigallocatechin gallate (101)
(-)-Epicatechin gallate (102)

Figure 3. Lipase inhibitors from *Camellia sinensis*.

Theaflavins and catechins were isolated from black tea residues and showed inhibitory activity *in vitro* against pancreatic lipase. The compounds were theaflavin-3-*O*-gallate **(97)**, theaflavin-3'-*O*-gallate **(98)**, theaflavin-3,3'-*O*-gallate **(99)**, theaflavin **(100)**, *(-)*-Epigallocatechin gallate **(101)**, *(-)*-Epicatechin gallate **(102)** (Figure 3). These compounds have IC_{50} values of 0.368, 0.320, 0.316, 0.679, 0.081, and 1.046 μg/mL, respectively [49].

Citrus unshiu

C. unshiu bark is a drug that has been used in East Asia to treat dyspepsia and vomiting since antiquity; it has also been used to treat nausea and tympanitis [50]. Two flavonoids extracted from the methanolic extract had inhibitory activity when tested against porcine pancreatic lipase and *Pseudomonas* spp., the hesperidin **(103)** and neohesperidine **(104)** with IC_{50} values of 32 and 46 µg/mL, respectively, against porcine pancreatic lipase [51] (Figure 4).

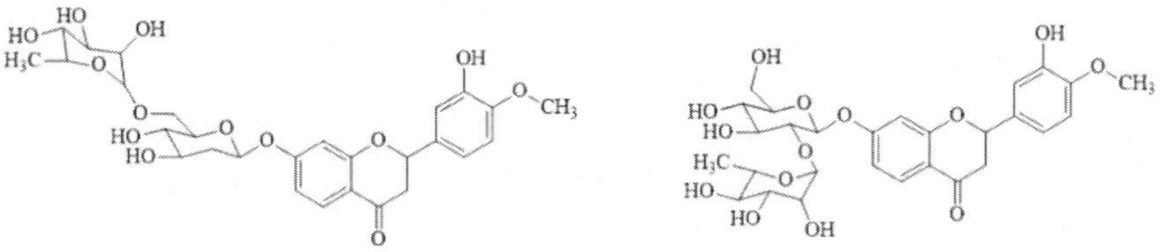

Hesperidin

(103)

Neohesperidin

(104)

Figure 4. Lipase inhibitors from *Citrus unshiu.*

Broussonetia kanzinoki

B. kanzinoki has been used in Chinese folk medicine as a tonic, diuretic, and an edema suppressant. Several prenylated flavonols, diphenyl propane, and flavans have been isolated from this plant. Antioxidant, anti-inflammatory, cytotoxic, and tyrosinase inhibitory effects have also been described [52, 53].

Phenolic compounds were isolated from B. kanzinoki stem bark. Among these, broussonone A **(105)**, broussonin A **(106)**, broussonin B **(107)**, and 7,4'-dihydroxyflavan **(108)** (Table 8) exhibit IC_{50} values ranging from 28.4 to 98.7 µM [1].

Table 8. Lipase inhibitors from *Broussonetia kanzinoki*

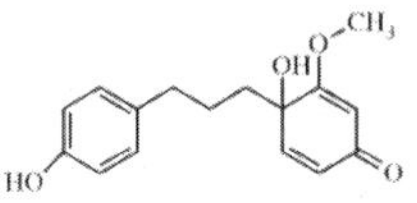

Broussonone A (105) | Broussonin A, Broussonin B (106) and (107) | 7,4'dihdroxiflavan (108)

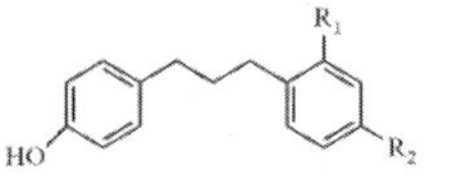

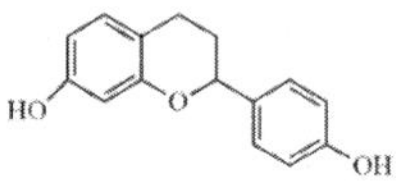

Compound	R1	R2	IC_{50} (µM)
(105)	-	-	28.4
(106)	OH	OCH_3	98.7
(107)	OCH_3	OH	75.4
(108)	-	-	85.1

Alpinia officinarum

A. Officinarum plant has been used since antiquity in European and Chinese medicine [54]. In China, it is used to treat colds, stomach pain, swelling, and as an agent to invigorate the circulatory system [55]. The dried root has also been described as having antioxidant, antiulcer, antidiabetic, antiemetic, antidiarrheal, anti-inflammatory, analgesic, and anticoagulant properties [55, 56].

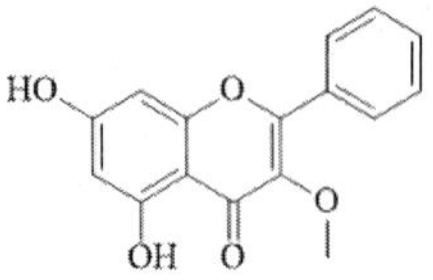

3-Methylethergalangin (109)

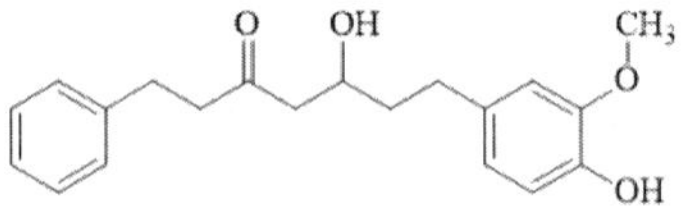

5-Hydroxy-7-(4'-hydroxy-3'-methoxyphenyl)-1-phenyl-3-heptanone (110)

Figure 5. Lipase inhibitors from *Alpinia officinarum*.

The aqueous extract from *A. officinarum* rhizome was tested against pancreatic lipase, and gave IC_{50} values of 3.7 mg/mL using tributyrin as substrate and 9.6 mg/mL using triolein as substrate. The bioactive compound 3-methylethergalangin **(109)** showed inhibitory activity against lipase, exhibiting IC_{50} values of 1.3 mg/mL using triolein as substrate and

3.3 mg/mL (using tributyrin as substrate). Another active compound isolated was 5-hydroxy-7-(4'-hydroxy-3'-methoxyphenyl)-1-phenyl-3-heptanone **(110),** which presented IC_{50} values of 1.5 mg/mL using triolein as substrate and 3.4 mg/mL using tributyrin as substrate [57] (Figure 5).

Filipendula kamtschatica

The plant is found in Japan. The Ainu people use *F. kamtschatica* infusion as an antidiarrheal agent, as well as a medication for eczema and urticaria [58, 59]. The hydroalcoholic extract of *F. kamtschatica* showed 95% inhibition of pancreatic lipase activity at 2.5 mg/mL. The flavonoid glycosides quercetin 3-*O*-*β*-xylopyranosyl-(1,2)-*O*-*β*-galactopyranoside **(111)**, 2-*O*-caffeoyl-4-*O*-galloyl-*L*-threonic acid **(112)**, 3-*O*-caffeoyl-4-*O*-galloyl-*L*-threonic acid **(113)**, gave IC_{50} values of 300, 246, and 26 μM, respectively [60] (Figure 6).

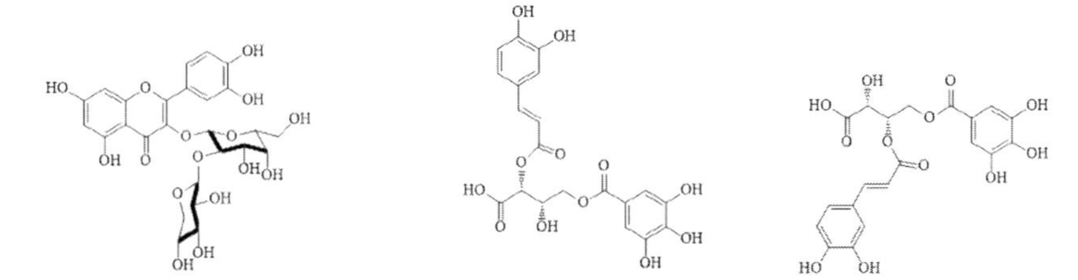

Quercetin 3-O-b-xylopyranosyl-(1'2)-O-b-galactopyranoside (111)

2-O-caffeoyl-4-O-galloyl-L-threonic acid (112)

3-O-caffeoyl-4-O-galloyl-L-threonic acid (113)

Figure 6. Lipase inhibitors from *Filipendula kamtschatica.*

Saponins

Saponins are found mainly in the rhizomes and roots of several plants. Their structures are composed of a sugar attached to a triterpene or steroid. Saponins mediate several biological activities, including lipase inhibition [61].

Acanthopanax sessiliflorus

The plants of the *Acanthopanax* species are found in China, Korea, and Japan. These plants were used historically to treat several disorders, including diabetes, rheumatoid arthritis, and tumors [62, 63]. A fraction abundant in saponins extracted from the leaves of *A. sessiliflorus* produced isolated lupano saponins-type containing activity. The author identified cessisolide **(114)** and chiisanoside **(115)**. *In vitro* evaluation against pancreatic lipase of these saponins gave IC_{50} values of 0.36 and 0.75 mg/mL, respectively [64, 65] (Figure 7).

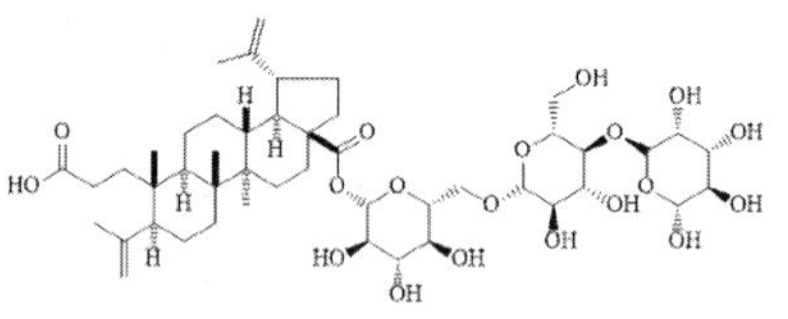

Cessiloside (114)

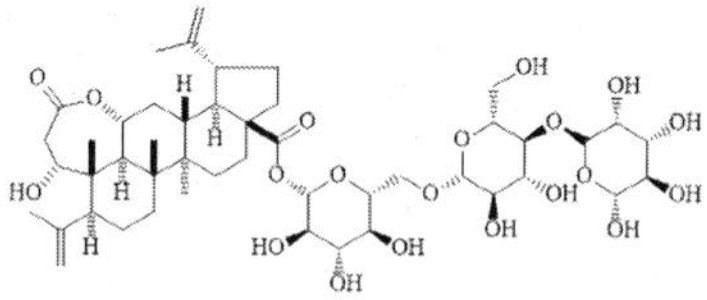

Chiisanoside (115)

Figure 7. Lipase inhibitors from *Acanthopanax sessiliflorus.*

Sapindus rarak

Sapindus rarak tree is found in Southern and Southeastern Asia. In Thailand, its pericarp is used as an antipruritic treatment [66].

Table 9. Lipase inhibitors from *Sapindus rarak*

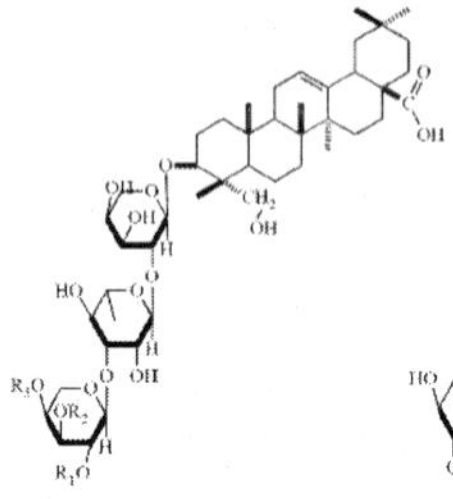

Rarasaponin I and II (116) and (117)

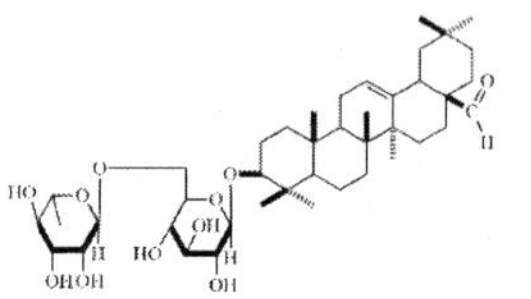

Raraoside A (118)

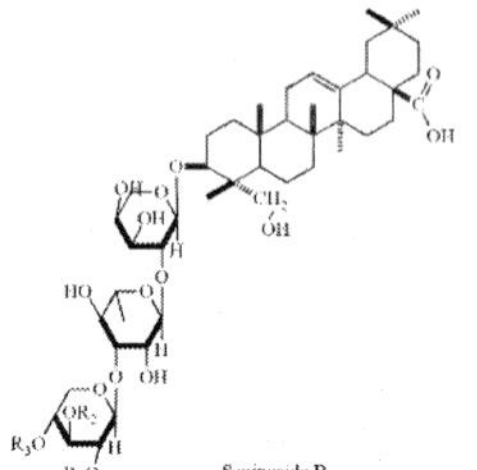

Sapinoside B
Hishoushi-saponin Ee
Mukurozi-saponin E1
(119) - (121)

Compound	R1	R2	R3	IC_{50} (μM)
(116)	Ac	H	H	131
(117)	H	Ac	H	172
(118)	-	-	-	151
(119)	H	H	H	125
(120)	H	Ac	H	117
(121)	H	H	Ac	100

The methanolic extract from the pericarp of *Sapindus rarak* was tested against pancreatic lipase. Its IC_{50} value was 614 µg/mL. The triterpene oligoglycosides of the oleanane-type were isolated from the active fraction, identifying the compounds rarasaponin I **(116)**, rarasaponin II **(117)**, raraoside A **(118)**. Furthermore, other saponins such as sapinoside B **(119)**, hishoushi saponin Ee **(120)**, mukurozi-saponin E1 **(121)** have been identified (Table 9) [67].

Acanthopanax senticosus

A. senticosus is a shrub found in Northeastern Asia, with various vegetal parts used in traditional Chinese medicine to treat various diseases, including hypertension, arthritis, tumors, ischemic heart disease, and neurasthenia. After fractionation of the ethanolic extract from *A. senticosus* fruits, the saponin fraction inhibited pancreatic lipase with an IC_{50} value of 3.63 mg/mL. Saponin triterpenes, including silphioside F **(122)**, copteroside B **(123)**, hederagenin 3-*O*-*β*-D-glucuronopyranoside-6'-*O*-methyl ester **(124)**, gypsogenin 3-*O*-*β*-D- glucuronopyranoside **(125)** gave activities with IC_{50} values ranging from 0.22 to 0.29 µM [65] (Table 10).

Table 10. Lipase inhibitors from *Acanthopanax senticosus*

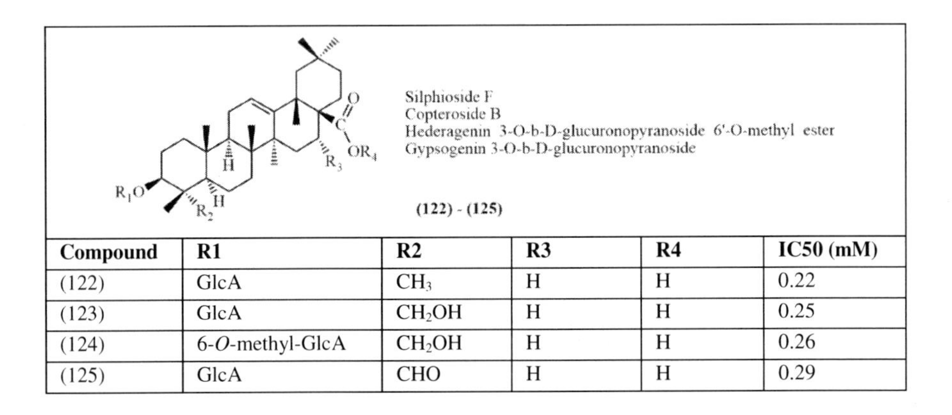

Compound	R1	R2	R3	R4	IC50 (mM)
(122)	GlcA	CH_3	H	H	0.22
(123)	GlcA	CH_2OH	H	H	0.25
(124)	6-*O*-methyl-GlcA	CH_2OH	H	H	0.26
(125)	GlcA	CHO	H	H	0.29

Terpenes

The terpenoids or isoprenoids consist of branched units of 5-carbons, similar to isoprene [68]. Their nomenclature depends on the number of isoprene units present in the structure, and are classified as monoterpenes, sesquiterpenes, diterpenes, triterpenes, tetraterpenes, and polyterpenes [69].

Ginkgo biloba

For centuries, *G. biloba* has been used in traditional Chinese medicine, as well as in Western countries. Crude extracts from *G. biloba* leaves are popularly used as phytotherapics or dietary supplements. Some clinical and experimental studies have used *G. biloba* extracts, investigating its neuroprotective, cerebrovascular, and cardiovascular effects [70, 71]. The most frequent clinical applications currently include Alzheimer's disease, memory improvement, vascular dementia, and others [72] (Table 11).

Table 11. Lipase inhibitors from *Ginkgo biloba*

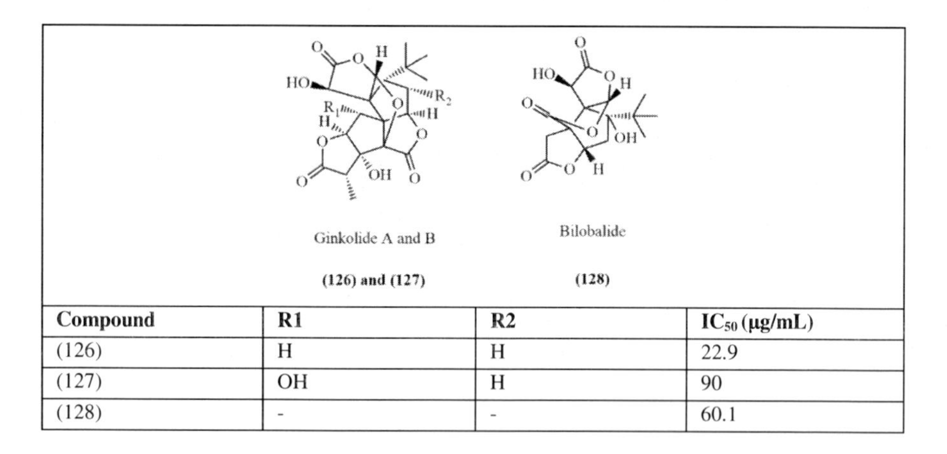

Ginkolide A and B

(126) and (127)

Bilobalide

(128)

Compound	**R1**	**R2**	**IC_{50} (μg/mL)**
(126)	H	H	22.9
(127)	OH	H	90
(128)	-	-	60.1

Actinidia arguta

A. arguta is a kiwi fruit from Japan, North China, Korea, and Siberia [73]. In traditional Chinese medicine, it is used to improve health [74]. It has been used as an additional treatment of obesity in Korea [75]. Nevertheless, the active compound from *A. arguta* roots and its mechanism of action has not been elucidated. Six bioactive triterpenes with pancreatic lipase inhibitory activity were isolated from *A. arguta* roots: 3-*O*-*trans*-*p*-coumaroyl actinidic acid **(129)**, ursolic acid **(130)**, 23-hydroxyursolic acid

(131), corosolic acid **(132)**, asiatic acid **(133)**, and betulinic acid **(134)**. These compounds gave IC_{50} values ranging from 15.83 to 76.45 μM (Table 12) [76].

Table 12. Lipase inhibitors from *Actinidia arguta*

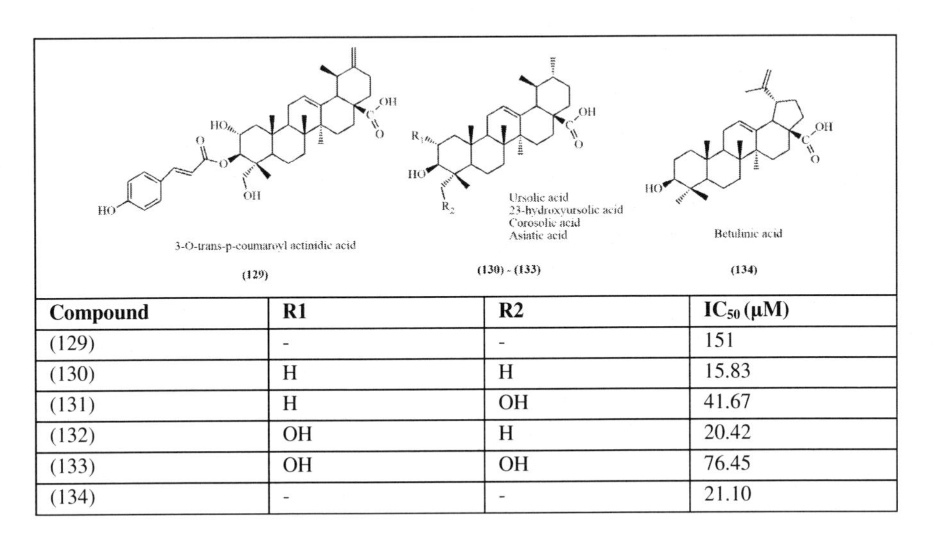

Compound	R1	R2	IC_{50} (μM)
(129)	-	-	151
(130)	H	H	15.83
(131)	H	OH	41.67
(132)	OH	H	20.42
(133)	OH	OH	76.45
(134)	-	-	21.10

Stilbenes

Stilbenes contain two phenyl moieties attached by a two-carbon methylene bridge. Resveratrol is one of the most studied natural polyphenols found in grapes [77].

Vitis vinifera

Vitis vinifera is a vine belonging to the *Vitaceae* family of approximately 60 species that are distributed in North America, Asia, and Europe. The highest cultivation purpose is that wine production, but it is also used to produce juices and fruit [78, 79, 80, 81]. Previous studies have demonstrated the antioxidant properties and health promoters of wine and grapes [82, 83]. From the *Vitis* species, the most relevant secondary metabolites from stilbene derivatives are a class of compounds containing polymerized resveratrol units. Several biological effects have been associated with these compounds, including anticancer, antibacterial, and anti-inflammatory effects [84]. Stilbenoids were isolated from an aqueous extract of *V. vinifera* roots, showing an IC_{50} value of 19.6 µg/mL against pancreatic lipase. The isolated compounds were wilsonol C **(135)**, ampelopsin A **(136)**, pallidol A **(137)**, *cis*-piceid **(138)**, *trans*-piceid **(139)**, with IC_{50} values of 6.7, 143.6, 144.2, 76.1, and 121.5 µM, respectively [85] (Figure 8).

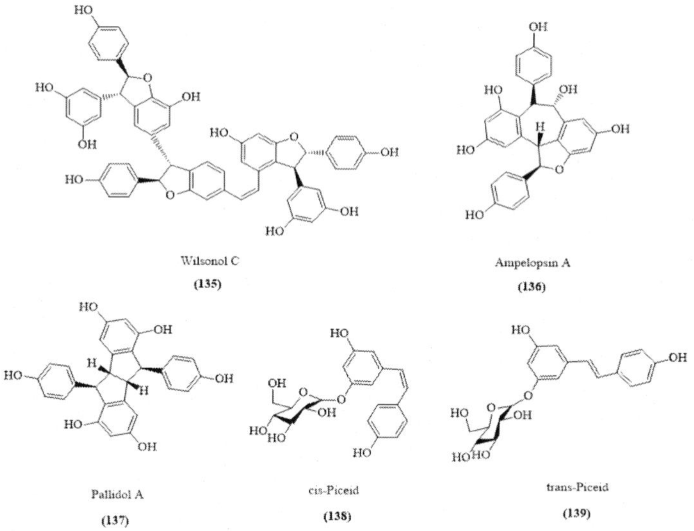

Wilsonol C
(135)

Ampelopsin A
(136)

Pallidol A
(137)

cis-Piceid
(138)

trans-Piceid
(139)

Figure 8. Lipase inhibitors from *Vitis vinifera*.

Microbial Sources

Streptomyces albolongus

Valylactone **(140)** was produced and isolated from strain MG147-CF2 from *Streptomyces albolongus*. Isolation occurred by shaken culture and fermented in the flask. Valylactone inhibited porcine pancreatic lipase with an IC_{50} value of 0.00014 µg/mL. Finally, it inhibited porcine esterase with an IC_{50} value of 0.029 µg/mL [86] (Figure 9).

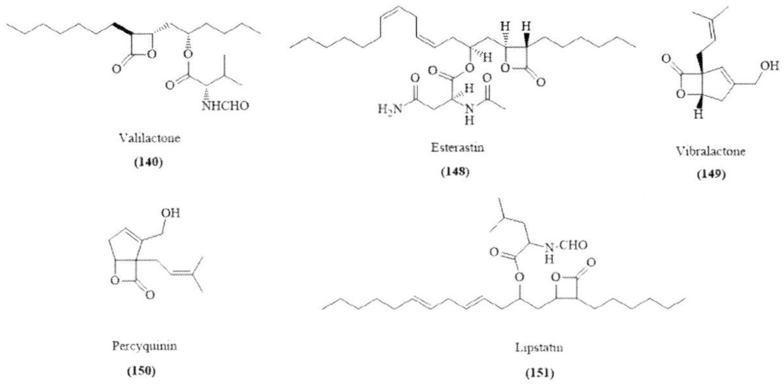

Valilactone (140) Esterastin (148) Vibralactone (149)

Percyquinin (150) Lipstatin (151)

Figure 9. Lipase inhibitors from diverse microbial sources.

Streptomyces spp.

Pancyclines A-E **(141-145)** were isolated from *Streptomyces* spp. NR 0619. These compounds have a *β*-lactone and two alkyl chains. The *N*-formamylalanyloxy substituent is contained in pancyclines A **(141)** and B **(142)** or the *N*-formylglycyloxy substituent present in panclicins C **(143)**, D **(144)**, and E **(145)** inhibited pancreatic lipase with IC_{50} values ranging from 0.62 to 2.9 μM [87] (Table 13).

Table 13. Pancyclin lipase inhibitors from *Streptomyces* spp.

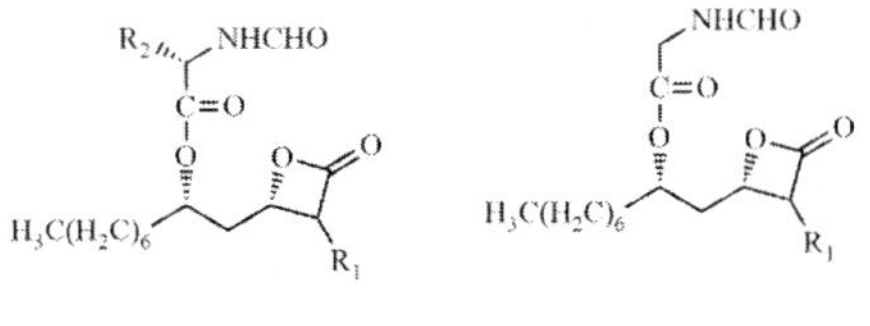

Pancilicin A and B (141) and (142) — Pancilicin C, D and E (143) - (145)

Compound	R1	R2	IC_{50} (μM)
(141)	$(CH_2)_7CH(CH_3)_2$	CH_3	2.9
(142)	$(CH_2)_9CH_3$	CH_3	2.6
(143)	$(CH_2)_7CH(CH_3)_2$	-	0.62
(144)	$(CH_2)_9CH_3$	-	0.66
(145)	$(CH_2)_{11}CH_3$	-	0.89

Streptomyces aburaviensis

Two ebelactones: A **(146)** and B **(147)** were isolated from the fermentation broth of *Actinomycetes* strain MG7-G1 that is closely related to *Streptomyces aburaviensis*. These two compounds inhibited porcine lipase with IC_{50} values of 0.003 and 0.0008 μg/mL, respectively. In addition, the ebelactones inhibited porcine esterase with IC_{50} values of 0.056 for ebelactone A, and 0.00035 μg/mL for ebelactone B [88] (Table 14).

Table 14. Ebelactone lipase inhibitors from *Streptomyces aburaviensis*

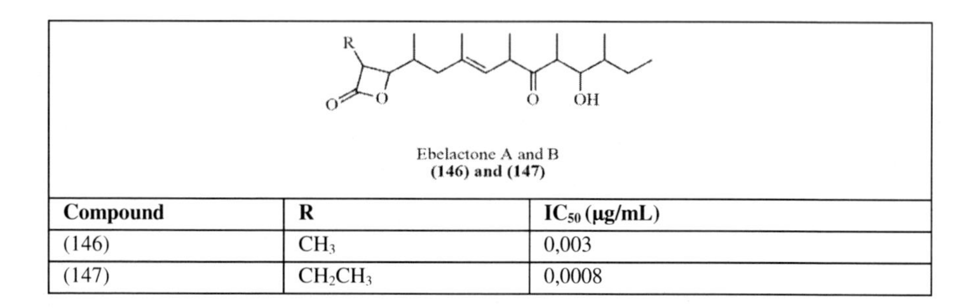

Ebelactone A and B
(146) and (147)

Compound	**R**	**IC_{50} (μg/mL)**
(146)	CH_3	0,003
(147)	CH_2CH_3	0,0008

Streptomyces lavendulae

Esterastine **(148)** inhibited porcine pancreatic lipase with an IC_{50} value of 0.0002 μg/mL. It was produced and isolated from *Streptomyces lavendulae* strain MD4-C1 through the fermentation broth [89] (see Figure 9).

Boreostereum virans

Vibralactone **(149)** is an atypical metabolite with inhibitory activity against pancreatic lipase (IC_{50} of 0.4 μg/mL). This compound was produced by isolation from the fungus basidiomycete and *Boreostereum virans* [90] (see Figure 9).

Stereum complicatum

Percyquinin **(150)** was produced by fungus basidiomycete *Stereum complicatum*, ST 001837, and presented an IC_{50} value of 2 μM for inhibition of pancreatic lipase [91] (see Figure 9).

LIPSTATIN: FROM DISCOVERY TO THE MARKET

Lipstatin **(151)** is a substance of microbial origin produced by *Streptomyces toxytricini*, isolated by scientists from the Hoffman-LaRoche company using a fermentative method. This substance is characterized by selective and potent irreversible inhibitory activity against pancreatic lipase, with an IC_{50} of 0.14 μM (Figure 9) [92, 93]. In the first experiments, the yield of lipstatin was 47.17 μg/g [94]. After improvement of the methods of production and obtaining lipstatin, researchers were able to obtain higher concentrations of lipstatin, ranging from 885 μg/g to 4208 μg/mL [94, 95, 96, 97]. The structure of lipstatin contains a *β*-lactone, two linear C6 and C13 alkyl chains in length. The C13 side chain contains two isolated unsaturations and a hydroxyl group esterified in *N*-formyl leucine [96, 98]. *β*-lactone present in the lipstatin structure is associated with irreversible inhibitory activity of lipase [92]. The opening of *β*-lactone resulted in almost total loss of activity, suggesting that this is an essential structure for irreversible inhibition of pancreatic lipase [99, 100]. Experiments in rats demonstrated their specificity when tested against other enzymes, phospholipase, and trypsin, where it did not show inhibition up to 200 μM concentration [92, 21].

Lipstatin has been only produced by *S. toxytricini* [92, 98, 101]; however, in a recent study, for the first time, lipstatin from *Streptomyces virginiae* was obtained. According to the current taxonomic norm, both bacterial species belong to the phenotypic cluster *Streptomyces lavendulae* [102]. Other compounds containing *β*-lactone have been reported in isolates from other species belonging to the *S. lavendulae* cluster, including valylactone [103] and the pancyclines [104]. Lipstatin is a key intermediate in the preparation of its hydrogenated derivative tetrahydrolipstatin

(Orlistat) [95]. Orlistat has an IC_{50} value of 0.34 μM for inhibition of pancreatic and gastric lipase via an irreversible action mode [92, 105, 106]. Orlistat covalently binds to the Ser^{153} in the active site, blocking triacylglycerols hydrolysis with subsequent excretion of fat in feces, thereby blocking approximately 1/3 of fat absorption [107, 108]. For more than 20 years, it has been studied regarding its side-effects, including diarrhea, oily stools, dyspepsia, abdominal pain, fecal stains, steatorrhea, and liver toxic effects [21, 109].

In 1998, FDA approved Orlistat, and it has been marketed as Xenical® for anti-obesity therapy. This drug reduces fat absorption by up to 30% and is given in combination with a low-calorie diet to decrease weight gain after treatment [110]. It is commercially available in 120 mg capsules. This amount is the recommended dose, three times per day [111]. Three randomized, double-blind, placebo-controlled trials using Orlistat over two years showed that 120 mg (three times/day), with an appropriate diet, resulted in clinically significant weight loss and reduction in weight recovery when compared to placebo. The Orlistat group lost (average) 10.2% versus placebo, which lost 6.1% of body weight ($p < 0.001$) from randomization to the end of the first year. In the second year, patients who continued treatment regained on average half the weight of placebo patients ($p < 0.001$) [10]. Currently, the drugs available for the treatment of obesity are only marketed under medical prescription [111]. At the end of the previous decade, the FDA and the EMA (European Medicines Agency) approved Orlistat 60 mg (Alli®), that does not require a medical prescription. However, it has a dubious efficacy [110, 112, 113, 10]. One of the adverse effects of the administration of Orlistat is reduced absorption of vitamins A, D, E, and *β*-carotene. Therefore, when it is prescribed, multivitamin supplementation is recommended for the reduction of these adverse effects [10]. Furthermore, Orlistat provides significant weight loss, maintenance, and improvement of the lipid profile. It also promotes control of glycemic levels in patients with type 2 diabetes mellitus [114, 115, 116]. One of the risk factors for type 2 diabetes mellitus is obesity, and weight loss is associated with the reduction of risk factors for cardiovascular diseases as well as better control of glycemic levels

[114]. In a meta-analysis that included patients with and without diabetes, Orlistat gave significantly more significant weight loss and decreased hemoglobin A1c levels in diabetic patients [10].

Another study involving 391 obese and diabetic adults of both genders showed that, after one year, the group treated with Orlistat 120 mg exhibited initial body weight loss of 6.2%, while the placebo group ($p < 0.001$) exhibited body weight loss of 4.3%. Treatment of Orlistat in combination with diet resulted in considerable improvement in glycemic control, fasting glycemia ($p < 0.001$), reduction of hemoglobin A1C ($p < 0.001$) and reduction in the dosage of sulfonylurea ($p < 0.01$). These results suggest that Orlistat may act as an effective therapy for both obese and diabetic patients [114].

The pancreatic lipase inhibition pathway has become a steadily growing market for anti-obesity drugs worldwide, where billing rates have been rising ever since its approval by the FDA. By 2003, Xenical® was already marketed in 93 countries. There were about 16 million individuals treated with Xenical® [117]. In England, from 1998 to 2005, there was a 36-fold increase in the number of prescriptions [118]. In 2005, sales of Orlistat reached US$ 450 million worldwide, with the United States accounting for 20% [119]. Roche reported that it earned $197 million from sales of Xenical® in the first half of 2009, and by July, more than 35 million people had purchased the drug [120]. GSK also reported that it earned $125 million from Alli® sales in the second half of 2009 [121]. Already in 2011, Alli® generated revenue of $154 million [122]. This steady increase in the market for Orlistat is also explained by the action of other antiobesity agents marketed, most of which act in the central nervous system, and for which there are increased risks of adverse effects. Reuptake inhibitors have been withdrawn from the market because of their cardiotoxicity, while neurotransmitter releasers are more likely to be withdrawn because of fatigue and dependence [37]. In this way, the anti-obesity market has created a tendency to expand to a range of pancreatic lipase inhibitors that already exist as other potential drugs in various phases of drug development.

Novel Anti-Obesity Drugs on the Market

The withdrawal of anti-obesity drugs by regulatory agencies has been attributable to their exasperating adverse effects, pertaining primarily to drugs that act on the central nervous system. This has increased the need for the development of novel drugs with acceptable adverse effects. Currently, the market points to a new drug called Cetilistat. Cetilistat inhibits pancreatic lipase, with an IC_{50} value of 5.95 nmol/L. It is the competitor molecule closest to the Orlistat market. Initially, Alizyme® (UK) developed Cetilistat, but it is currently being developed by Takeda® (Japan) and Norgine® (Amsterdam) [123].

In a 12-week phase II clinical trial, 612 obese diabetic patients with BMI 28–45 kg/m^2 were enrolled. When compared with placebo, Cetilistat (120 mg and 80 mg) provided superior weight loss (4.32 kg and 3.85 kg) versus 2.86 kg placebo. Weight loss promoted by Orlistat (3.78 kg) was similar to weight loss promoted by Cetilistat. In the groups treated with Orlistat and placebo, there was a more significant number of discontinuations due to gastrointestinal adverse effects [124]. Phase III clinical trials on obese patients with diabetes mellitus 2 and dyslipidemia resulted in the approval of Cetilistat in September 2013 in Japan for the therapy of obese patients with type 2 diabetes mellitus and dyslipidemia [125]. Cetilistat is marketed as Oblean® 120 mg [126, 127]. Experimental results are still required to demonstrate its efficacy for anti-obesity treatment.

Final Considerations

Currently, Orlistat and Cetilistat are the drugs used for the treatment of obesity that have acceptable adverse effect profiles. Knowledge of the structure and function of lipases is essential for developing of new inhibitors. In the literature, there is has been challenging to generate a PL experimental model, because PL activities may have different properties, even those with similar amino acid sequences. On the other hand, most of the compounds followed the physiologic substrate characteristics, primarily with the presence of hydrophobic groups. The interfacial composition of lipases presents a promising alternative for new binding sites. In this way, the amphipathic molecules are effective strategies for inhibition of PL.

This chapter presented several synthetic and natural molecules with high inhibitory activity against lipases with the purpose of inspiring researchers and introducing alternative paths for rational studies in the development of new antiobesity prototypes.

REFERENCES

[1] Hoon J, Liu Q, Lee C, Ahn M, Yoo H, Yeon B, Kyeong M. A new pancreatic lipase inhibitor from Broussonetia kanzinoki. *Bioorg Med Chem Letters* 2012;22:2760–2763. doi:10.1016/j.bmcl.2012.02.088.

[2] Downey M, Atkinson RL, Billington CJ, Bray GA, Eckel RH, Finkelstein EA, Tremblay A. Obesity as a Disease: A White Paper on Evidence and Arguments Commissioned by the Council of The Obesity Society. *Obesity* 2008; 16:1161–1177. doi:10.1038/oby.2008.231.

[3] WHO – World Heath Organization. *Obesity and overweight.* Avaliable in: who.int/en/news-room/fact-sheets/detail/obesity-and-overweight. Access in: 10/01/2019. 2019.

[4] Wong E, Kaur N, Ma N, Patel K. Obesity: A Focus on Pharmacotherapy. *The Journal for Nurse Practitioners* 2013;9:387–395. doi:10.1016/j.nurpra.2013.03.015.

[5] Han TS, Sattar N, Lean M. Assessment of obesity and its clinical implications. *BMJ* 2006;333: 695–698. doi:10.1136/bmj.333.7570.695.

[6] O'Donovan G, Thomas EL, McCarthy JP, Fitzpatrick J, Durighel G, Mehta S, Bell JD. Fat distribution in men of different waist girth,

fitness level and exercise habit. *Int J Obesity* 2009;33:1356–1362. doi:10.1038/ijo.2009.189.

[7] Porter SA, Massaro JM, Hoffmann U, Vasan RS, O'Donnel CJ, Fox CS. Abdominal subcutaneous adipose tissue: A protective fat depot?. *Diabetes Care* 2009;32:1068–1075. doi:10.2337/dc08-2280.

[8] Mccowen C, Blackburn L. Obesity and Its Comorbid Conditions. *Clinical Cornestone,* 1980, 2(3).

[9] Pi-Sunyer, X. The Medical Risks of Obesity. *Postgrad Med* 2009;121:20–33. doi:10.3810/pgm.2009.11.2074.

[10] Glandt M, Raz I. Present and Future: Pharmacologic Treatment of Obesity. *J Obesity* 2011;1-13. doi:10.1155/2011/636181.

[11] Popkin BM. Symposium: Obesity in Developing Countries: Biological and Ecological Factors The Nutrition Transition and Obesity in the Developing World 1. *The Journal of Nutrition* 2001;131:871–873. Available in: http://jn.nutrition.org/content/131/3/871S.short.

[12] Health A, Gordon-Larsen P, Adair LS, Nelson MC, Popkin BM. Five-year obesity inicidence in the transition period between adolescence and adulthood: the National Longitudinal Study of Adolescent Health. *AJCN* 2004;569–575. DOI: 10.1093/ajcn/80.3.569.

[13] Abete I, Parra MD, Zulet MA, Martínez JA. Different dietary strategies for weight loss in obesity: role of energy and macronutrient content. *Nutr Res Reviews* 2006;19:5–17. doi:10.1079/nrr2006112.

[14] Schrauwen P, Westerterp KR. The role of high-fat diets and physical activity in the regulation of body weight. *British J Nutr* 2000;84:417–427. doi:10.1017/s0007114500001720.

[15] Marti A, Moreno-Aliaga MJ, Hebebrand J, Martínez JA. Genes, lifestyles, and obesity. *International J Obesity* 2004;28:29–36. doi:10.1038/sj.ijo.0802808.

[16] Kemper HC, Stasse-Wolthuis M, Bosman W. The prevention and treatment of overweight and obesity. Summary of the advisory report by the Health Council of the Netherlands. *Netherlands J Med*

2004;62:10–17. Available in: http://ovidsp.ovid.com/ovidweb.cgi?T=JS&PAGE=reference&D=emed6&NEWS=N&AN=2004350818.

[17] Chakhtoura M, Rahme M, Fuleihan GE. (2017). Vitamin D Metabolism in Bariatric Surgery Vitamin D Obesity Bariatric surgery RYGB SG Guidelines. *Endocrinology and Metabolism Clinics of NA* 2017;46: 947–982. doi:10.1016/j.ecl.2017.07.006.

[18] Bal BS, Finelli FC, Shope TR, Koch TR. Nutritional deficiencies after bariatric surgery. *Nat Rev Endocrinol* 2012;8:544–556. doi:10.1038/nrendo.2012.48.

[19] Snow V, Barry P, Fitterman N, Qaseem A, Weiss K. Pharmacologic and Surgical Management of Obesity in Primary Care : A Clinical Practice Guideline from the American College of Physicians. *Ann Intern Med* 2005;142:525-531. doi:10.7326/0003-4819-142-7-200504050-00011.

[20] Umashanker D, Igel LI, Aronne LJ. Current and Future Medical Treatment of Obesity. *Gastrointest Endoscopy Clin N Am* 2017;27:181–190. doi:10.1016/j.giec.2016.12.008.

[21] Kumar P, Dubey KK. (2015). Current trends and future prospects of lipstatin: A lipase inhibitor and pro-drug for obesity. *RSC Advances* 2015;5:86954–86966. doi:10.1039/c5ra14892h.

[22] Lowe ME. The triglyceride lipases of the pancreas. *J Lipid Res* 2002;43:2007–2016. doi:10.1194/jlr.r200012-jlr200.

[23] De La Garza AL, Milagro FI, Boque N, Campión J, Martínez JA. Natural inhibitors of pancreatic lipase as new players in obesity treatment. *Planta Med* 2011;77:773–785. doi:10.1055/s-0030-1270924.

[24] Lowe E. Pancreatic Triglyceride Lipase and Colipase : Insights Into dietary fat digestion. *Gastroenterology* 1994;52;1524–1536.

[25] Liu MS, Ma Y, Hayden MR, Brunzell JD. Mapping of the epitope on lipoprotein lipase recognized by a monoclonal antibody (5D2) which inhibits lipase activity. *Biochim Biophys* 1992;1128:113–115. doi:10.1016/0005-2760(92)90264-V.

[26] Belle V, Fournel A, Woudstra M, Ranaldi S, Prieri F, Thomé V, Carrière F. Probing the opening of the pancreatic lipase lid using

site-directed spin labeling and EPR spectroscopy. *Biochem* 2007; 46:2205–2214. doi:10.1021/bi0616089.

[27] Lookene A, Bengtsson-Olivecrona G. Chymotryptic cleavage of lipoprotein lipase: Identification of cleavage sites and functional studies of the truncated molecule. *Eur J Biochem* 1993;213:185–194. doi:10.1111/j.1432-1033.1993.tb17747.x.

[28] Ellington AD, Szostak JW. Nature Publishing Group. *Lett Nat* 1990;346:818–822. doi:10.1038/346183a0.

[29] Brzozowski AM, Derewenda U, Derewenda ZS, Dodson GG, Lawson DM, Turkenburg JP, Thim L. A model for interfacial activation in lipases from the structure of a fungal lipase-inhibitor complex. *Nature* 1991;351:491–494. doi:10.1038/351491a0.

[30] Petersen MTN, Fojan P, Petersen SB. How do lipases and esterases work: the electrostatic contribution. *J Biotechnol* 2001;85:115-147. doi:10.1016/S0168-1656(00)00360-6.

[31] Sridhar SNC, Bhurta D, Kantiwal D, George G, Monga V, Paul AT. Design, synthesis, biological evaluation and molecular modelling studies of novel diaryl substituted pyrazolyl thiazolidinediones as potent pancreatic lipase inhibitors. *Bioorg Med Chem Lett* 2017;27:3749–3754. doi:10.1016/j.bmcl.2017.06.069.

[32] Magrioti V, Verger R, Constantinou-Kokotou V. Triacylglycerols Based on 2-(N-tert-Butoxycarbonylamino) Oleic Acid Are Potent Inhibitors of Pancreatic Lipase. *J Med Chem* 2004;47:288–291. doi:10.1021/jm034202s.

[33] Vu D, Vu T, Nguyen P, Tran D. Virtual Screening, Oriented-synthesis and Evaluation of Lipase Inhibitory Activity of Benzyl Amino Chalcone Derivatives. *Med Pharm Res* 2018;1:26–36. doi:10.32895/ump.mpr.1.1.26/suffix.

[34] Buduma K, Chinde S, Kumar A, Sharma P, Shukla A, Srinivas KVNS, Kumar K. Synthesis and evaluation of anticancer and antiobesity activity of analogs. *Bioorg Med Chem Lett* 2016;26: 1633–1638. doi:10.1016/j.bmcl.2016.01.073.

[35] Sridhar SNC, Ginson G, Reddy POV, Tantak MP, Kumar D, Paul AT. Synthesis, evaluation and molecular modelling studies of 2-

(carbazol-3-yl)-2-oxoacetamide analogues as a new class of potential pancreatic lipase inhibitors. *Bioorg Med Chem* 2017;25:609–620. doi:10.1016/j.bmc.2016.11.031.

[36] Sridhar SNC, Palawat S, Paul AT. Design, synthesis, biological evaluation and molecular modelling studies of indole glyoxylamides as a new class of potential pancreatic lipase inhibitors. *Bioorg Chem* 2019;85:373–381. doi:/10.1016/j.bioorg.2019.01.012.

[37] Sankar V, Maida Engels SE. Synthesis, biological evaluation, molecular docking and in silico ADME studies of phenacyl esters of N-Phthaloyl amino acids as pancreatic lipase inhibitors. *Future J Pharm Sci* 2018;4:276–283. doi:10.1016/j.fjps.2018.10.004.

[38] Sridhar SNC, Bhurta D, Kantiwal D, George G, Monga V, Paul AT. Design, synthesis, biological evaluation and molecular modelling studies of novel diaryl substituted pyrazolyl thiazolidinediones as potent pancreatic lipase inhibitors. *Bioorg Med Chem Lett* 2017; 27:3749–3754. doi:10.1016/j.bmcl.2017.06.069.

[39] Menteşe E, Yılmaz F, Emirik M, Ülker S, Kahveci B. Synthesis, molecular docking and biological evaluation of some benzimidazole derivatives as potent pancreatic lipase inhibitors. *Bioorg Chem* 2017;76:478–486. doi:10.1016/j.bioorg.2017.12.023.

[40] Liu TT, He XR, Xu RX, Wu XB, Qi YX, Huang JZ, Chen QX. Inhibitory mechanism and molecular analysis of furoic acid and oxalic acid on lipase. *Int J Biol Macromol* 2018;120:1925–1934. doi:10.1016/j.ijbiomac.2018.09.150.

[41] Hu B, Cui F, Yin F, Zeng X, Sun Y, Li Y. Caffeoylquinic acids competitively inhibit pancreatic lipase through binding to the catalytic triad. *Int J Biol Macromol* 2015;80:529–535. doi:10.1016/ j.ijbiomac.2015.07.031.

[42] Ashihara H, Deng WW, Mullen W, Crozier A. Distribution and biosynthesis of flavan-3-ols in Camellia sinensis seedlings and expression of genes encoding biosynthetic enzymes. *Phytochem* 2010;71:559-66. doi: 10.1016/j.phytochem.2010.01.010.

[43] Coggon P, Moss GA, Graham HN, Sanderson GW. Biochemistry of tea fermentation. Oxidative degallation and epimerization of the tea

flavanol gallates. *J Agr Food Chem* 1973;21:727–733. doi:10.1021/jf60188a025.

[44] Hilal Y, Engelhardt U. Characterisation of white tea Comparison to green and black tea. *Journal für Verbraucherschutz und Lebensmittelsicherheit* 2007;2:414 - 421, 2007. doi:10.1007/s00003-009-0485-2.

[45] Cheng TO. All teas are not created equal. *Int J Cardiol* 2006;108:301–308. doi:10.1016/j.ijcard.2005.05.038.

[46] Catterall F, Copeland E, Clifford MN, Ioannides C. Effects of black tea theafulvins on aflatoxin B1 mutagenesis in the Ames test. *Mutagenesis* 2003;18:145–150. doi:10.1093/mutage/18.2.14566.2010.525509.

[47] Liang YC, Chen YC, Lin YL, Lin-Shiau SY, Ho CT, Lin JK. Suppression of extracellular signals and cell proliferation by the black tea polyphenol, theaflavin-3,3′-digallate. *Carcinogenesis* 1999;20:733–736. doi:10.1093/carcin/20.4.733.

[48] Duffy SJ, Keaney JF, Holbrook M, Gokce N, Swerdloff PL, Frei B, Vita JA. Short- and Long-Term Black Tea Consumption Reverses Endothelial Dysfunction in Patients With Coronary Artery Disease. *Circulation* 2001;104:151–156. doi:10.1161/01.cir.104.2.151.

[49] Yuda N, Tanaka M, Suzuki M, Asano Y, Ochi H, Iwatsuki K. Polyphenols Extracted from Black Tea (Camellia sinensis) Residue by Hot-Compressed Water and Their Inhibitory Effect on Pancreatic Lipasein vitro. *J Food Sci* 2012;77:H254–H261. doi:10.1111/j.1750-3841.2012.02967.x.

[50] Oh YC, Cho WK, Jeong YH, Im GY, Yang MC, Hwang YH, Ma JY. Anti-Inflammatory Effect of Citrus Unshiu Peel in LPS-Stimulated RAW 264.7 Macrophage Cells. *Am J Chinese Med* 2012;40:611–629. doi:10.1142/s0192415x12500462.

[51] Kawaguchi K, Mizuno T, Aida K, Uchino K. Hesperidin as an Inhibitor of Lipases from Porcine Pancreas and Pseudomonas. *Biosci Biotech Biochem* 1997;61:102–104. doi:10.1271/bbb.61.102.

[52] Baek YS, Ryu YB, Curtis-Long MJ, Ha TJ, Rengasamy R, Yang MS, Park KH. Tyrosinase inhibitory effects of 1,3-diphenylpropanes

from Broussonetia kazinoki. *Bioorg Med Chem* 2009;17:35–41. doi:10.1016/j.bmc.2008.11.022.

[53] Zhang PC, Wang S, Wu Y, Chen RY, Yu DQ. Five New Diprenylated Flavonols from the Leaves of Broussonetia kazinoki. *J Nat Prod* 2001;64:1206–1209. doi:10.1021/np010283o.

[54] Bown D. *Encyclopedia of Herbs and Their Uses*. Dorling Kindersley 1995:424.

[55] Basri AM, Taha H, Ahmad N. A review on the pharmacological activities and phytochemicals of Alpinia officinarum (Galangal) extracts derived from bioassay-guided fractionation and isolation. *Phcog Rev* 2017;22:11:43-56. Available in http://www.phcogrev.com/text.asp?2017/11/21/43/204369.

[56] Lee J, Kim KA, Jeong S, Lee S, Park HJ, Kim NJ, Lim S. Anti-inflammatory, anti-nociceptive, and anti-psychiatric effects by the rhizomes of Alpinia officinarum on complete Freund's adjuvant-induced arthritis in rats. *J Ethnopharmacol* 2009;126:258–264. doi:10.1016/j.jep.2009.08.033.

[57] Shin JE, Han MJ, Song MC, Baek NI., Kim DH. 5-Hydroxy-7-(4′-hydroxy-3′-methoxyphenyl)-1-phenyl-3-heptanone: A Pancreatic Lipase Inhibitor Isolated from Alpinia officinarum. *Biol Pharm Bull* 2004;27:138–140. doi:10.1248/bpb.27.138.

[58] Hashimoto Y, Kawanishi K, Tomita H, Moriyasu M, Uhara Y, Kato A. Enfleurage Chromatography: A New Technique for Identifying Volatile Components in a Small Amount of Samples from Natural Occurrence. *Anal Lett* 1983;16:317–322. doi:10.1080/000327183 080 64469.

[59] Kanetoshi A, Inoue S, Anetai M, Fujimoto T, Aoyagi M, Sato M. An In Vitro Screening Test for Anti-cancer Components of Wild Plants in Hokkaido. *Rep Hokkaido Inst Pub Health* 2005;55:49.

[60] KATO E, Yama M, Nakagomi R, Shibata T, Hosokawa K, Kawabata J. Substrate-like water soluble lipase inhibitors from Filipendula kamtschatica. *Bioorg Med Chem Lett* 2012;22:6410–6412. doi:10.1016/j.bmcl.2012.08.055.

[61] Birari RB, Bhutani KK. Pancreatic lipase inhibitors from natural sources: unexplored potential. *Drug Discov Today* 2007;12:879–889. doi:10.1016/j.drudis.2007.07.024.

[62] Fujikawa T, Yamaguchi A, Morita I, Takeda H, Nishibe S. Protective effects of Acanthopanax senticosus Harms from Hokkaido and its components on gastric ulcer in restrained cold water stressed rats. *Biol Pharm Bull* 1996;19:1227-30.

[63] Jung BS, Shin MK, “Hyang Yak Dae Sa Jeon” 3rd ed. Young Lim Sa Publisher, Seoul, Korea, 2003; 432-425.

[64] Yoshizumi K, Hirano K, Ando H, Hirai Y, Ida Y, Tsuji T, Terao J. Lupane-Type Saponins from Leaves of Acanthopanax sessiliflorus and Their Inhibitory Activity on Pancreatic Lipase. *J Agr Food Chem* 2006;54:335–341. doi:10.1021/jf052047f.

[65] Li F, Li W, Fu H, Zhang Q, Koike K. Pancreatic Lipase-Inhibiting Triterpenoid Saponins from Fruits of Acanthopanax senticosus. *Chem Pharm Bull* 2007;55:1087–1089. doi:10.1248/cpb.55.1087.

[66] Chung MS, Kim NC, Long L, Shamon L, Wan Ahmad WY, Sagrero-Nieves L, Kinghorn AD. Dereplication of saccharide and polyol constituents of candidate sweet-tasting plants: Isolation of the sesquiterpene glycoside mukurozioside IIb as a sweet principle of Sapindus rarak. *Phytochem Anal* 1997;8:49-54. doi:10.1002/(SICI) 10991565(199703)8:2<49::AID-PCA339>3.0.CO;2-C.

[67] Morikawa T, Xie Y, Asao Y, Okamoto M, Yamashita C, Muraoka O, Yoshikawa M. Oleanane-type triterpene oligoglycosides with pancreatic lipase inhibitory activity from the pericarps of Sapindus rarak. *Phytochem* 2009;70:1166–1172. doi:10.1016/j.phytochem. 2009.06.015.

[68] Sharkey TD, Yeh S. Isoprene emission from plants. *Annu Rev Plant Physiol Plant Mol Biol* 2001;52:407-436.

[69] Santos MRV, Moreira FV, Fraga BP, Souza DP de, Bonjardim LR, Quintans-Junior LJ. Cardiovascular effects of monoterpenes: a review. *Revista Brasileira de Farmacognosia* 2011;21:764–771. doi:10.1590/s0102-695x2011005000119.

[70] Mahady G. Ginkgo Biloba for the Prevention and Treatment of Cardiovascular Disease: A Review of the Literature. *J Cardiov Nursing* 2002;16:21-32.

[71] Yao P, Song F, Li K, Zhou S, Liu S, Sun X, Liu L. Ginkgo biloba Extract Prevents Ethanol Induced Dyslipidemia. *Am J Chinese Med* 2007;35:643–652. doi:10.1142/s0192415x07005132.

[72] Singh B, Kaur P, Gopichand Singh RD, Ahuja PS. Biology and chemistry of Ginkgo biloba. *Fitoterapia* 2008;79:401–418. doi:10.1016/j.fitote.2008.05.007.

[73] Ferguson R. Actubudua arguta - the hardy kiwifruit. New Zealand Kiwifruit 1991;81:23-24.

[74] Kim JG, Xiao PG. Traditional Drugs of the East Color Edition. Young-Rim Publishing Co 1995:207.

[75] Webby RF, Markham KR. Flavonol 3-O-triglycosides from Actinidia species. *Phytochem* 1990;29:289–292. doi:10.1016/0031-9422(90)89052-b.

[76] JANG DS, Lee GY, Kim J, Lee YM, Kim JM, Kim YS, Kim JS. A new pancreatic lipase inhibitor isolated from the roots of Actinidia arguta. *Archiv Pharm Res* 2008;31:666–670. doi:10.1007/s12272-001-1210-9.

[77] Pandey KB, Rizvi SI. Plant Polyphenols as Dietary Antioxidants in Human Health and Disease. *Oxid Med Cell Longev* 2009;2:270–278. doi:10.4161/oxim.2.5.9498.

[78] Rossetto M, McNally J, Henry RJ. Evaluating the potential of SSR flanking regions for examining taxonomic relationships in the Vitaceae. *TAG Theor Appl Geneti* 2002;104:61–66. doi:10.1007/s00122020000.

[79] Sefc KM, Steinkellner H, Lefort F, Botta R, Machado AC, Borrego J, Maletić E, Glössl J. Evaluation of the genetic contribution of local wild vines to European grapevine cultivars. *Am J Enol Viticult* 2003;54:15–21.

[80] Crespan M. Evidence on the evolution of polymorphism of microsatellite markers in varieties of Vitis vinifera L. *Theo Appl Gen* 2004;108:231–237.

[81] This P, Jung A, Boccacci P, Borrego J, Botta R, Costantini L, Maul E. Development of a standard set of microsatellite reference alleles for identification of grape cultivars. *Theo Appl Gen* 2004;109:1448–1458. doi:10.1007/s00122-004-17603.

[82] Frankel EN, Waterhouse AL, Teissedre PL. Principal Phenolic Phytochemicals in Selected California Wines and Their Antioxidant Activity in Inhibiting Oxidation of Human Low-Density Lipoproteins. *J Agr Food Chem* 1995;43:890–894. doi:10.1021/jf00052a008.

[83] Monagas M, Hernández-Ledesma B, Gómez-Cordovés C, Bartolomé B. Commercial Dietary Ingredients fromVitis viniferaL. Leaves and Grape Skins: Antioxidant and Chemical Characterization. *J Agr Food Chem* 2006;54:319–327. doi:10.1021/jf051807j.

[84] Peng SC, Cheng CY, Sheu F, Su CH. (2008). The antimicrobial activity of heyneanol A extracted from the root of taiwanese wild grape. *J Appl Microbiol* 2008;105:485–491. doi:10.1111/j.1365-2672.2008.03766.x.

[85] KIM YM, Lee EW, Eom SH, Kim TH. Pancreatic lipase inhibitory stilbenoids from the roots of Vitis vinifera. *Int J Food Sci Nutri* 2013;65:97–100. doi:10.3109/09637486.2013.832172.

[86] Kitahara M, Asano M, Naganawa H, Maeda K, Hamada M, Aoyagi T, Nakamura H. Valilactone, an inhibitor of esterase, produced by actinomycetes. *J Antibiotics* 1987;40:1647–1650. doi:10.7164/antibiotics.40.1647.

[87] Mutoh M, Nakada N, Matsukuma S, Ohshima S, Yoshinri K, Watanabe J, Arisawa M. Panclicins, novel pancreatic lipase inhibitors. I. Taxonomy, fermentation, isolation and biological activity. *J Antibiotics* 1994;47:1369–1375. doi:10.7164/antibiotics.47.1369.

[88] Umezawa H, Aoyagi T, Uotani K, Hamada M, Takeuchi T, Takahashi S. Ebelactone, an inhibitor of esterase, produced by actinomycetes. *J Antibiotics* 1980;33:1594–1596. doi:10.7164/antibiotics.33.1594.

[89] Umezawa, H., Aoyagi, T., Hazato, T., Uotani, K., Kojima, F., Hamada, M., & Takeuchi, T. Esterastin, an inhibitor of esterase, produced by Actinomycetes. *The Journal of Antibiotics* 1978;31:639–641. doi:10.7164/antibiotics.31.639.
[90] Liu DZ, Wang F, Liao TG, Tang JG, Steglich W, Zhu HJ, Liu JK. Vibralactone: A Lipase Inhibitor with an Unusual Fused β-Lactone Produced by Cultures of the BasidiomyceteBoreostereum vibrans. *Org Lett* 2006;8:5749–5752. doi:10.1021/ol062307u.
[91] Hopmann C, Kurs M, Mueller G, Toti L. Percyquinnin, a process for its production and its use as a pharmaceutical. US 6596518B2.
[92] Weibel EK, Hadvary P, Hochuli E, Kupfer E, Lengsfeld H. Lipstatin, an inhibitor of pancreatic lipase, produced by Streptomyces toxytricini. I. Producing organism, fermentation, isolation and biological activity. *J Antibiotics* 1987;40:1081–1085. doi:10.7164/antibiotics.40.1081.
[93] Hadváry P, Hochuli E, Kupfer E, Lengsfeld H, Weibel EK. Leucine derivatives 1986. US patent 4,598,089.
[94] Luthra RC, Umesh RC, Dubey RC. Medium optimization of lipstatin from Streptomyces toxytricini ATCC 19813 by shake flask study. *Int J Microbiol* 2012;4:266-269.
[95] Luthra U, Kumar H, Kulshreshtha N, Tripathi A, Trivedi A, Khadpekar S, Chaturvedi A, Dubey RC. Medium optimization for the production of lipstatin by Streptomyces toxytricini using full factorial design of experiment 2013;11:73–76.
[96] Kumar MS, Verma V, Soni S, Rao ASP. Identification and Process Development for enhanced Production of Lipstatin from Streptomyces Toxytricini (Nrrl 15443) and further Downstream Processing. *Adv Biotechnol* 2012;11:06–11.
[97] Zhu T, Wang L, Wang W, Hu Z, Yu M, Wang K and Cui Z, Enhanced production of lipstatin from Streptomyces toxytricini by optimizing fermentation conditions and medium. *J Gen Appl Microbiol* 2014;60:106–111.
[98] Hochuli E, Kupfer E, Maurer R, Meister W, Mercadal Y, Schmidt K. Lipstatin, an inhibitor of pancreatic lipase, produced by

Streptomyces toxytricini: II Chemistry and structure elucidation. *J Antibiotics* 1987;40:1086–1091. doi:10.7164/antibiotics.40.1086.

[99] Stalder H, Oesterhelt G, Borgström B. Tetrahydrolipstatin: Degradation products produced by human carboxyl-ester lipase. *Helvetica Chim Acta* 1992;75:1593–1603. doi:10.1002/hlca.19920750513.

[100] Borgström B. Mode of action of tetrahydrolipstatin: a derivative of the naturally occurring lipase inhibitor lipstatin. *Biochim Biophys Acta* 1988;962:308–316. doi:10.1016/0005-2760(88)90260-3.

[101] Yoshinari K, Aoki M, Ohtsuka T, Nakayama N, Itezono Y, Mutoh M, Yokose K. Panclicins, novel pancreatic lipase inhibitors: II Structural elucidation. *The Journal of Antibiotics* 1994;47:1376–1384. doi:10.7164/antibiotics.47.1376

[102] Sladič G, Urukalo M, Kirn M, Lešnik U, Magdevska V, Benički N, Pelko M, Gasparič A, Raspor P, Polak T, Fujs S, Hoskisson PA, Petković H. New Lipstatin-Producing Streptomyces Strain. *Food Technol Biotechnol* 2014;52:276–284.

[103] Kitahara M, Asano M, Naganawa H, Maeda K, Hamada M, Aoyagi T, Nakamura H. Valilactone, an inhibitor of esterase, produced by actinomycetes. *J Antibiotics* 1987;40:1647–1650. doi:10.7164/antibiotics.40.1647.

[104] Mutoh M, Nakada N, Matsukuma S, Ohshima S, Yoshinri K, Watanabe J, Arisawa M. Panclicins, novel pancreatic lipase inhibitors: I Taxonomy, fermentation, isolation and biological activity. *J Antibiotics* 1994;47:1369–1375. doi:10.7164/antibiotics.47.1369.

[105] Borgström B. Mode of action of tetrahydrolipstatin: a derivative of the naturally occurring lipase inhibitor lipstatin. *Biochim Biophys Acta* 1988;962:308–316. doi:10.1016/0005-2760(88)90260-3.

[106] Hadváry P, Lengsfeld H, Wolfer H. Inhibition of pancreatic lipasein vitroby the covalent inhibitor tetrahydrolipstatin. *Biochem J* 1988;256:357–361. doi:10.1042/bj2560357.

[107] Henness S, Perry CM. *Orlistat. Drugs* 2006;66:1625–1656. doi:10.2165/00003495-200666120-00012.

[108] Curran MP, Scott LJ. *Orlistat. Drugs* 2004;64:2845–2864. doi:10.2165/00003495-200464240-00010.

[109] Hartmann D, Hussain Y, Guzelhan C, Odink J. Effect on dietary fat absorption of orlistat, administered at different times relative to meal intake. *Br J Clin Pharmac* 1993;36:266–270.

[110] FDA – Food Drug Administration - Orlistat (marketed as Alli and Xenical) *Information,* 2015. Availabre in: <https://www.fda.gov/drugs/postmarket-drug-safety-information-patients-and-providers/orlistat-marketed-alli-and-xenical-information>. Access in: may 9, 2019.

[111] Khan LK, Serdula MK, Bowman BA, Williamson DF. (2001). Use of Prescription Weight Loss Pills among US Adults in 1996–1998. *Ann Internal Med* 2001;134:282. doi:10.7326/0003-4819-134-4-200102200-00.

[112] EMA – European Medicines Agency, Alli (previously Orlistat GSK). Available in: <https://www.ema.europa.eu/en/medicines/human/EPAR/alli-previously-orlistat-gsk >. Acesso em: 10 mai 2019.

[113] Anderson JW. Orlistat for the management of overweight individuals and obesity: a review of potential for the 60-mg, over-the-counter dosage. *Exp Op Pharmacother* 2007;8:1733–1742. doi:10.1517/14656566.8.11.1733.

[114] Hollander PA, Elbein SC, Hirsch IB, Kelley D, McGill J, Taylor T, Weiss SR, Crockett SE, Kaplan RA, Comstock J, Lucas CP, Lodewick PA, Canovatchel W, Chung J, Hauptman J. Role of orlistat in the treatment of obese patients with type 2 diabetes: A 1-year randomized double-blind study. *Diabetes Care* 1998;21(8):1288-94. doi:10.2337/diacare.21.8.1288.

[115] Miles JM, Leiter L, Hollander P, Wadden T, Anderson JW, Doyle M, Klein S. Effect of Orlistat in Overweight and Obese Patients With Type 2 Diabetes Treated With Metformin. *Diabetes Care* 2002;25:1123–1128. doi:10.2337/diacare.25.7.1123.

[116] Padwal RS, Majumdar SR. Drug treatments for obesity: orlistat, sibutramine, and rimonabant. *The Lancet* 2007;369:71–77. doi:10.1016/s0140-6736(07)60033-6.

[117] Roche. *Submission for Reclassification of Xenical® (Orlistat capsules 120mg) To Pharmacist Only Medicine for Weight Control in Adults*, 2003. Available in: <https://medsafe.govt.nz/profs/class/Agendas/agen31-Xenical.pdf>. Access: may 12, 2019.

[118] Srishanmuganathan J, Patel H, Car J, Majeed A. National trends in the use and costs of anti-obesity medications in England 1998–2005. *J Public Health* 2007;29:199–202. doi:10.1093/pubmed/fdm013.

[119] Bray G. Obesity and the Metabolic Syndrome. *Humana Press* 2007:290. doi:10.1007/978-1-59745-431-5.

[120] Taylor PW, Arnet I, Fischer A, Simpson IN. Pharmaceutical Quality of Nine Generic Orlistat Products Compared with Xenical®. *Obesity Facts* 2010;3:231–237. doi:10.1159/000319450.

[121] Richwine L, *US probes Roche, Glaxo diet drug over liver injury*. Disponível em:< https://www.reuters.com/article/us-roche-xenical/u-s-probes-roche-glaxo-diet-drug-over-liver-injury-idUSTRE57N4U320090825 >. Acesso em: 14 mai 2019.

[122] Hirschler B, GSK recalls weight-loss drug Alli in US on tampering concerns, 2014. Disponível em: <https://www.reuters.com/article/us- gsk- alli/ gsk- recalls- weight- loss- drug- alli- in- u- s- on-tampering-concerns-idUSBREA2Q12K20140327>. Acesso em: 14 mai 2019.

[123] Yamada Y, Kato T, Ogino H, Ashina S, Kato K. Cetilistat (ATL-962), a Novel Pancreatic Lipase Inhibitor, Ameliorates Body Weight Gain and Improves Lipid Profiles in Rats. *Horm Metabol Res* 2008;40:539–543. doi:10.1055/s-2008-1076699.

[124] Kopelman P, de Groot HG, Rissanen A, Rossner S, Toubro S, Palmer R, Hickling RI. Weight Loss, HbA1c Reduction, and Tolerability of Cetilistat in a Randomized, Placebo-controlled Phase 2 Trial in Obese Diabetics: Comparison With Orlistat (Xenical). *Obesity* 2010;18:108–115. doi:10.1038/oby.2009.155.

[125] Hainer V. Overview of new antiobesity drugs. *Exp Op Pharmacother* 2014;15:1975–1978. doi:10.1517/14656566.2014.946904.

[126] PMDA, *New Drugs Approved in FY 2013*, 2013. Available in: https://www.pmda.go.jp/files/000153463.pdf. Access in: may 17, 2019.

[127] Norgine, *Norgine and Taked announce the new drug application approval ofoblean®(cetilistat) tablets 120mgin japan for the treatment of obesity with complications 2013.*

In: Advances in Medicinal Chemistry … ISBN: 978-1-53616-368-1
Editor: E. Ferreira da Silva-Júnior

Chapter 2

MEDICINAL CHEMISTRY OF MODULATION OF BACTERIAL RESISTANCE BY INHIBITION OF EFFLUX SYSTEMS

Pedro Gregório Vieira Aquino**, PhD
Federal Rural University of Pernambuco, Garanhuns, Brazil

ABSTRACT

Since the advent of the use of antimicrobials to control diseases in humans and animals, we are forced to deal with the loss of their efficiency and with the appearance of species of microorganisms resistant to the action of these substances. The factors leading to this problem are numerous, ranging from antimicrobial selective pressure to indiscriminate and non-rational use. There are several mechanisms that bacteria have developed to resist the action of antimicrobials, such as the production of enzymes that inactivate the antimicrobial molecules; development of systems that protect the molecular target of the drug, making it difficult for the molecule to reach the target or modifying the active site to reduce binding affinity; decreased cellular permeability to the antimicrobial and the development of active efflux systems. The latter acts by forcing the

* Corresponding Author's E-mail: pgvaquino@gmail.com.

movement of substances against a concentration gradient, decreasing the intracellular concentration of the drug, leading to a decrease in its efficiency. The existence of these efflux systems has been described since the 1980s, both of which are systems specific to certain substrates and broad spectrum systems. There are five main families of efflux pumps, whose classification is based on the amino acid sequences homology: major facilitator superfamily (MFS), small multidrug resistance family (SMR), resistance-nodulation-cell-division family (RND), ATP-binding cassette family (ABC) and multidrug and toxic compound extrusion family (MATE). All these families present structural differences, in terms of energy source, substrates that can carry any type of bacterial organism from which they are distributed. In view of the above, it is intended to show the recent advances in Medicinal Chemistry in the field of inhibition of efflux pumps, analyzing the development of new tools based on natural products chemistry, synthetic medicinal chemistry, molecular biology tools and in silico techniques, that may bring new alternatives for the modulation of antimicrobial resistance.

Keywords: antimicrobial resistance, efflux pumps, medicinal chemistry

Introduction

Several substances have been reported as inhibitors of antimicrobial efflux. An important example is the flavonoid pinostrobin (**1,** Figure 1), which at a dose of 56 μM is able to reduce the minimal inhibitory concentration (MIC) of ciprofloxacin 128-fold against *Staphylococcus aureus*, by inhibiting other systems, other than NorA (one of the most studied efflux systems of *S. aureus*), in addition to inhibiting the formation of biofilms by *Escherichia coli* and *Pseudomonas aeruginosa*, which some studies have reported that depend on efflux pumps for their formation [1].

It is common for some bacteria to have efflux pumps belonging to different families, such as *Mycobacterium avium*, which has the MAV_1406 and MAV_1695 pumps, the first being the Major Facilitator Superfamily (MFS) and the second belonging to the ATP-Binding Cassette (ABC) Family. In view of this, it is interesting to develop substrates capable of inhibiting pumps of different families, such as the quinoline (**2,**

Figure 1), capable of exerting synergism with clarithromycin (**3**, CLA, Figure 1) in a *M. avium* resistant strain [2].

Some researches have been directed at the development of alternatives for the control of bacteria dispersed in the environment, such as the disinfectant baptized as HLE, an association of EDTA (**4**, Figure 1), lactic acid (**5**, Figure 1) and hydrogen peroxide (H_2O_2), capable of exerting an antimicrobial effect and preventing the formation of biofilms when applied at the dose of 0.15-0.4% and also to inhibit the expression of the EfrAB (ABC family), NorE (MFS superfamily) and MexCD (RND family) efflux systems, and may perform synergism with other antimicrobial substances [3].

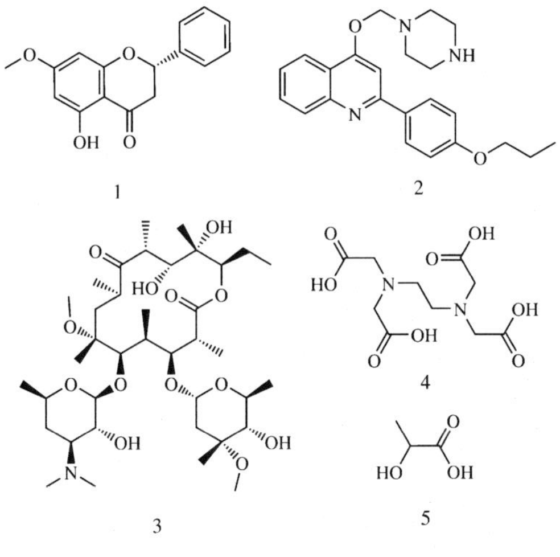

Figure 1. Chemical structures of resistance modulators and clarithromycin.

There are five main families of efflux pumps: Major Facilitator Superfamily (MFS), Small Multidrug Resistance Family (SMR), Resistance-Nodulation-Cell-Division Family (RND), ATP-Binding Cassette Family (ABC) and Multidrug and Toxic Compound Extrusion Family (MATE), each presenting differences in structural characteristics and diversity of substrates which they can transport, but having in common the ability to cause the movement of antimicrobial substances against a concentration gradient across bacterial cell membranes [4]. Based on this division of families of pumps, we will discuss the substances (or mixtures of substances) that have been described over the past 28 years as capable of modulating the action of these pumps, reversing the bacterial resistance.

MFS SUPERFAMILY

This is one of the largest known family of transporters and a good review of the properties of this family can be found at the ref. [5].

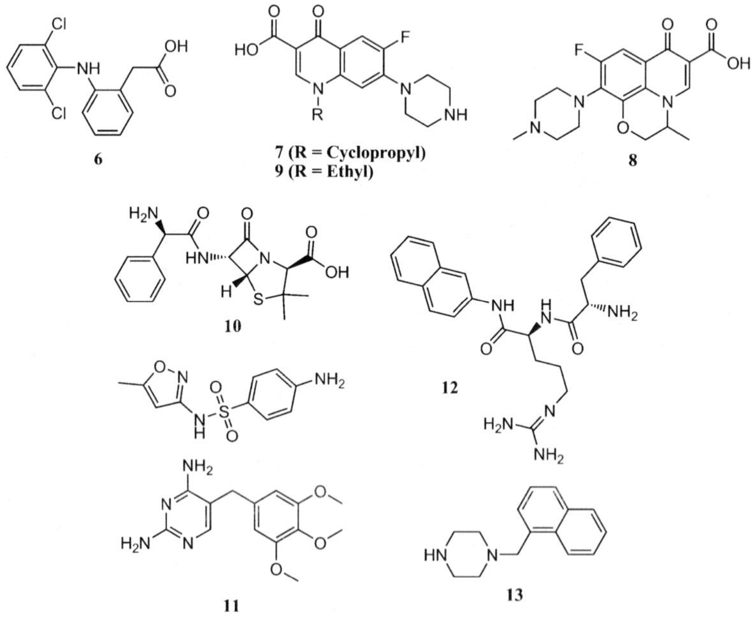

7 (R = Cyclopropyl)
9 (R = Ethyl)

Figure 2. Chemical structures of modulators of MFS superfamily efflux pumps.

Several substances have been associated with alteration in the expression of transporters of this family. For example, it was observed that when *S. aureus* was grown in a medium containing 80 μg/mL diclofenac (**6**, Figure 2) there was a subregulation of the *emrAB/qacA* family genes, indicating that the bacterium would have an increased susceptibility to antimicrobials such as ciprofloxacin (**7**, CIP, Figure 2), ofloxacin (**8**, OFL, Figure 2), and norfloxacin (**9**, NOR, Figure 2) [6]. Another study showed that when *E. coli* isolated from bovine mastitis was grown in medium containing ampicillin (**10**, AMP, Figure 2), sulfamethoxazole-trimethoprim (**11**, SMT, Figure 2), PAβN (**12**, phenylalanine-arginine β-naphthylamide, Figure 2), or NMP (**13**, 1-(1-naphthylmethyl)-piperazine, Figure 2), the expression of the *emrA* and *emrB* genes was increased, suggesting that these antimicrobial and efflux pump inhibitors could exert an increased pressure of expression of resistance factors [7].

MFS Inhibitors from Natural Origin

A study with *Allium sativum* and its main constituent, allyl sulfide (**14**, Figure 3), were able to inhibit the antibiotic efflux mediated by EmrD-3 (Multidrug Resistance Protein D) from *Vibrio cholerae*, both at a dose of 20 mg/mL, increasing its susceptibility to linezolid (**15**, LIN, Figure 3) (4-fold), erythromycin (**16**, ERY, Figure 3) (4-fold), tetracycline (**17**, TET, Figure 3) (16-fold), chloramphenicol (**18**, CLM, Figure 3) (4-fold), kanamycin (**19**, KAN, Figure 3) (33-fold), lincomycin (**20**, LNC, Figure 3) (260-fold), and vancomycin (**21**, VAN, Figure 3) (521-fold) [8].

NorA is a protein that confers resistance mainly to the action of the quinolones in *S. aureus* through their efflux. A description of its role in quinolone resistance has been studied since the early 1990s, where evidence suggested that the *S. aureus* bacterium had two mechanisms of resistance, one involving mutation in DNA gyrase (the molecular target of the antibiotic) and another involving an energy-dependent process that could be blocked by CCCP (**22**, carbonyl cyanide *m*-chlorophenyl hydrazone, Figure 4), an enhancer at a dose of 50 mM [9]. One of the first inhibitors described for this efflux pump was the alkaloid reserpine (**23**, Figure 4), a competitive inhibitor of the efflux pump [10–12]. The first *in vitro* studies aimed at potentiating the effect of antibiotics performed with this alkaloid showed that it was able to reduce the IC_{50} of NOR (**9**, Figure 2) in 4-13 times when inoculated in the concentration of 33 μM, besides suppressing the emergence of bacterial colonies resistant to antibiotics [13,

14]. Further studies demonstrated that potentiation was less pronounced when more lipophilic quinolones, such as sparfloxacin (**24**, SPR, Figure 4) and moxifloxacin (**25**, MOX, Figure 4) were used, both having a 1-2 fold reduction in the minimum inhibitory concentration (MIC), about half of the reduction observed in MIC of CIP (**7**, Figure 2), suggesting that the efflux pump has a higher affinity for more hydrophilic drugs [15].

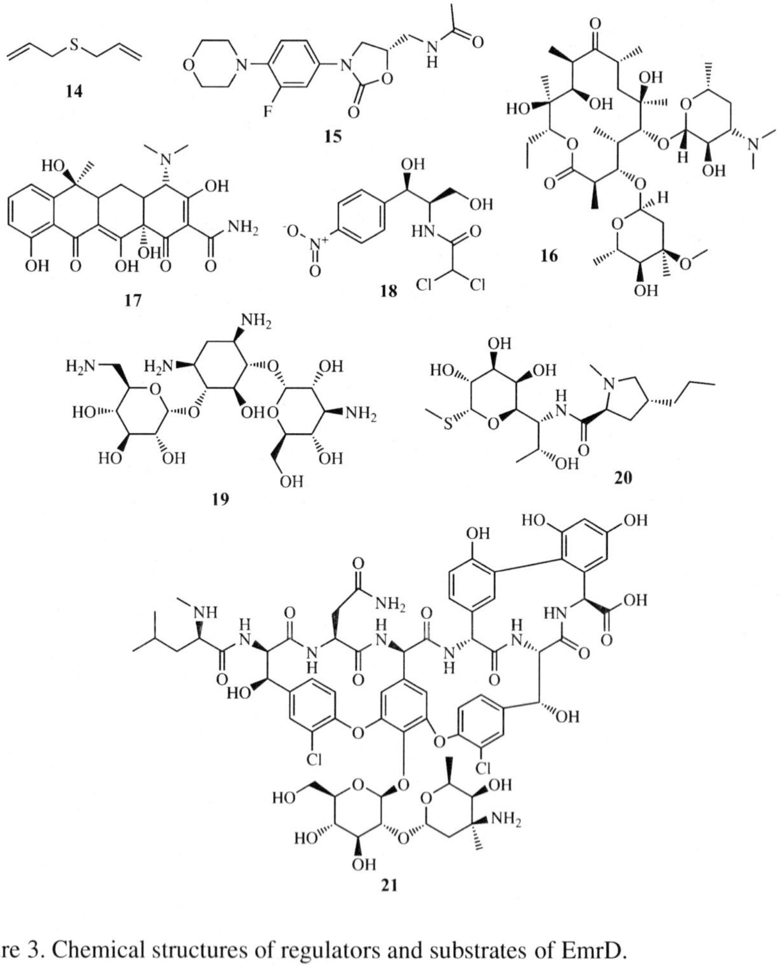

Figure 3. Chemical structures of regulators and substrates of EmrD.

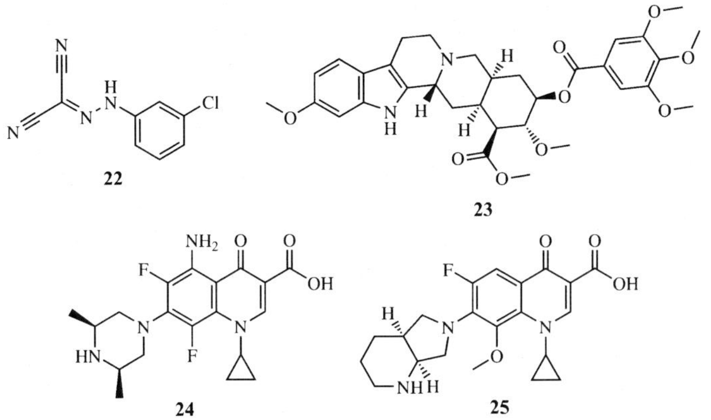

Figure 4. Chemical structures of modulators and substrates of NorA.

Another substrate for the NorA protein are the alkaloids similar to berberine (**26**, Figure 5), which suffer immediate efflux upon entering the bacterial cell. Some plants that produce this type of alkaloid also produce a flavolignan called 5'-methoxyhydnocarpin (**27**, Figure 5), which has no intrinsic antimicrobial activity but is able to reduce the MIC of NOR (**9**, Figure 2) 4-fold and berberine (**26**, Figure 5) 16-fold at the dose of 20 μM when administered to a wild-type *S. aureus* strain. When tested in a strain that did not produce the NorA pump, the substance did not change the MIC of NOR (**9**, Figure 2) but reduced the MIC of berberine (**26**, Figure 5) by 16 times, suggesting that it may also act on systems other than NorA [16].

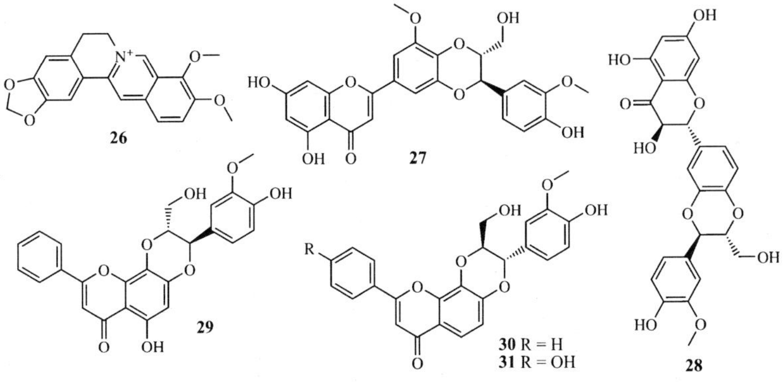

Figure 5. Chemical structures of berberine and flavonoids that interact with NorA.

The studies with compound **27** (Figure 5) continued with the intention of establishing the relations between chemical structure and inhibitory activity of NorA, being synthesized a series of flavones and flavonignanes with structures analogous to **27** (Figure 5) [17]. The study showed that free phenolic groups at positions 5 and 7 lead to compounds of activity comparable to **27** (Figure 5), but that the most active compound has the ring A completely free of hydroxyls. The presence of a 3-hydroxy group drastically decreases or completely abolishes activity in flavone-derived systems, except in the case of silybin (**28**, Figure 5), which showed high activity. Flavolignans with 3-methoxy-4-hydroxy groups showed some activity, but compounds with the 3,5-dimethoxy-4-hydroxy pattern were somewhat less active. On the other hand, compounds with only the 4-hydroxy group were inactive. The analogs of scutellaprostin (**29**, Figure 5) **30** and **31** (Figure 5) were equally active, the hydroxyls in ring B being important in this system. Simple flavones, especially those with free hydroxyl groups, were less active and those with no substitution in ring B or A had moderate activity. Alkylation at the 4-position appears to be important for activity, especially with short lipophilic chains. However, as lipophilicity increases, activity is completely lost [17]. A few years later it was demonstrated that silybin (**28**, Figure 5) was able to reduce the expression of the NorA and qacA/B genes, both coding for the expression of efflux bombs, in Methicillin-Resistant *Staphylococcus aureus* (MRSA), reestablishing their sensitivity to antimicrobials [18].

Still from the perspective of NorA inhibition by natural products, chalcone **32** (Figure 6), obtained from organic extracts of *Dalea versicolor*, was identified as capable of increasing the sensitivity of *S. aureus* to berberine (**26**, Figure 5) and to TET (**17**, Figure 3) in four times at the dose of 34 μM [19]. From another species of the genus, *D. spinosa*, compound **33** (Figure 6) was isolated, which had no direct antimicrobial activity but was able to potentiate the activity of berberine (**26**, Figure 5) strongly in *S. aureus* that dit not expressed NorA than in wild-type *S. aureus* and did not show any effect against NorA-overproducing *S. aureus*, when tested in the concentration range from 42 to 56 μM. Its acetylated analog **34** (Figure 6) was synthesized and tested, showing potentiation of berberine activity (**26**,

Figure 5) in 62-fold in the wild strain, 2-fold in the non-NorA producing strain and 33-fold in the super-productive strain of NorA, demonstrating that the acetylation of the compound directed its activity to the NorA pump [20]. Another substance identified as resistance modulator was baicalein (**35**, Figure 6), a structural analog of compound **32** (Figure 6) isolated from *Scutellaria baicalensis*, which restored sensitivity to CIP (**7**, Figure 2) in *S. aureus* when administered *in vitro* at 59 μM [21].

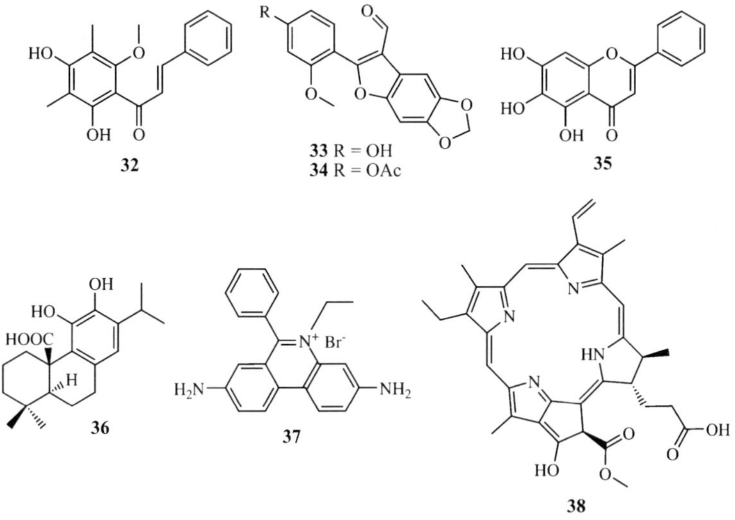

Figure 6. Chemical structures of resistance modulators of natural origin.

In a bioguided fractionation study of chloroform extract of the aerial parts of *Rosmarinus officinalis*, the abietane-type diterpene carnosic acid (**36**, Figure 6) was isolated, capable of inhibiting the efflux of ethidium bromide (EtdBr, **37**, Figure 6) in *S. aureus* with an IC_{50} of 50 μM, enabling it as a possible inhibitor of NorA [22]. Following the methodology of bioguided fractionation by the inhibition of *S. aureus* efflux pump from *Berberis fremontii*, *B. repens* and *B. aquifolium* extracts resulted in the isolation of 5'-methoxyhydnocarpin-D (**27**, Figure 5), and pheophorbide A (**38**, Figure 6), which were able to completely inhibit *S. aureus* growth in the presence of berberine at doses of 2 and 0.8 μM [23].

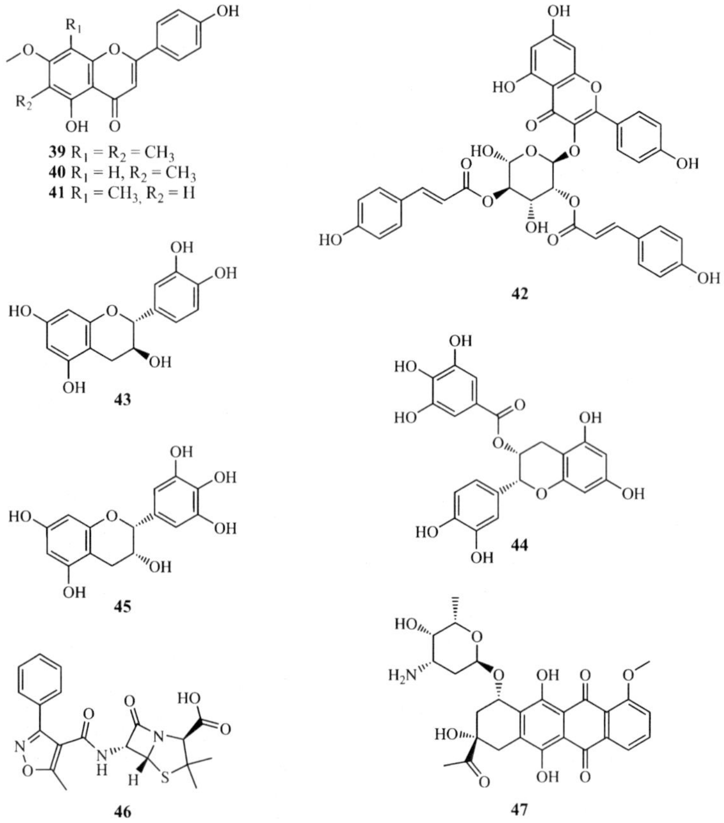

$39\ R_1 = R_2 = CH_3$
$40\ R_1 = H, R_2 = CH_3$
$41\ R_1 = CH_3, R_2 = H$

Figure 7. Chemical structures of oxacillin and resistance modulators of natural origin.

A similar strategy was the synergism-guided fractionation, combining mass spectrometry characterization and synergistic activity with the alkaloid berberine (**26**, Figure 5). This strategy was applied in the fractionation of the ethanolic extract of the leaves and roots of *Hydrastis canadensis*, resulting in the isolation of sideroxylin (**39**, Figure 7), 8-desmethylsideroxylin (**40**, Figure 7), and 6-desmethylsideroxylin (**41**, Figure 7), none of which had intrinsic antimicrobial activity but were capable of decreasing the MIC of berberine (**26**, Figure 5) [24]. Using as a guide the inhibition of EtdBr (**37**, Figure 6) efflux, the bioguided fractionation strategy was applied to the fractionation of the ethanolic extract of *Persea lingue* leaves, resulting in the isolation of dicoumaroyl kaempferol rhamnoside **42** (Figure 7), exhibiting an IC_{50} of inhibition of ethidium bromide efflux and a capacity to increase by eight times the activity of ciprofloxacin against *S. aureus* both at the dose of 2 μM [25]. Using the technique of fractionation guided by increased sensitivity it was possible to identify the substances *(+)*-Catechin (**43**, Figure 7), *(-)*-Epicatechin gallate (**44**, Figure 7), and *(-)*-Epigallocatechin (**45**, Figure 7) in the extract of *Fructus crataegi* (hawthorn, *Crataegus* sp.) as responsible for increasing the sensitivity of MRSA to oxacillin (**46**, Figure 7). However, the compounds in combination were shown to be more efficient than either alone, especially the mixture of catechin (**43**, Figure 7) (441 μM) with Epicatechin gallate (**44**, Figure 7) (36 μM), also capable of increasing the intracellular accumulation of daunomycin (**47**, Figure 7) and decreasing expression of genes *norA*, *norC* and *abcA*, responsible for the expression of efflux pumps [26].

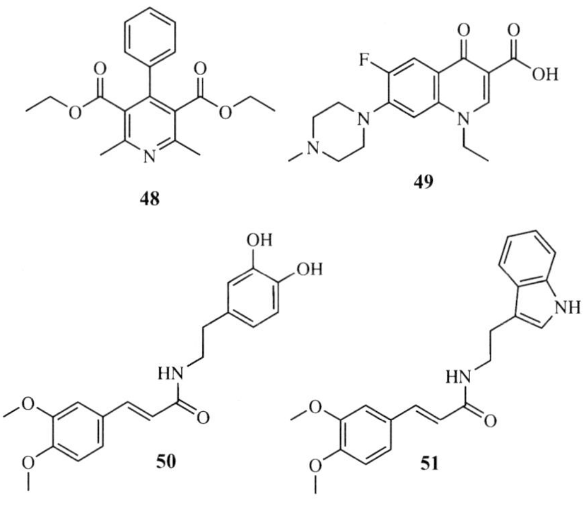

48 **49** **50** **51**

Figure 8. Chemical structures of perfloxacin and natural resistance modulators.

Another inhibitory substance of NorA identified was the pentasubstituted pyridine **48** (Figure 8), isolated by the technique of bioguided extract fractionation of the plant *Jatropha ellipitica* and able to increase the sensitivity of *S. aureus* to CIP (**7**, Figure 2), and NOR (**9**, Figure 2) in 5.5 times and to perfloxacin (PER, **49**, Figure 8) in 1.5 times, at a dose of 305 μM [27]. Using the same technique but with the methanolic extract of *Mirabilis jalapa* led to the isolation of amide **50** (Figure 8), capable of increasing sensitivity to NOR (**9**, Figure 2) in 8-fold at 292 μM in a strain of *S. aureus* overproductive of NorA, what led the researchers to synthesize substances derived from it for a structure-activity relationship (SAR) study, being tryptamine **51** (Figure 8) identified as being able to increase sensitivity to NOR (**9**, Figure 2) 4-fold at the 29 μM dose, eight times at the dose of 57 μM, and at sixteen times at the dose of 286 μM [28].

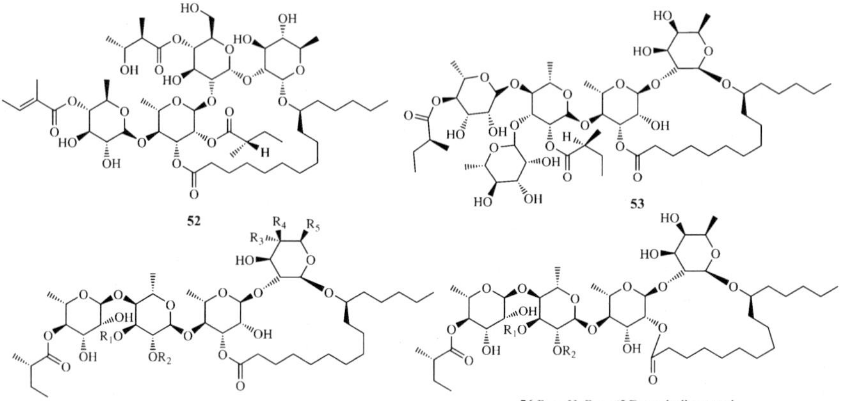

54 $R_1 = R_4 = H$, $R_2 = n$-dodecanoyl, $R_3 = OH$, $R_5 = CH_2OH$
55 $R_1 = n$-dodecanoyl, $R_2 = R_4 = H$, $R_3 = OH$, $R_5 = CH_2OH$
58 $R_1 = n$-dodecanoyl, $R_2 = R_3 = H$, $R_4 = OH$, $R_5 = CH_3$

56 $R_1 = H$, $R_2 = (2S)$-methylbutanoyl
57 $R_1 = (2S)$-methylbutanoyl, $R_2 = H$

Figure 9. *Ipomoea* resistance modulators.

Twenty-two convolvulaceous oligosaccharides of the series tricolorin, scammonin, and orizabin were evaluated in a study with Mexican species of the plant known as Morning Glory (*Ipomoea sp.*). Several were identified that could be classified as bactericidal against strains of *S. aureus* producing tetracycline efflux pump (TetK) and NorA. Among the compounds tested, two of the orizabin series had no antibacterial activity, but were able to inhibit the efflux of EtdBr (**37**, Figure 6) more potently than reserpine (**23**, Figure 4) and increase susceptibility to NOR (**9**, Figure 2), especially compound **52** (Figure 9), which increased it by 16-fold, at a dose of 0.87 μM, against a strain of *S. aureus* overproducing NorA [29]. A further study with the chloroform extract of the flowers of *I. murucoides* identified the substance stoloniferin I (**53**, Figure 9) as capable of increasing sensitivity to NOR (**9**, Figure 2) at eight times the dose of 4 μM [30]. Further study with the same extract identified the murucoidins **54**, **55**, **56**, and **57** (Figure 9) as resistance modulators via inhibition of NorA, without intrinsic antibacterial activity, increasing the sensitivity to NOR (**9**, Figure 2) four times in the dose of 22-25 μM and the murucoidin **59** (Figure 9), as an antimicrobial (MIC 29 μM) capable of increasing sensitivity to NOR (**9**, Figure 2) 4-fold at the dose of 4 μM [31].

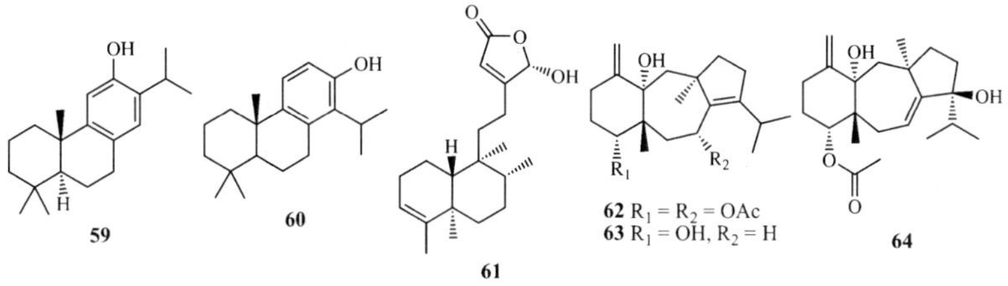

Figure 10. Diterpenes modulators of resistance.

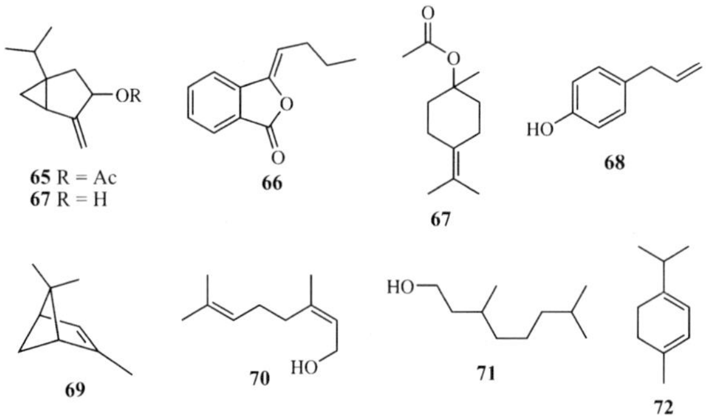

Figure 11. Terpenes from resistance modulators essential oils.

In a study with *Chamaecyparis lawsoniana*, eight compounds were isolated, being identified diterpenes with antibacterial activity in concentrations ranging from 4 to 128 μg/mL. Among them, ferruginol (**59**, Figure 10) was identified as capable of increasing MRSA sensitivity 80-fold at 28 μM and to reduce efflux of EtdBr (**37**, Figure 6) by 40% at the dose of 10 μM in the NorA-overproductive *S. aureus* strain [32]. From another plant of this genus, *C. nootkatensis* was isolated the phenolic diterpene totarol (**60**, Figure 10), capable of reducing the efflux of EtdBr (**37**, Figure 6) with IC_{50} of 15 μM and able to decrease the MIC of TET (**17**, Figure 3), and NOR (**9**, Figure 2) four times and of ERY (**16**, Figure 3) in eight times at a dose of 3 μM [33]. Another diterpene inhibitor of NorA is **61** (Figure 10), isolated from leaves of *Polyalthia longifolia*, capable of reducing the MIC of fluoroquinolones by 16-fold and reducing the microbial load in the systemic swiss albino mice infection model when given in combination with NOR (**9**, Figure 2) [34]. Diterpenes **62**, **63**, and **64** (Figure 10), found in *Canistrocarpus cervicornis* brown alga, are capable of reducing the MIC of NOR (**9**, Figure 2) and TET (**17**, Figure 3) in four times (4 alone or a mixture of 2 and 3) against *S. aureus* expressing NorA and TetK [35].

Another work, using the essential oil and the crude extract in dichloromethane obtained from the roots of *Ligusticum porteri*, increased the activity of NOR (**9**, Figure 2) in NorA-producing *S. aureus* in 4 times at the dose of 100 μg/mL, being this activity associated with the presence of the substances sabinyl acetate (**65**, Figure 11), (*Z*)-ligustilide (**66**, Figure 11), sabinol (**67**, Figure 11), and 4-terpinyl acetate (**68**, Figure 11) [36]. Another crude extract with the potential to supply NorA inhibitory substances is that from roots of *Heracleum lanatum*, capable of synergism with CIP (**7**, Figure 2) and NOR (**9**, Figure 2) [37]. In a study of the crude extract in ether from *Syzygium aromaticum*, it was shown that it was able to reduce the MIC of CIP (**7**, Figure 2) in clinical isolates of multi-drug resistant *S. aureus* [38]. The crude extract from the leaves of *Ocimum americanum* was also able to modulate the NorA pump, increasing by 2-4 times the sensitivity to NOR (**9**, Figure 2) of NorA overproducing *S. aureus* and MRSA strains [39].

The essential oil obtained from the leaves of *Croton zehntneri*, whose main component was estragole (**68**, Figure 11), showed potential in the modulation of *S. aureus* resistance via inhibition of NorA, being shown that it was able to increase the zone of inhibition of norfloxacin in 39.5%, in the *in vitro* evaluation by gas contact [40]. The essential oil obtained from the leaves of another species, *C. grewioides*, also presented modulating activity of NorA, reducing the MIC of NOR (**9**, Figure 2) in 64 times and TET (**17**, Figure 3) in 4 times, while its main constituent, *α*-pinene (**69**, Figure 11), was able to reduce MIC of TET (**17**, Figure 3) by 32-fold, without affecting MIC of NOR (**9**, Figure 2) [41]. The monoterpenes found in diverse essential oils nerol (**70**, Figure 11), dimethyl-octanol (**71**, Figure 11), and estragole (**68**, Figure 11) were able to halve the MIC of EtdBr (**37**, Figure 6) at doses of 1.6-1.7 mM and nerol (**70**, Figure 11) was able to reduce the MIC of NOR (**9**, Figure 2) in 16-fold against NorA producing *S. aureus* at a dose of 3,3 mM [42]. The effects of *Chenopodium abrosioides* essential oil and its main constituent, *α*-terpinene (**72**, Figure 11), on the modulation of *S. aureus* resistance were evaluated. It was identified that the essential oil, but not the terpene, was able to decrease the MIC of the NOR (**9**, Figure 2) and EtdBr (**37**, Figure 6) [43].

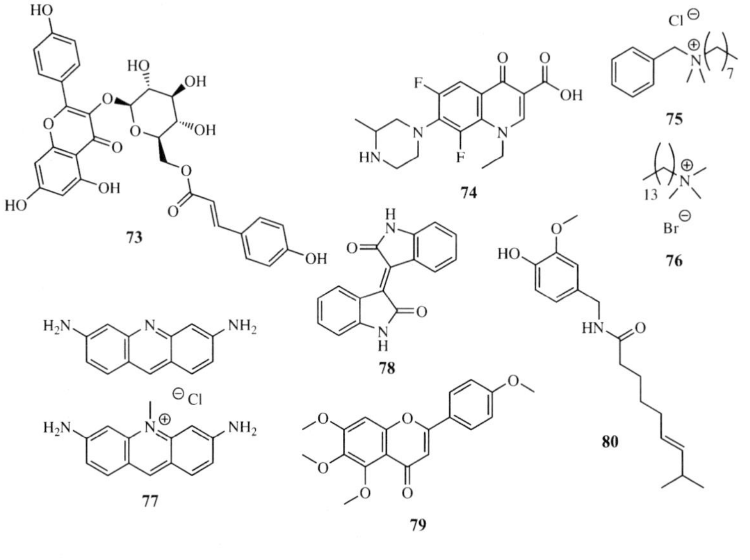

Figure 12. Structures of lomefloxacin, biocides and resistance modulators.

In a study of resistance modulators in Brazilian biodiversity, the substance tiliroside (**73**, Figure 12) was isolated from *Herissantia tiubae*, capable of increasing the sensitivity of NorA-producing *S. aureus* to NOR (**9**, Figure 2) and CIP (**7**, Figure 2) in 16-fold, to lomefloxacin (LOM, **74**,

Figure 12) in four-fold, twice to OFL (**8**, Figure 2) and 128 times to benzalkonium chloride (BCL, **75**, Figure 12), cetrimide (CTR, **76**, Figure 12), acriflavine (ACR, **77**, Figure 12) and EtdBr (**37**, Figure 6) biocides at the dose of 108 μM [44].

Another substance with modulation potential of the activity of NorA is indirubin (**78**, Figure 12), isolated from the chloroform extract of the leaves of *Wrightia tinctoria*, able to reduce the MIC of CIP (**7**, Figure 2) four times in the dose of 9,5 μM [45]. A study with ethanolic extract of the aerial parts of *Praxelis clematidea* resulted in the isolation of six flavones, none of which had antibacterial activity, but all were able to modulate the activity of NorA, especially 4',5,6,7-tetramethoxyflavone (**79**, Figure 12), capable of reducing the MIC of NOR (**9**, Figure 2) in 16-fold at 0.2 mM [46]. Another natural NorA inhibitor is capsaicin (**80**, Figure 12), which is able to halve the MIC of CIP (**7**, Figure 2) against NorA overproducing *S. aureus* at a dose of 41 μM, in addition to increasing the post-antibiotic effect by 1.1 hours and decreasing the invasive capacity of the bacterium to macrophages, thus reducing its virulence [47].

In a work with *Mesua ferrea* extracts seven coumarins were isolated, five of which were able to halve the efflux of EtdBr (**37**, Figure 6) and reduce the MIC of NOR (**9**, Figure 2) by eight times being compounds **81** and **82** (Figure 13) chosen as promising pump inhibitors of NorA-mediated efflux [48]. Another natural product with the potential for inhibition of NorA is the acylphloroglucinol **83** present in lipophilic extracts from the aerial parts of *Hypericum olympicum* L. cf. *uniflorum*, capable of increasing the intracellular accumulation of enoxacin (ENO, **84**, Figure 13) at a dose of 50 μM [49].

Also of natural origin, flavonoid artonin I (**85**, Figure 13), isolated from *Morus mesozygia*, showed antibacterial and efflux inhibitory activity in *S. aureus* [50]. Tannic acid (**86**, Figure 13) was also shown to modulate NorA activity by reducing the MIC of NOR (**9**, Figure 2) in 32-fold at 75 μM [51,52]. Phytochemical study of rhizome extract of *Alpinia cacarata* led to the isolation of five flavonoids, of which galangin (**87**, Figure 13) and kaempferol (**88**, Figure 13) were good modulators of NorA activity, decreasing the efflux of EtdBr (**37**, Figure 6) twice in the doses of 29 and

54 μM, respectively, and decreasing MIC of NOR (**9**, Figure 2) four times at the same doses [53]. From *Boerhavia diffusa* the substance boeravinone B (**89**, Figure 13) was isolated, which was able to reduce the MIC of the EtdBr (**37**, Figure 6) in 16-fold, of NOR (**9**, Figure 2) in 8-fold, of CIP (**7**, Figure 2), OFL (**8**, Figure 2), and MOX (**25**, Figure 4) in 2-fold and of gatifloxacin (GAT, **90**, Figure 13) in 4-fold, when administered at the dose of 25 μM to NorA-producing *S. aureus*, suggesting it is a potential NorA [54].

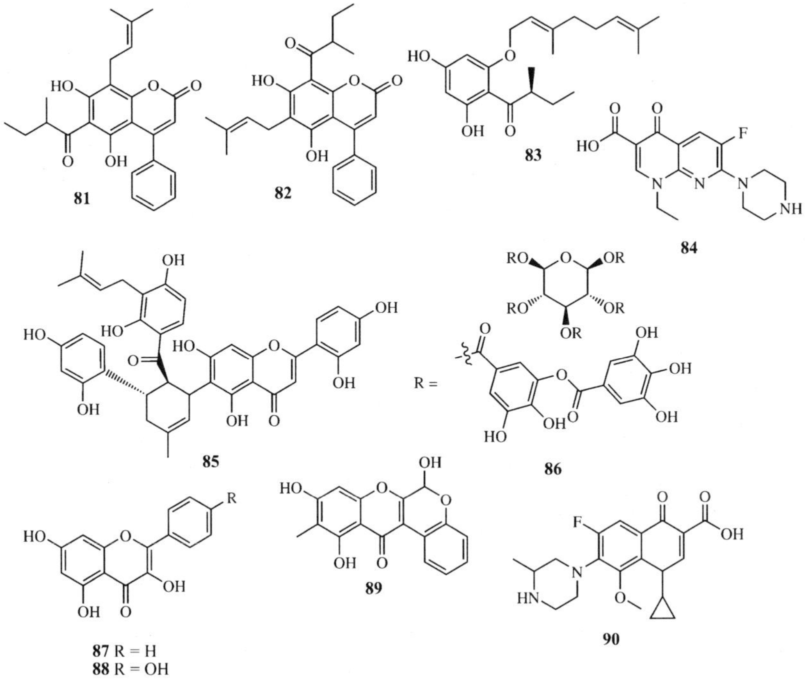

Figure 13. Aromatic modulators of resistance and gatifloxacin.

A study of extracts of the shells and seeds of the fruits of *Vaccinium corymbosum* and *Rubus fruticosus*, as well as their main constituents, protocatechuic (**91**), *p*-coumaric (**92**), vanillic (**93**), caffeic (**94**), and gallic acids (**95**, Figure 14) showed their capacity to reduce methicillin MIC by 128-fold at the dose of 20 μg gallic acid equivalents and to decrease the expression of *mecA*, *norA*, *norB*, *norC*, *mdeA*, *sdrM,* and *sepA* genes, pointing out that these side products of the processing of such fruits may have utility as antimicrobial resistance modulators [55], and this effect was confirmed and better characterized by *in silico* and *in vitro* studies in a subsequent study with caffeic (**94**) and gallic acids (**95**, Figure 14) [56].

A library of 117 chalcones was evaluated for its ability to inhibit EtdBr (**37**, Figure 6) efflux mediated by NorA, among which chalcones **95** and **96** (Figure 14) were the best, showing IC_{50} of 9.0 and 7.7 μM, respectively. Of these, **95** induced a four-fold increase in sensitivity to CIP (**7**, Figure 2) at a dose of 8 μM [57]. A virtual screening technique applied to a library of 182 flavonoids from Chinese traditional medicine plants identified 33 substances as potential candidates to NorA inhibitors [58].

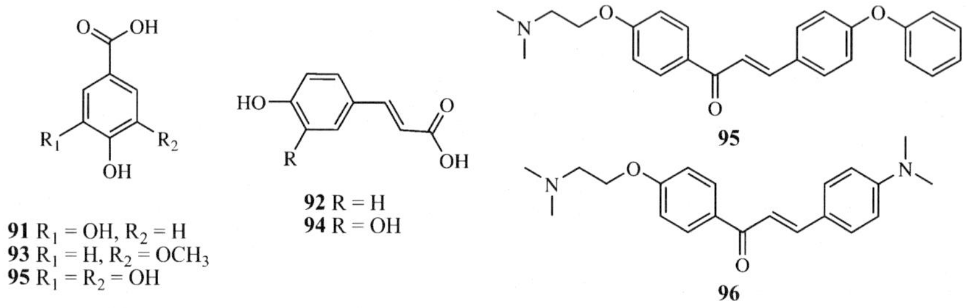

91 R_1 = OH, R_2 = H
93 R_1 = H, R_2 = OCH_3
95 R_1 = R_2 = OH

92 R = H
94 R = OH

95

96

Figure 14. Phenolic acids and chalcones modulators of efflux.

Within the arsenal of natural products, the substances osthol (**97**, Figure 15) and curcumin (**98,** Figure 15) show the potential of inhibiting both P-glycoprotein (P-gp) and NorA, achieving 4- and 8-fold reductions in ciprofloxacin MIC at 25 μM [59]. In a study with 100 clinical isolates of *S. aureus* curcumin (**98,** Figure 15) was shown to decrease mRNA expression to NorA, proving its potentiating effect of CIP (**7**, Figure 2) activity through modulation of the NorA pump [60].

Rv1258c has been associated with the resistance of *Mycobacterium tuberculosis* to drugs such as rifampicin (RIF, **99**, Figure 15). A study with piperine (**100**, Figure 15) showed that it was able to reduce the MIC of rifampicin by 4-8 times at 88 μM, besides reducing the frequency of mutations in the bacterium and increasing the post-antibiotic effect of RIF (**99**, Figure 15), becoming an important resistance modulator. Docking study revealed that piperine has more affinity binding to the protein than reserpine, with hydrogen bonding of the amide oxygen to Arg[141] and the 1,3-dioxole system being important with Leu[55], Leu[56], Gln[342], and His[343], in addition to a π-π interaction of the aromatic ring with His[343] [61].

Tet is a group of efflux pumps responsible for TET (**17**, Figure 3) and analogous molecules resistance. It was demonstrated that the association of tetracycline with cholecalciferol (**101**, Figure 15), when administered to TetK-producing *S. aureus*, halved the MIC of TET (**17**, Figure 3), even cholecalciferol (**101**, Figure 15) not having intrinsic antimicrobial [62].

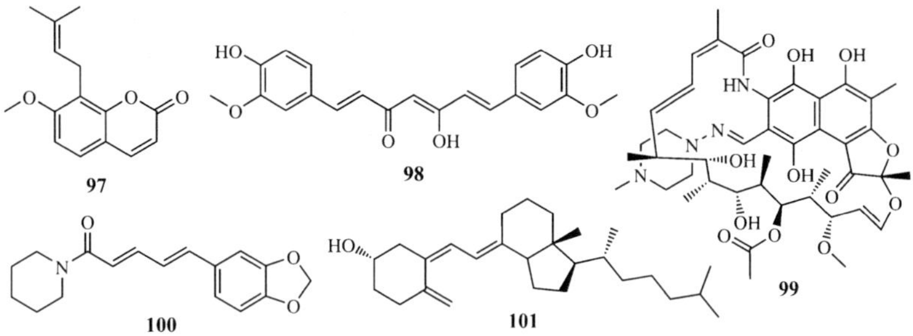

Figure 15. Structures of rifampicin and resistance modulators.

Synthetic MFS Inhibitors

To address the problem with reserpine toxicity to humans at the dose required for NorA blockade, one of the first efforts studied a library of 9600 structurally diverse compounds and identified compounds INF-392 (**102**, Figure 16), INF-55 (**103**, Figure 16) and INF-271 (**104**, Figure 16) as having promising frameworks for the development of inhibitory substances of NorA, capable of reducing MIC of CIP (**7**, Figure 2) four times at 0.5, 6, and 5 μM doses, respectively [63]. Another work has sought to establish the relationships between structure and activity in 2-aryl-1*H*-indoles derivatives based on the inhibitor INF-55 (**103**, Figure 16), describing that the C-5 substituent is important for the activity and that the presence of electron donor groups in this region reduces the activity, **105** (Figure 16) being identified as a promising candidate capable of increasing sensitivity to berberine (**26**, Figure 5) 13.3 times at the dose of 2 μM [64]. Molecular modeling study with 2-aryl-1*H*-indoles identified compound **106** (Figure 16) as a probable inhibitor of NorA [65]. Among a series of 37 derivatives 1-(1*H*-indol-3-yl)ethanamines, which had weak antibacterial activity but many were able to reestablish sensitivity to CIP (**7**, Figure 2) in a dose-dependent manner, **107** (Figure 16) was able to reduce the MIC of CIP (**7**, Figure 2) 4-fold at 1 μM [66]. A series of 48 compounds were created, among which **108-111** (Figure 16) had low IC_{50} of NorA inhibition and high MIC [67]. Still exploring the indolic system, compounds **112** and **113** (Figure 16) were identified as potent inhibitors of EtdBr (**37**, Figure 6), both having IC_{50} of 2 μM [68].

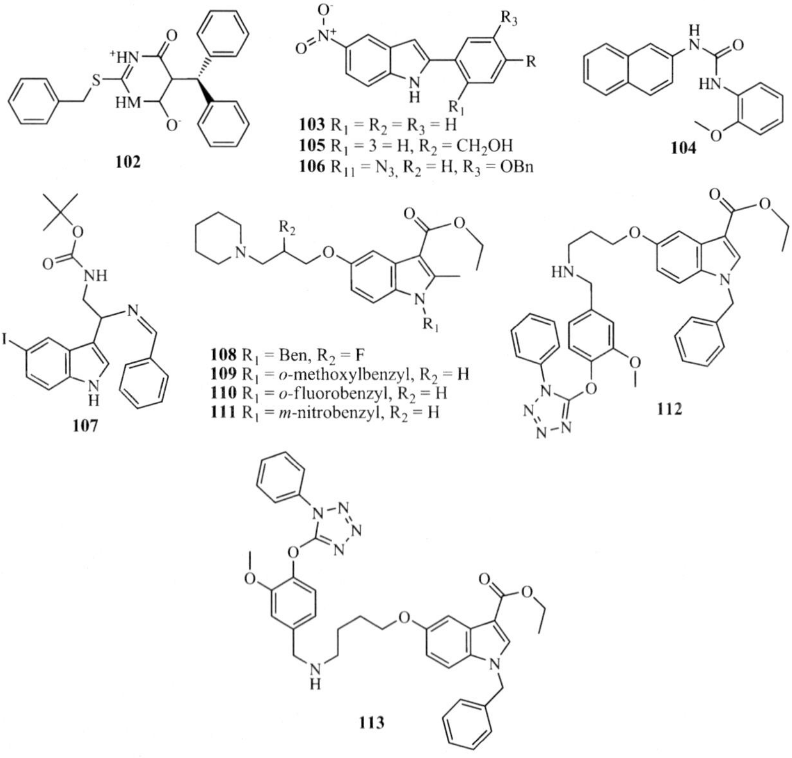

103 $R_1 = R_2 = R_3 = H$
105 $R_1 = 3 = H, R_2 = CH_2OH$
106 $R_{11} = N_3, R_2 = H, R_3 = OBn$

108 R_1 = Ben, R_2 = F
109 R_1 = *o*-methoxylbenzyl, R_2 = H
110 R_1 = *o*-fluorobenzyl, R_2 = H
111 R_1 = *m*-nitrobenzyl, R_2 = H

Figure 16. Resistance modulators of INF series and its derivatives.

Further study with the synthesis of a series of pyrrolo[1,2-a]quinoxalines and their bioisosterosyl pyrrolo[1,2-a]thieno[3,2-e]pyrazine showed the importance of substitution with methoxy groups and the exchange of the *N,N*-dimethylamino group by the pyrrolidine group (**114**, Figure 17) for the first framework and also that the second framework (**115**, Figure 17) had an interesting activity, being the compound **114** able to reduce the MIC of NOR (**9**, Figure 2) 16 times at the dose of 40 μM and **115** at the dose of 169 μM [69].

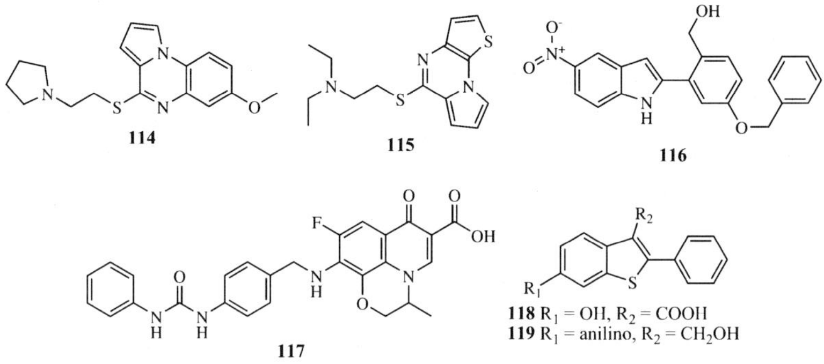

114 **115** **116** **117**

118 R_1 = OH, R_2 = COOH
119 R_1 = anilino, R_2 = CH_2OH

Figure 17. Synthetic modulators of bacterial resistance.

A series of 2-aryl-5-nitro-1*H*-indoles was studied in the search for synthetic NorA-inhibitor scaffolds, compound **116** (Figure 17) being identified as capable of increasing the sensitivity of NorA-producing *S. aureus* to CIP (**7**, Figure 2) in 8-fold and to NOR (**9**, Figure 2) in 16-fold at 13 μM [70]. A different strategy was the design of compounds based on the structure of fluoroquinolones, substrates of NorA, **117** (Figure 17) being identified as capable of reducing the efflux of EtdBr in 96.4% at 10 μM [71].

In an organic synthesis work, a series of 38 new arylated benzo[*b*]thiophenes or were produced, among which most had no antimicrobial activity against *S. aureus*, but were capable of selectively inhibiting NorA pump [72]. A few years later a series of 34 2-arylbenzo[*b*]thiophenes were synthesized, of which the two most prominent, **118** and **119** (Figure 17), decreased the MIC of CIP (**7**, Figure 2) 16-fold at doses of 1 and 1.5 μM, respectively [73].

Another group of substances studied as potential inhibitors of NorA is aminoguanidine hydrazones, the quinoline derivative **120** (Figure 18) being identified as the most potent in the series tested, binding to NorA in the same site as NOR (**9**, Figure 2), decreasing MIC of NOR 16-fold and EtdBr (**37**, Figure 6) in 32 times [74].

One of the strategies used to discover new compounds targeting NorA is the virtual screening strategy. An example was the screening of a library of 300,000 compounds, among which three substances (**121-123**, Figure 18) were capable of inhibiting the efflux of EtdBr (**37**, Figure 6) by NorA-producing *S. aureus* by more than 80%. Among these compounds, **121** and **122** had no intrinsic antibacterial activity and **121** was able to reduce the MIC of CIP (**7**, Figure 2) against NorA-producing *S. aureus* four-fold at the dose of 23 μM and eight-fold at the dose of 68 μM [75].

Another strategy used in the development of substances capable of escaping the efflux by the NorA pump was molecular hybridization, and a hybrid was constructed between berberine (**26**, Figure 5, substrate for NorA) and INF-55 (**103**, Figure 16, NorA blocker), joined by a methylene ether bond (**124**, Figure 18), obtaining a substance with MIC 1.7 μM, 382 times more potent than berberine [76]. Subsequent study synthesized isomers where the binding between the two compounds occurred at the 2', 3' and 4'-positions of 5-nitro-2-phenyl-1*H*-indole, demonstrating that the effect of the change was negligible with respect to the antimicrobial effect, but only **125** and **126** (Figure 18) were able to effectively inhibit NorA [77]. In an attempt to elucidate the mechanism of action of such hybrids a series of berberine-INF-55 hybrid analogs were prepared where INF-55 was replaced by weaker inhibitors than it, leading researchers to conclude that the mechanism of action of the hybrid should be different from the association between the two chemically separated substances [78]. Hybridization was also used in the synthesis of chalcone and glabridin hybrids, and it was shown that compound **127** (Figure 18) was able to reduce NOR (**9**, Figure 2) MIC 16-fold against MRSA and also to reduce microbial burden in blood, liver, kidney, lung, and spleen in the systemic infection pattern of Swiss albino mice [79]. Hybridization was also used as a strategy to increase the accumulation of methylene blue (**128**, Figure 18), increasing the photosensitizing power of this molecule in photodynamic antimicrobial therapy, this goal is achieved by synthesis of compound **129** (Figure 18), methylene blue and INF-55 hybrid [80]. The compounds INF-55 and INF-271 were also used in hybridization with methylene blue (**128**) and showed to be able to increase the sensitivity of *Escherichia coli* and *Acinetobacter baumannii* to photodynamic therapy, surprising therefore that these bacteria do not express NorA, which shows that such inhibitors may show cross-inhibition with other classes of efflux system [81].

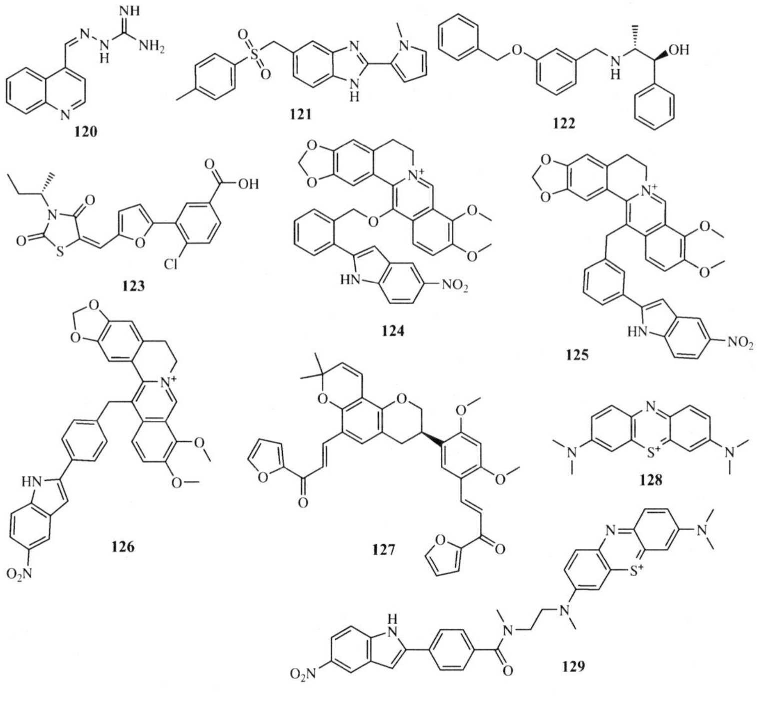

Figure 18. Synthetic modulators of bacterial resistance.

Modification of quinolones with a thiopyranopyridinyl group at the C-7 position generated a series of substances that did not present intrinsic antibacterial activity but that were able to reestablish sensitivity to CIP (**7**, Figure 2) through the modulation of NorA, drawing attention to compounds **130** and **131** (Figure 19), capable of inhibiting the efflux of EtdBr (**37**, Figure 6) in 91 and 90%, respectively, at a dose of 50 μM and of increasing susceptibility to CIP [82]. A continuation of this work proposed to work on the 2-phenyl-4*H*-chromen-4-one nucleus, common to naturally occurring NorA inhibitory flavones and flavolignans, resulting in the 2-phenyl-4-hydroxyquinoline derivatives **132** and **133** (Figure 19), capable of completely restoring the sensitivity to CIP in NorA overproducing S. *aureus* strains [83].

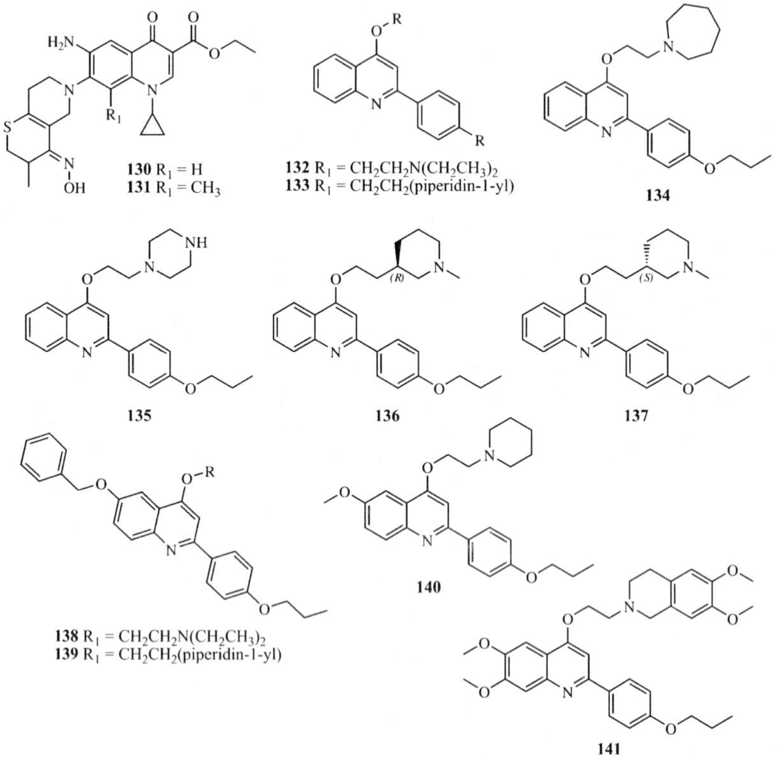

Figure 19. Modifications on the quinolone nucleus.

Subsequent studies by the same group synthesized a series of 2-phenylquinolines designed by pharmacophoric modeling based on ligand directed to NorA, arriving at the compounds **134** and **135** (Figure 19), able to inhibit the efflux of EtdBr with IC_{50} of 6.4 and 8.3 μM (86 and 93.4%, respectively), besides completely re-establishing the sensitivity to CIP [84]. Another concern of the group was regarding the effect of the different enantiomers of these compounds, which led them to separate the (*R*) and (*S*) enantiomers (**136** and **137**, Figure 19) and to test them separately, demonstrating that the racemic mixture had an IC_{50} of 15.9 μM in the inhibition of EtdBr efflux, whereas the *R* and *S* enantiomers presented 16.5 and >50 μM, respectively, proving the importance of the geometry of the molecule for the inhibition of NorA [85]. Additional studies aimed to explore the importance of the C-6 position, leading to the synthesis of compounds **138** and **139** (Figure 19), with IC_{50} of EtdBr efflux inhibition of 3 and 2 μM, respectively and potentiation of CIP activity four times at 2 μM [86]. The introduction of methoxy groups was also evaluated, and a series of 35 compounds were synthesized, of which **140** and **141** were the most active, presenting IC_{50} of inhibition of EtdBr of 4 and 1 μM, respectively, and potentiation of the effect of CIP in eight times at the dose of 2 μM [87].

Inspired by the quinolones framework, a series of 2,1-benzothiazine-2,2-dioxide was synthesized, none of which had intrinsic antibacterial activity, but **142** and **143** (Figure 20) were capable of dose-dependently re-establishing the sensitivity of NorA-overproductive *S. aureus* to CIP (**7**, Figure 2) [88]. Based on the structure of fluoroquinolones, an attempt was made to synthesize substances with antimicrobial activity at the same time capable of inhibiting NorA. These efforts led to the synthesis of a series of 3-aryl-4-methyl-2-quinolin-2-ones, especially **144** (Figure 20), capable of reducing MIC of CIP against *S. aureus* twice in the dose of 4 μM and in four times at the dose of 31 μM [89].

Another important group for the inhibitory activity of NorA is boronic acid. In a study of 150 heterocycles carrying the boronic group, 24 hits were reported, capable of increasing sensitivity to CIP (**7**, Figure 2) four times at doses ranging from 0.5 to 8 μg/mL, with no intrinsic antibacterial activity and none of which had the activity boron group was removed, drawing attention to compound **145** (Figure 20), which increased sensitivity to CIP at the dose of 2 μM [90]. As a result of this first work, a series of 6-(aryl)alkoxypyridine-3-boronic acids were synthesized, having as main substances**146** and **147**, capable of potentiating the action of CIP in four doses at 16 and 15 μM, respectively [91].

Using the *in vivo* model of zebrafish infection it was identified the compound **148** (Figure 20) as a potentiator of CIP (**7**, Figure 2) effect in the range 4- to 256-fold, depending on the bacterial strain, and capable of decreasing the microbial load on zebrafish muscle and skin, and it was also described that it functioned as a competitive inhibitor of NorA [92].

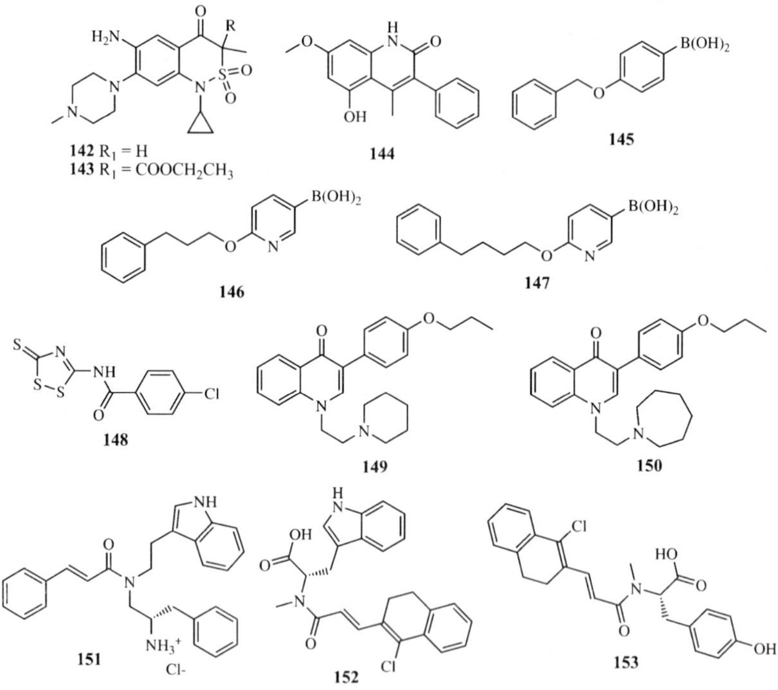

142 $R_1 = H$
143 $R_1 = COOCH_2CH_3$

Figure 20. Synthetic resistance modulators.

Another class of inhibitors of NorA is 3-phenylquinolones, derivatives **149** and **150** (Figure 20) being the most potent in a series of eight different compounds, by reducing the MIC of fluoroquinolones and macrolides in the range of 4- to 128-fold against *Mycobacterium avium* [93]. The *N,N'*-disubstituted cinnamamides were also evaluated as inhibitors of NorA, and **151** (Figure 20) was identified as the most promising, capable of reducing ciprofloxacin MIC in the 4-8-fold range against NorA-producing *S. aureus*

[94]. Another strategy used was the synthesis of acrylic acid amides with amino acids as inhibitors of NorA, which led to the identification of compounds **152** and **153** as being able to reduce the MIC of ciprofloxacin by 16-fold at doses of 3.4 and 3.6 μM, respectively [95].

Another study evaluated the importance of the association of different drugs in the control of *S. aureus*, producer of NorA and NorB, reporting that the product named Vancoplus (a combination of VAN (**21**, Figure 3) and ceftriaxone (CFT, **154**, Figure 21) and VRP1020) was able to control MRSA and NorA and NorB with MIC of 0.25-0.5 μg/mL, in addition to reducing the expression of the coding genes of the efflux pumps [96].

The use of nanocomposites was also a strategy adopted in the modulation of resistance and it was demonstrated that Fe_3O_4 and silver nanocomposites biosynthesized by *Spirulina platensis* were able to reduce the MIC of CIP (**7**, Figure 2) in resistant S. *aureus* and to decrease the expression of the *norA* and *norB* genes by this bacterium [97].

Other groups have been dedicated to understanding, from the molecular point of view, how to block the activity of the NorA protein. An effort in this direction made use of computational tools to show that inhibition of the efflux of antibacterials by the substance was due to a direct interaction between it and the antimicrobial molecule and not by the blocking of the protein itself, as well as interaction between small molecules could alter the antimicrobial logP, facilitating its uptake by gram-negative bacteria [98]. A docking study with a number of substances known to inhibit NorA has shown that the major binding site thereof is a large hydrophobic cleft present in the protein. With such data in hand, the researchers submitted a library of 150 compounds, identifying 14 as potential leads [99].

Natural Products-Inspired Synthetic MFS Inhibitors

From citral (**155**, Figure 21) and citronellal (**156**, Figure 21) monoterpenes a series of amides were synthesized, calling attention **157**, **158** and **159**, which were able to reduce the MIC of CIP (**7**, Figure 2) four times in the dose of 78, 79, and 78 μM, respectively [100]. On the basis of flavanone naringenin (**160**, Figure 21), a series of halogenated substances were synthesized, all of which had antimicrobial activity and some of them capable of reducing the MIC of CIP, the most potent compounds being fluorinated derivatives **161** and **162** (Figure 21) [101]. Another study proposed the synthesis of 36 heterocyclic chalcone derivatives, identifying five compounds (**163-167**, Figure 21) as potent inhibitors of NorA and capable of reducing the MIC of CIP four times at a dose of 52 μM [102].

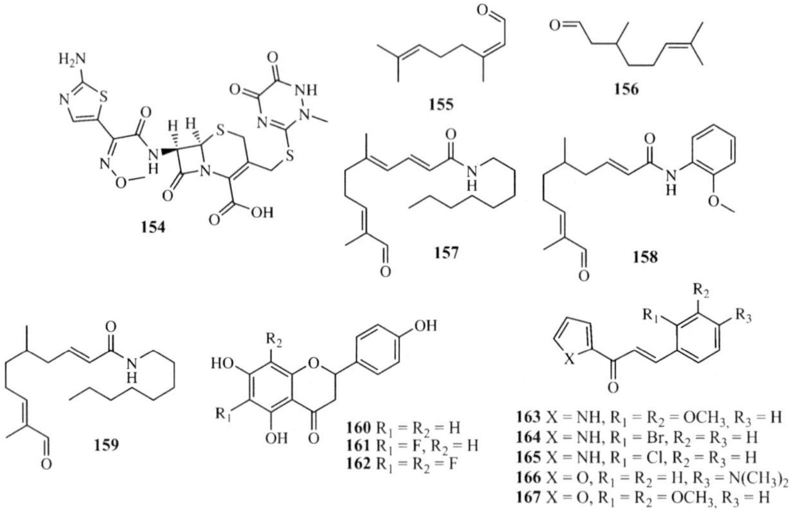

160 $R_1 = R_2 = H$
161 $R_1 = F, R_2 = H$
162 $R_1 = R_2 = F$

163 $X = NH, R_1 = R_2 = OCH_3, R_3 = H$
164 $X = NH, R_1 = Br, R_2 = R_3 = H$
165 $X = NH, R_1 = Cl, R_2 = R_3 = H$
166 $X = O, R_1 = R_2 = H, R_3 = N(CH_3)_2$
167 $X = O, R_1 = R_2 = OCH_3, R_3 = H$

Figure 21. Ceftriaxone and natural products-inspired synthetic derivatives.

A series of synthetic riparins, compounds derived from alkamides isolated from *Aniba riparia*, were evaluated for their antimicrobial and NorA inhibitory activity, showing that the antimicrobial activity increases with the lipophilicity of the compound and that **168** (Figure 22) showed the best NorA modulating effect, reducing the MIC of NOR (**9**, Figure 2), CIP (**7**, Figure 2), and EtdBr (**37**, Figure 6) by a factor of eight, at a dose of 448 μM [103]. Another natural product-inspired synthetic product capable of modulating NorA activity was benzochromene **169** (Figure 22), able to inhibit the pump with minimum effective concentration (MEC) of 4 μM, reestablishing sensitivity to CIP in MRSA [104].

Based on lamellarins structure, polysubstituted pyrroles of natural origin, a series of analogs were synthesized, among which compound **170** (Figure 22) was shown to be capable of inhibiting human P-gp and *S. aureus* NorA, decreasing the MIC of CIP (**7**, Figure 2) in 4-fold at the dose of 6 μM [105]. Ferulic acid derivatives were also synthesized to find new NorA inhibitory substances against MRSA. Among them, **171** and **172** (Figure 22) were identified as the most potent, both of which performed synergism with CIP, and **172** probably inhibited NorA and **171** acting to inhibit a different efflux system [106].

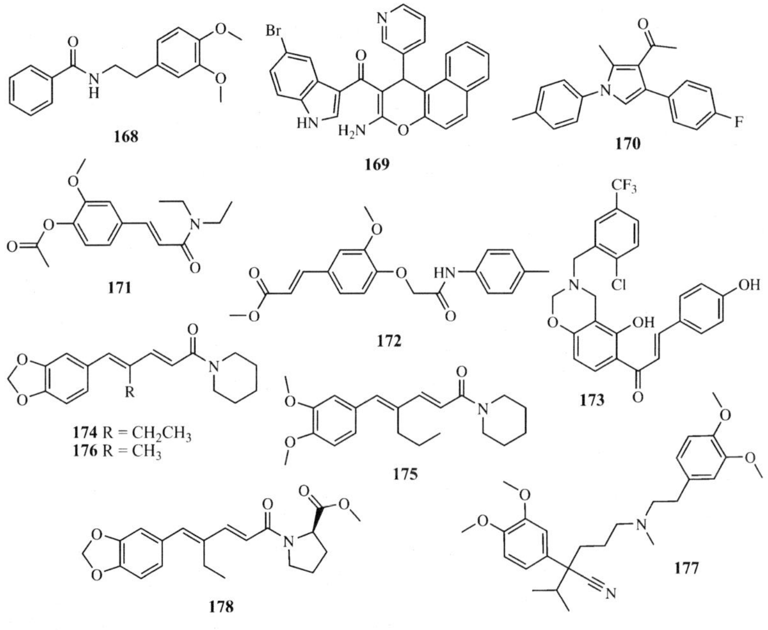

168 169 170 171 172 173

174 R = CH_2CH_3
176 R = CH_3

175 177 178

Figure 22. Verapamil and NorA modulators.

Based on the structure of the antibacterial flavonoid isoliquiritigenin, the bi-functional chalcone **173** (Figure 22) was synthesized and shown to be capable of inhibiting NorA-mediated efflux and reducing the MIC of NOR (**9**, Figure 2) 4-fold in a series of *S. aureus* clinical isolates, including Vancomycin Intermediate *S. aureus* [107].

Study with a library of 200 piperine derivative compounds using as template *S. aureus* SA1199B (NorA overproducer) identified compounds **174** and **175** (Figure 22) as capable of reducing the MIC of CIP (**7**, Figure 2) in 8-fold at 20 and 21 μM, respectively, and **176** (Figure 22) as capable of reducing the MIC of CIP 4-fold at 36 μM [108]. Another study evaluated a series of 38 piperine analogs, identifying 25 with activity levels similar to reserpine (**23**, Figure 4), carnosic acid (**36**, Figure 6) and verapamil (**177**, Figure 22), confirming the potential of this framework as a source of NorA inhibitory substances [109]. Finally, in the field of molecular modeling, a model was developed capable of predicting the bioactivity of new piperins designed as inhibitors of NorA, paving the way for the faster development of new bioactive molecules [110]. Based on the similarity with piperine, 18 amides derived from piperic acid and 4-ethylpiperic acid with α-, β-, and γ-amino acids were synthesized, **178** (Figure 22) being identified as the most potent inhibitor, decreasing MIC of ciprofloxacin in 16 times at the dose of 35 μM [111].

MFS Inhibitors from Repositioning of Drugs

In an attempt to understand the effect of inhibition of NorA on the pharmacodynamics of quinolones, tests were performed on three genetically related strains of *S. aureus* (wild, SA1199; NorA hyperproductive with a mutation in the *grlA* gene, SA1199B; and inducible hyperproductive of NorA, SA1199-3). Cyclosporine (calcineurin inhibitor, 8 μM), reserpine (**23**, Figure 4, 33 μM), omeprazole (**179**, Figure 23, inhibitor of H^+ and K^+ ATPase pump, 290 μM), lansoprazole (**180**, Figure 23, inhibitor of H^+ and K^+ ATPase pump, 270 μM), verapamil (**177**, Figure 22, calcium channel blocker, 220 μM) and diltiazem (**181**, Figure 23, calcium channel blocker, 241 μM) and as antibiotics the substances NOR (**9**, Figure 2), CIP (**7**, Figure 2) and levofloxacin (LEV, **182**, Figure 23). Results indicated that cyclosporin and calcium channel blockers had negligible effects, while membrane gradient blockers and reserpine improved the activity of the more hydrophilic quinolones (2-16-fold decrease in MIC of CIP and 2-4 times of NOR), with reserpine being the substance with the highest levels of potentiation. In addition, reserpine and omeprazole increased bacterial growth suppression time (post-antibiotic effect) in the three strains, SA1199, SA1199B and SA1199-3, at 3-5, 2 and 2-6 times, respectively [112].

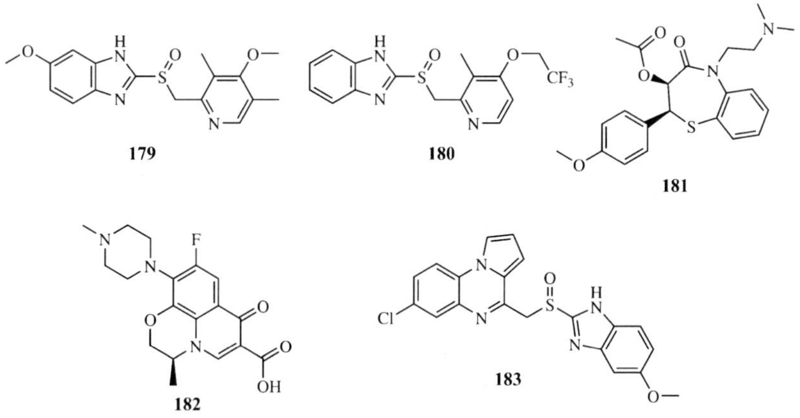

Figure 23. Pump and channel blockers.

Further study with omeprazole (**179**, Figure 23) simulated its clinical effect in an *in vitro* model using infected fibrin-platelet matrices, showing that it (290 μM) reduced the MIC of NOR (**9**, Figure 2) by 4-8 times and that of CIP (**7**, Figure 2) in 2-8-fold, with no significant effect on LEV (**182**, Figure 23) MIC, and having the time it takes ciprofloxacin to kill 99.9% of bacteria and decreasing the frequency of bacterial resistance by 100 times [113]. Years later a whole series of omeprazole analogs were synthesized in order to identify compounds capable of inhibiting NorA, and compound **183** (Figure 23) was identified as capable of reducing the MIC of NOR by 16-fold when tested at the dose of 312 μM [114].

A synthetic inhibitor of P-gp-mediated tumor multidrug resistance (**184**, Figure 24) was evaluated for its potential for inhibition of NorA, demonstrating a reduction of 4 to 8 fold in the MIC of fluoroquinolones at the dose of 18 μM [115]. An important caveat was made some years later, and it was shown that of 32 structurally diverse compounds, only 4 were able to potently inhibit both human P-gp and *S. aureus* NorA, three efficiently inhibited the first and seven potently inhibited only the second, indicating that there is an open pathway for the development of substances that can be safely used in humans. Among the possible relationships to be established, it has been described that a certain degree of lipophilicity is required for both pumps to be inhibited; small substances are not inhibitory, large substances are strong inhibitors and intermediates, generally preferentially inhibit NorA; and that the volume of polar groups relative to the total volume of the molecule is essential to direct selectivity to NorA [116]. Another study that attempted to assess what is required for the same molecule to bind at the same time to NorA and P-gp used the pharmacophoric modeling associated with machine learning to demonstrate that hydrophobicity is essential for a molecule to bind to both pumps; that aromatic rings are essential for inhibition of NorA, but not of P-gp; and that a hydrogen-bonding acceptor is required to inhibit the two pumps, but that a donor is needed to inhibit only NorA [117].

In the search for NorA inhibitors based on substances already used in the therapy, a series of substances analogous to selective serotonin reuptake inhibitors (SSRI) were synthesized, and the compound **185** (Figure 24), the enantiomer of paroxetine, was identified as capable of increasing the susceptibility to NOR (**9**, Figure 2) at 8-fold against *S. aureus* with overexpression of NorA and mutant with overexpression of non-NorA efflux system at doses of 70 and 35 μM, respectively [118]. Based on the SSRI, a series of phenylpiperidines were synthesized, which were evaluated for their ability to block NorA and MepA, with 3-arylpiperidines, in particular compound **186** (Figure 24) being identified as capable of to reduce the MIC of EtdBr (**37**, Figure 6) 64-fold at the dose of 8 μM against a NorA-overproducing strain [119]. Sertraline (**187**, Figure 24) was another SSRI that demonstrated the ability to potentiate the antimicrobial activity of CIP (**7**, Figure 2), LEV (**182**, Figure 23), NOR, MOX (**25**, Figure 4), and gentamicin (GEN, **188**, Figure 24) against *S. aureus* [120].

Following the same repositioning line, phenothiazine prochlorperazine (**189**, Figure 24) and thioxanthene *trans*-flupentixol (**190**, Figure 24) were identified as capable of increasing susceptibility to NOR (**9**, Figure 2) by 64- and 8-fold, respectively, at doses of 67 and 29 μM [121]. Based on phenothiazines, a series of 1,4-benzothiazines were synthesized, where **191** and **192** (Figure 24) were shown to increase sensitivity to CIP (**7**, Figure 2) in four and 32-fold at 88 and 74 μM, respectively [122]. The continuation of this work led to the synthesis of compound **193**, capable of inhibiting in 49% the efflux of EtdBr (**37**, Figure 6) and to reduce the MIC of CIP in 8-fold at 94 μM [123].

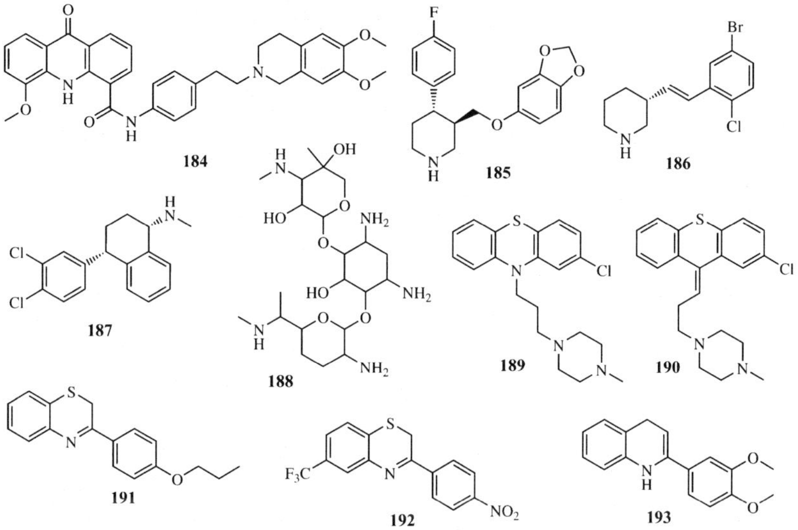

Figure 24. Gentamicin and NorA modulators.

Using a library of approved drugs, *in silico* screening was performed in an attempt to discover new inhibitors of NorA, and gefitinib (**194**, Figure 25) and dasatinib (**195**, Figure 25) compounds were identified as being the most promising, with an IC_{50} of EtdBr (**37**, Figure 6) efflux inhibition of 9.7 and 11.5 μM, respectively [124]. From the perspective of drug repositioning, it was thought that the possibility of analogs to celecoxib (**196**, Figure 25, COX-2 selective inhibitor) could serve as inhibitors of NorA. Evaluation of a library of 150 analogs, all bearing the 1,4-dihydropyrazolo[4,3-c]benzothiazine-5,5-dioxide system led to the identification of compound **197** (Figure 25), capable of inhibiting the efflux of EtdBr in 77 % at a dose of 50 μM, with a MIC five times greater than that of CIP (**7**, Figure 2) and capable of completely reversing the resistance to the latter in a dose-dependent manner [125].

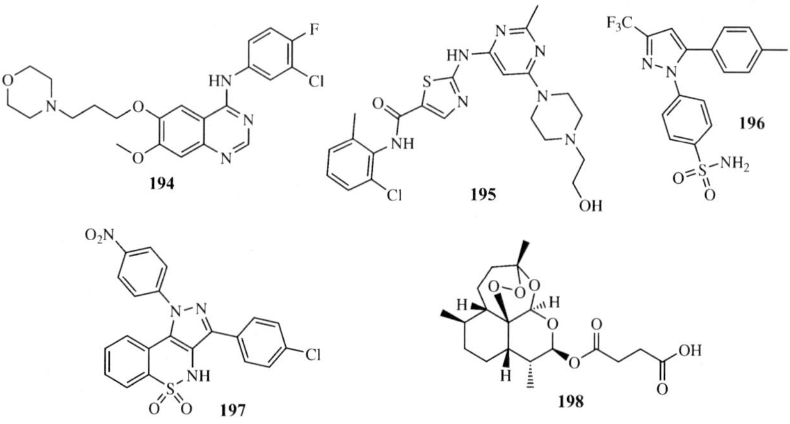

Figure 25. Chemical structures of NorA modulators.

Artesunate (**198**, Figure 25) was another substance that had its proven efficacy in the inhibition of efflux pumps, and it was shown that it was capable of causing an increase in the accumulation of daunorubicin and oxacillin antibiotics in MRSA and also of inhibiting gene expression *norA*, norB, and *norC* genes, reducing the expression of efflux pumps [126].

RND SUPERFAMILY

Members of this family are of particular importance to antimicrobial resistance in gram-negative bacteria and are composed by three proteins that work together in a tripartite MFP-RND-OMF system, being the first located at the periplasmic space, making the connection between the other two: the RND located at the inner membrane and the OMF located at the outer membrane. Several reviews are available in the literature speaking of the properties of the members of this family [127–130].

It is interesting to note that efflux pumps from this family are associated with another virulence factor in gram-negative bacteria, the formation of biofilms. An assay with a series of *Salmonella typhimurium* mutant strains, unable to produce multiple efflux pumps showed that they were unable to efficiently produce biofilm, as well as those in which the pumps were inhibited by the administration of blockers such as PA*β*N (**12**, Figure 2), CCCP (**22**, Figure 4), and chlorpromazine (**199**, Figure 26) [131, 132].

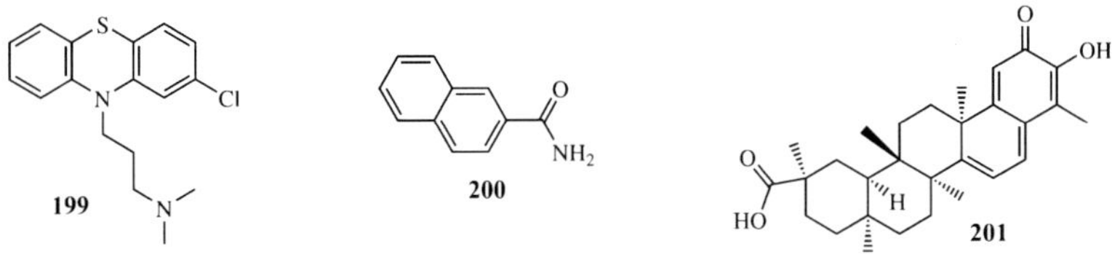

Figure 26. RND systems modulators.

Other bacteria that present efflux pumps of this family are *Campylobacter jejuni* and *C. coli*, which produces pumps known as CmeABC and CmeDEF, among others. The importance of these systems was demonstrated by the administration of inhibitors *β*-naphthylamide (**200**, Figure 26) and NMP (**13**, Figure 2), which were able to reverse resistance to a number of antimicrobials, similar to mutation in *cmeB* and *cmeF* genes, coding for the pump, as well as mutation of the repressor gene the pump, *cmeR* [133].

The MexXY-OprM system has been reported as one of the resistance factors in *Pseudomonas aeruginosa*. One of the first attempts to understand how it worked showed that its inhibition could decrease the inhibition of divalent cations on the activity of aminoglycosides [134]. In *Neisseria gonorrhoeae*, MtrCDE is the most well-described efflux system, one of its inhibitors being reserpine (**23**, Figure 4), which has been shown to increase the activity of TET (**17**, Figure 3) against clinical isolates up to 62 times at the dose of 41 μM [135]. *In vitro* studies with the structure of this protein have shown that the binding interface between MtrC and MtrE is an important modulator of its function, residing there a promising region where new ligands can be developed [136].

One of the first demonstrations of the existence and importance of Sme efflux system was in the gram-negative bacterium *Stenotrophomonas maltophilia*, intrinsically resistant to a series of structurally diverse antimicrobials and implicated in nosocomial infections. The researchers demonstrated by reverse transcription PCR that the wild-type bacterium expressed the *smeDEF* genes and that their deletion led to increased susceptibility to various antibiotics [137]. Another efflux system expressed by *S. maltophilia* is SmeYZ, and it has been shown that its expression can be reduced by administration of the naturally occurring product celastrol (**201**, Figure 26), a pentacyclic triterpenoid isolated from the roots of *Tripterygium wilfordii* [138].

RND Inhibitors from Natural Origin

A strategy used in the search of inhibitors of the systems of this family was the virtual screening of a library of compounds of natural origin. In this screening the researchers were able to exclude those substances that presented structural similarities with others that were recognized substrates for the efflux pumps AcrAB-TolC and MexAP-OprM, identifying as good candidates for inhibitors the lanatoside C (**202**, Figure 27) and daidzein (**202**, Figure 27), which were validated *in vitro*, where they were able to inhibit efflux of EtdBr (**37**, Figure 6) in *E. coli* and *P. aeruginosa* at doses of 16 and 63 μM, respectively [139].

Working with the hypothesis that natural marine products could be a good source of compounds with activity against terrestrial bacteria due to the absence of the previous contact with these substances, a research was carried out with extracts of marine microorganisms to identify inhibitors of RND family efflux pumps. The substance **203** (Figure 27) was isolated from *Pseudoalteromonas piscida* extract and described as being able to reduce the MIC of fluoroquinolones, aminoglycosides, macrolides, *β*-lactams, and chloramphenicol in the range of 2-16-fold in AcrAB-TolC,

MexAB-OprM and MexXY expressing strains [140]. Natural terrestrial products have also shown efficacy against gram-negative bacteria producing RND family pumps, such as the methanolic extract from the bark of *Nauclea pobeguiinii* and resveratrol (**204**, Figure 27), isolated from this extract, the first demonstrating synergism with CLM (**18**, Figure 3) and KAN (**19**, Figure 3) the second with streptomycin (STR, **205**, Figure 27) and CIP (**7**, Figure 2), against resistant bacteria [141].

Following the natural product line, the total alkaloid extract from the seeds of *Sophorea alopecuroides* was shown to be able to revert *E. coli* resistance to CIP (**7**, Figure 2) via inhibition of the AcrAB-TolC pump at the dose of 8 mg/mL [142]. In *E. aerogenes* it was demonstrated that gerylamine (**206**, Figure 27) and geraniol (**207**, Figure 27) were able to increase sensitivity to CLM (**18**, Figure 3) via non-competitive inhibition of AcrAB-TolC [143]. Still in the search for compounds of natural origin, a virtual screening identified the substances plumbagin (**208**, Figure 27), nordihydroguaretic acid (**209**, Figure 27) and shikonin (**210**, Figure 27) as possible inhibitors of AcrB, increasing the sensitivity of *E. coli* to a series of antimicrobials *in vitro* [144]. Extracts of the species *Allium sativum*, *Syzygium aromaticum*, *Berberis aristata*, *Rhus cotinus*, and *Phyllanthus emblica* showed synergism with the TET (**17**, Figure 3) against *Salmonella enterica* expressing AcrAB-TolC, indicating that such species may serve as sources of inhibitors of this protein [145]. Using the docking strategy it was identified that a potential inhibitor of AcrB is punigratane (**211**, Figure 27), present in *Punica granatum*, capable of binding to the same site as the inhibitor MBX2319 (**212**, Figure 27), which suggests that it can be used as a resistance modulator [146]. Resveratrol (**204**, Figure 27) was also able to modulate the action of AcrAB-TolC, causing its inhibition at the dose of 250 μM in *E. coli* [147].

In an attempt to find adjuvants for the therapy of *Acinetobacter baumanii* infections, a series of phenolic compounds produced by plants were tested, and the ellagic (**213**, Figure 28) and tannic (**86**, Figure 13) acids were identified as capable of increasing the activities of novobiocin (**214**, Figure 28), coumermycin (**215**, Figure 28), chlorobiocin (**216**, Figure

28), RIF (**99**, Figure 15), and fusidic acid (**217**, Figure 28) at the dose of 40 μM, probably by inhibition of the AdeIJK system [148].

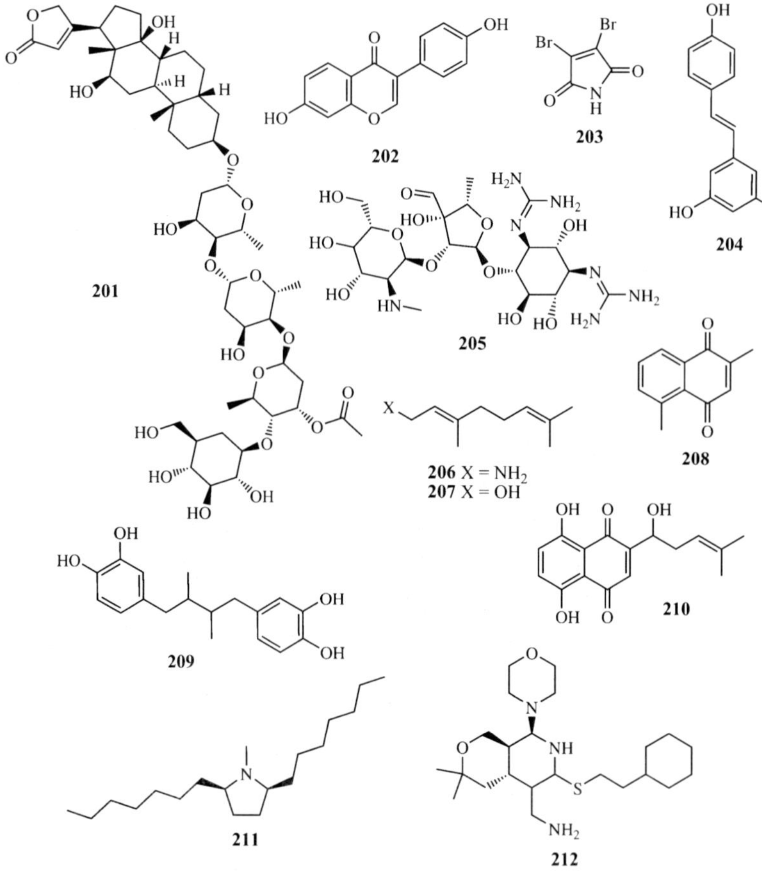

Figure 27. RND inhibitors from natural origin.

In an evaluation of a library of 85,000 microbial fermentation extracts, the compounds **218** (Figure 28) and **219** (Figure 28) were identified as potent and specific inhibitors of the MexAB-OprM system of *P. aeruginosa*, increasing sensitivity to LEV (**182**, Figure 23) in 4 times at 1 μM and in 8 times at 4 and 2 μM, respectively [149].

Using berberine, the main component of the methanolic extracts of *Coptis japonica* rhizome and the bark of *Phellodendron chinense*, it was shown that it was able to reduce the MIC of aminoglycosides against *P. aeruginosa*, *Achromobacter xylosoxidans* and *Burkholderia cepacia,* besides other classes of antibiotics, such as cephalosporins, macrolides, and lincosamides against *P. aeruginosa*, all mediated by MexXY [150]. Curcumin, when encapsulated in nanoparticles and administered at a dose of 400 μg/mL, caused a half reduction in MIC of CIP (**7**, Figure 2) in clinical isolates of *P. aeruginosa* that were not sensitive to this antimicrobial when administered alone, in addition to decreasing expression of the genes *mexX* and *oprM*, which suggests that it can act in the modulation of the resistance via system MexXY-OprM [151].

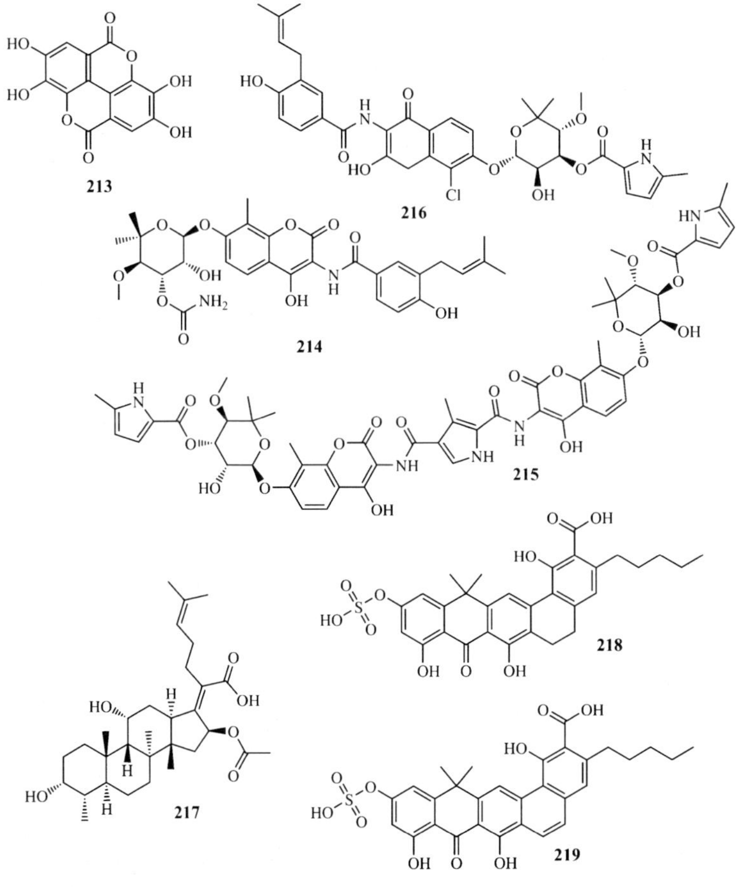

Figure 28. RND modulators of natural origin.

Synthetic RND Inhibitors

Mex-Opr systems have been related to resistance to fluoroquinolones in *P. aeruginosa*. In an assay using whole cells it was possible to identify the substance PAβN (**12**, Figure 2) at a dose of 45 μM as a reducer of the intrinsic resistance of the bacterium to LEV (**182**, Figure 23) in eight times as well as the acquired resistance due to expression of pumps of efflux 32-64 times for the same antibiotic, in addition to reducing the frequency with which new resistant bacteria appeared [152, 153]. From the rationalization that PAβN was a substrate of cathepsin C, it was proposed that other substances that are also substrates of this enzyme would be potential inhibitors of efflux pump, and PA4MβN (**220**, Figure 29) was identified as being able to reduce the MIC of LEV against *P. aeruginosa* in 16 times, in addition to preventing the appearance of quinolone-resistant mutants [154].

A series of arylpiperazines capable of reversing the multidrug resistance of *E. coli* overproducing acrAB and acrEF were selected from a library of *N*-heterocyclic organic compounds. Among them, the most potent was **221** (Figure 29) compound, which was able to decrease the MIC of LEV (**182**, Figure 23) 4-fold at the dose of 93 μM, and later demonstrated that it also exerted its effect in several clinical isolates of the bacterium [155, 156].

In a study with *Enterobacter aerogenes* strains (agent that causes respiratory tract infections in a hospital environment), clinical isolates with resistance induced by plasmid containing *marA* (a stimulant of the expression of the AcrAB-TolC system and porins) were assayed and the pyridoquinoline **222** (Figure 29) was identified as a competitive inhibitor of NOR (**9**, Figure 2) efflux, increasing its activity by 8 times in the concentration of 37 μM [157]. Another class of substances that proved to be inhibitory to the AcrAB-TolC system were the alkylaminoquinolines, especially compound **223** (Figure 29), capable of reducing the MIC of CLM (**18**, Figure 3) 16-fold at a dose of 200 μM, also against *E. aerogenes* [158]. In another study using clinical isolates of *E. aerogenes* and *Klebsiella pneumoniae*, alkoxyquinoline **224** (Figure 29) was identified as capable of reestablishing susceptibility to CLM, decreasing its MIC 8-fold in a dose of 500 μM and16-fold at 1 mM [159].

Another promising framework is indole, with **225** and **226** (Figure 29) substances being identified as capable of reducing MICs of CLM (**18**, Figure 3) (32 and 64 times), TET (**17**, Figure 3) (4 times), ERY (**16**, Figure 3) (2 and 16-fold), CIP (**7**, Figure 2) (16 and 8-fold), thiamphenicol (8-fold), and florfenicol (4-fold and 8-fold) in a clinical isolate of TolC overproducing *E. coli* [160]. Later another group reported that the framework served as a basis for substances capable of decreasing the MIC of several antibiotics against *E. coli* between 2-64 times, having been shown from experience with mutant bacteria that these compounds bind in the TolC portion of the efflux pump [161].

Based on the biological activity profile of 5-arylhydantoins, a series of 5,5-diphenylhydantoins were proposed as substances capable of influencing the effect of nalidixic acid on the AcrAB-TolC producing *E. aerogenes* strain. It was reported that none of the synthesized compounds had antibacterial activity and that the compounds containing the 2-methoxyphenylpiperazine group on the N^1-terminal fragment and methylcarboxyl acid at the N^3-position of hydantoin were most promising for future studies, especially compounds **227** and **228** (Figure 29), capable potentiate the effect of the antimicrobial 4-fold at the dose of 62 [162]. A later study within the class described the synthesis of a series of amine-alkyl derivatives, of which **229** and **230** (Figure 29) were able to reduce the MIC of all antibiotics tested in an AcrAB-TolC overproducing *E. aerogenes* strain at 63 μM [163].

Another synthetic compound capable of blocking AcrAB is pyranopyridine MBX2319 (**212**, Figure 27), which at a dose of 3 μM, has been shown to increase the susceptibility of *E. coli* to CIP (**7**, Figure 2) (twice), LEV (**182**, Figure 23) (four times), and piperacillin (eight times), and it was later demonstrated that it bound to the AcrB portion, preventing binding of the substrate [164, 165]. Later it was demonstrated cross-studies using X-ray crystallography and molecular dynamics simulations that this class of substances binds in a cage rich in phenylalanine in AcrB through hydrophobic interactions, in addition to the formation of a network of hydrogen bonds involving the protein and water molecules [166]. Tetrahydropyridines were also evaluated for their potential for inhibition of efflux pump in *E. coli*, and NUNL02 was identified as a specific inhibitor of AcrB, acting through competitive inhibition [167].

In the assay with a series of 14 2-substituted benzothiazoles, none of them had intrinsic antibacterial activity, but some were able to reverse resistance to CIP (**7**, Figure 2), drawing attention to the compound **231** (Figure 30), capable of reducing the MIC of CIP by 31 times. A docking study with this series of substances showed that they interacted with the phenylalanine-rich region of the AcrB protein, with more intensity than CIP, thus functioning as substrates instead [168].

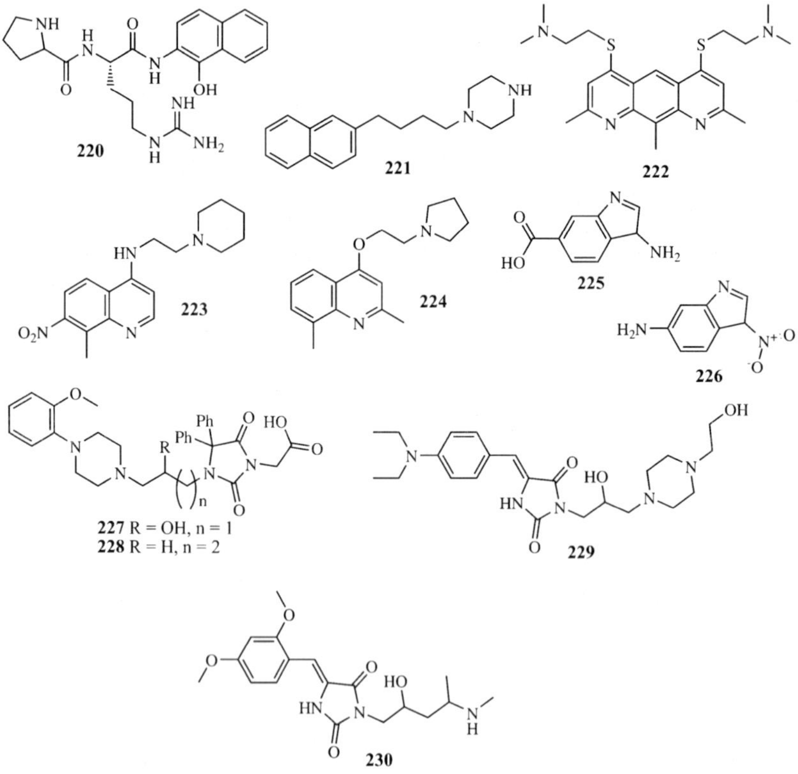

220 **221** **222** **223** **224** **225** **226**
227 R = OH, n = 1
228 R = H, n = 2
229 **230**

Figure 29. Synthetic RND modulators.

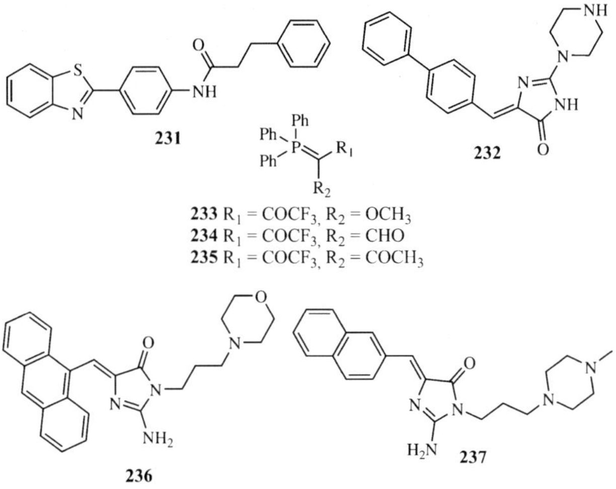

233 $R_1 = COCF_3$, $R_2 = OCH_3$
234 $R_1 = COCF_3$, $R_2 = CHO$
235 $R_1 = COCF_3$, $R_2 = COCH_3$

Figure 30. Synthetic RND modulators.

Piperazine arylideneimidazolones are another class of AcrAB-TolC inhibitors and **232** (Figure 30) biphenyl was shown to be able to double the sensitivity of *E. coli* to oxacillin and LIN (**15**, Figure 3) at the 25 μM dose and to decrease rifampicin MIC by 32-fold, being suggested that in addition to an efflux pump inhibitor, the substance acts as a permeabilizer of the outer membrane of the bacterium [169]. Phosphorous-containing compounds were also inhibitors of this system in *E. coli*, as is the case of phosphorus ylides, with **233**, **234**, and **235** (Figure 30) being the most important compounds capable of inhibiting AcrAB-TolC and influencing the expression of *acrA* and *acrB* genes [170]. Another series of compounds synthesized was the series of 5-arylideneimidazolones loading amine group at position 3, of which **236** (Figure 30) showed the ability to reduce the MIC of oxacillin versus an MRSA at a dose of 125 μM by 32 times, and compound **237** (Figure 30) showed the most potent duality of action, acting as an MRSA therapy adjuvant (16-fold increase in oxacillin activity at 125 μM) and *E. aerogenes* AcrAB-TolC inhibitor (80% reduction in activity at 250 μM) [171].

In a rational planning work, a series of 2-naphthamides directed against AcrB from *E. coli* were designed, and none of the compounds were identified as having antimicrobial activity, but most were able to inhibit

Nile Red efflux, and compound **238** (Figure 31) reduced the MIC of CLM (**18,** Figure 3), and ERY (**16**, Figure 3) at 1 mM against resistant strains up to a level of activity equal to a sensitive strain, with no effect on RIF (**99**, Figure 15) MIC, which is not a substrate of AcrB [172]. After this work, a series of 4-substituted 2-naphthamide derivatives of compound **238** were designed as specific AcrB ligands in *E. coli*, compound **239** (Figure 31) being identified as capable of eight-fold reduction of ERY MIC at the dose of 470 μM [173]. Another synthetic compound synthesized was the 4-substituted 2-naphthamide series, the most potent compounds being shown to be **240** (Figure 31) (decreased MIC of ERY and CLM 8-fold at 377 μM), **241** (Figure 31) (decrease twice in MIC ERY and CLM at 319 μM) and **242** (Figure 31) (2-fold decrease in MIC of CLM at 341 μM) [174]. Following this same line of frameworks a series of 5-methoxy-2,3-naphthalimide was synthesized, among which compound **243** (Figure 31) was shown to be the most potent AcrAB-TolC modulator, performing synergism with LEV (**182**, Figure 23), increasing its antimicrobial activity in 4 times at 23 μM and 16 times at 46 μM, in addition to abolishing the efflux of the Nile Red at 100 μM [175].

Another modulation strategy was based on the antisense technique, where the administration of the phosphorothioate oligodeoxynucleotide 831, directed against the *acrB* gene, caused a strong decrease in mRNA expression associated with this gene and consequently a reestablishment of sensitivity to fluoroquinolones in *E coli* [176].

A study with *E. coli* isolated from cases of bovine mastitis showed that the expression of AcrA and AcrB could be decreased during the administration of the NMP (**13**, Figure 2) and PA*β*N (**12**, Figure 2) efflux pump inhibitors, suggesting that these substances can be used in the mastitis therapy [7]. An attempt to understand the mechanism of action of PA*β*N showed that it was able to inhibit efflux of other molecules by attaching to the bottom of the distal cuff pocket (hydrophobic trap) and also through interaction at the top of the binding pocket [177].

Based on the structure of compound **244** (Figure 31), the first specific inhibitor of the MexAB-OprM system, a series of compounds was synthesized in order to improve its activity profile *in vivo*, which was not active in murine models, even though there was a reduction in the MIC of levofloxacin 8 times in the dose of 1.6 μM against *P. aeruginosa in vitro.*

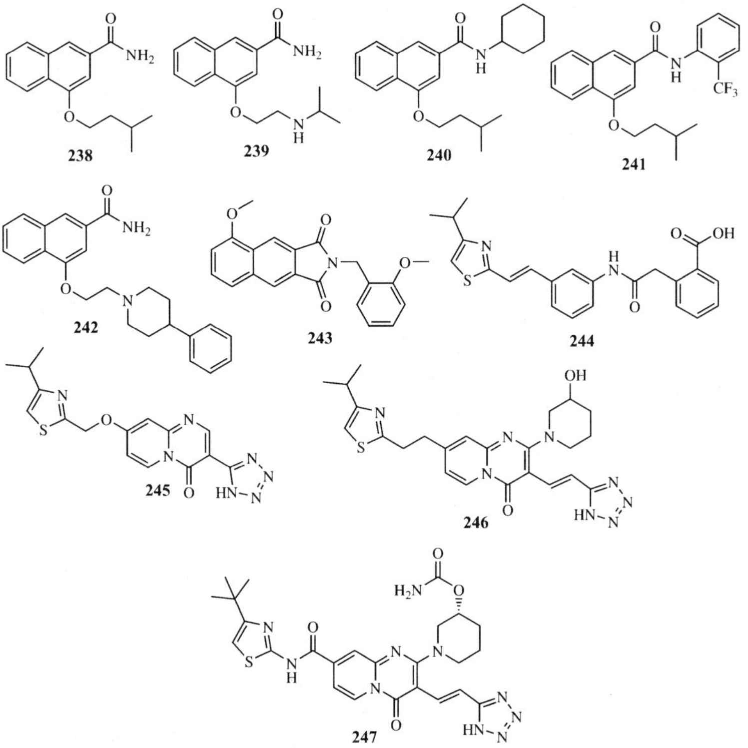

Figure 31. Synthetic RND modulators.

SAR studies showed that pyridine variants with ethers or ethylene thioethers in place of the styrene system had a promising activity profile, but were not active *in vivo*, whereas the incorporation of more polar groups in order to increase the solubility profile completely abolished the activity [178]. Further studies led to the synthesis of compound **245** (Figure 31), which is capable of increasing the sensitivity of *P. aeruginosa in vivo* to levofloxacin and sitafloxacin at a dose of 50 mg/kg, with the pyridopyrimidine framework being identified as important for this activity profile [179]. Proceeding with the project, it was described that the incorporation of a hydrophobic group at position 2 of the pyridopyrimidine framework increased activity and that the incorporated polar groups in order to improve the pharmacokinetic profile of the substance class did not cause a decrease in activity. In addition, it was shown that inclusion of an olefin spacer between the tetrazole and pyridopyrimidine nuclei also caused an increase in activity, drawing attention to compound **246** (Figure 31), capable of increasing the sensitivity to azithromycin 8-fold at the dose of 8 μM in the presence or absence of human serum albumin and levofloxacin 4 times at 1 μM and 8 times at 8 μM [180]. In order to solve the problem of the low stability of the compounds by photoisomerization, a new battery of compounds was synthesized and it was identified that the replacement of the ethylene bridge between the thiazole and pyridopyrimidine groups by an amide function solved this problem and led to an increase of the potency of the compounds, although the novel compounds have low solubility, with attention being drawn to compound **247** (Figure 31), capable of increasing the sensitivity to azithromycin 8-fold at the dose of 0.45 μM in the absence and at 4 μM in the presence of human serum albumin and levofloxacin in 4-fold at 0.45 μM and in 8-fold at 0.91 μM [181].

Another mode of substitution explored was through carbon-linked analogs at the 2-position, which provided compounds with good potency, especially when a cyclic substituent was introduced, attention being drawn to compound **248** (Figure 32), by charging a hydrophilic substituent and potentiating the activity of 8-fold levofloxacin at the dose of 8 μM absence of albumin and 33 μM in its presence and at 8-fold the activity of azithromycin at the dose of 4 μM [182]. Aromatic substituents at the 2-position where also evaluated, where derivatives containing benzylamine were identified as having a good solubility profile, with substance **249** (Figure 32) being identified, containing the morpholine substituent on the amino group as a potent and low toxicity substance, causing a potentiation of the activity of 8-fold levofloxacin in the dose of 7 μM in the absence of albumin and 28 μM in its presence and of azithromycin also in 8-fold but at the dose of 3 μM [183].

Finally, the research group reached a series of 4-oxo-4*H*-pyrido [1,2-a] pyrimidines, loading at the 2-position piperidines with quaternary ammonium salt side chains, electing compound D13-9001 (**250**, Figure 32) as a candidate for preclinical development, which had a potentiating effect of levofloxacin activity eight times at the dose of 2.5 μM when in the absence of albumin and 5 μM in its presence, in addition to increasing the potency of azithromycin in eight times at the dose 2.5 μM, with a solubility of 747 μg/mL at pH 6.8 [184]. It was also described that the inhibition of the MexAB-OprM system by this substance was able to reduce the invasive capacity of *P. aeruginosa*, being a second mechanism useful in the control of the infection by this microorganism [185]. A study using molecular dynamics simulations suggested that its effect comes from the high amount of energy that is required to make this molecule pass through the extrusion channel and its binding affinity with the distal binding pocket of the protein [186].

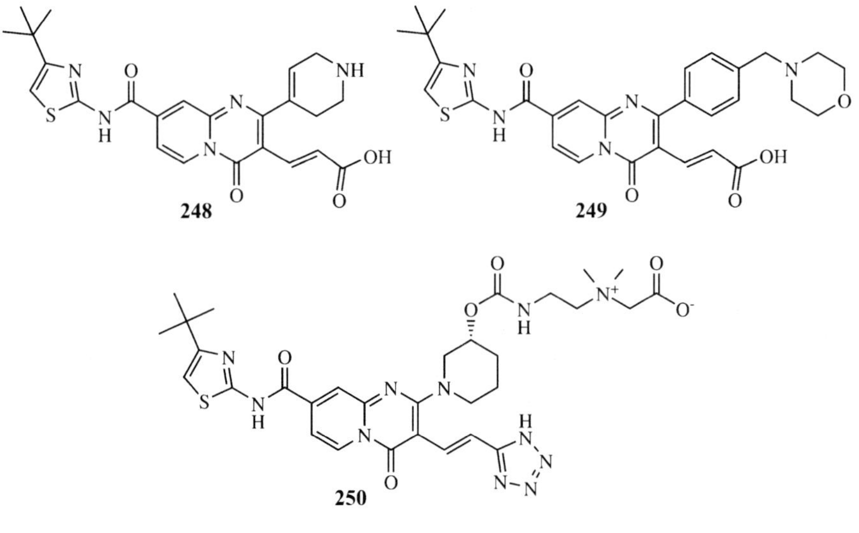

248 **249** **250**

Figure 32. Synthetic RND modulators.

Another strategy developed for resistance modulation in *P. aeruginosa* was the antisense strategy, by the administration of a phosphorothioate oligodeoxynucleotide capable of inhibiting the expression of OprM, leaving the system dysfunctional and reestablishing the sensitivity to antimicrobials [187].

Natural Products-Inspired Synthetic RND Inhibitors

Synthetic inhibitors based on natural products have also been proposed as pump modulators of the RND family, such as the indanone derived from gallic acid (**95**, Figure 14), capable of inhibiting the efflux of EtdBr and also ATPase, of extending the post-antibiotic effect of tetracycline and still being tolerated at a dose of 300 mg/kg in mice [188].

RND Inhibitors from Repositioning of Drugs and Other Alternatives

One of the strategies used was the search for compounds already used in therapeutics, but for other purposes. In an assay with several strains of *E. coli* producing the efflux pumps AcrAB, AcrEF, MdtEF and MexAB in an attempt to reposition SSRI as pump inhibitors of the RND family identified sertraline (**187**, Figure 24) as capable of performing synergism with tetracycline, oxacillin, linezolid, and clarithromycin, but induced the expression of *marA* and *acrB*, acting as an inducer of the expression of the pumps it inhibited, limiting its clinical utility [189]. Other antibiotics with which sertraline (**187**, Figure 24) performed synergism were ciprofloxacin, levofloxacin, norfloxacin, moxifloxacin, cefixime, and gentamicin, and it was shown to have an intrinsic antibacterial activity against *E. coli* and *P. aeruginosa* [120].

A study with *Salmonella enterica* described that phenothiazines, such as chlorpromazine (**199**, Figure 26), were not directly inhibitory to the pump, but they were involved in the expression of genes, including *acrB*, thus presenting synergism with antimicrobials [190]. The phenothiazines chlorpromazine (**199**, Figure 26), thiethylperazine (**251**, Figure33) and promethazine (**252**, Figure 33) were also able to perform synergism with vancomycin, reducing its MIC against *Enterococcus faecalis* and *E. faecium* in 24-32 fold [191]. In a study with phenothiazine derivatives, *N*-hydroxyalkyl-2-aminophenotiazines, the question arose whether there would be a difference in activity between pairs of enantiomers. In an assay where the inhibitory capacity of AcrAB-TolC from *E. coli* was assessed by the efflux of EtdBr, it was identified that compound **253** (Figure 33) was the most active, both enantiomers having equal potency each isolated and equal to the racemic [192].

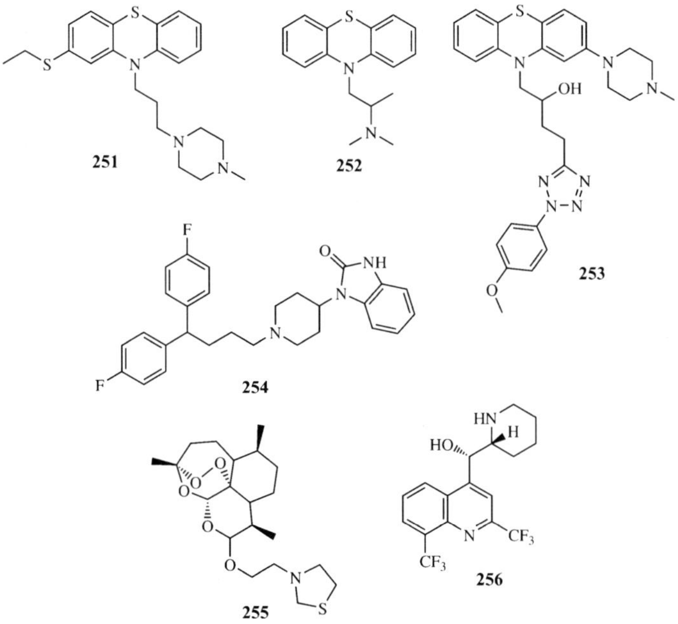

Figure 33. Reposition drugs against RND.

Another psychotropic substance capable of modulating the AcrAB-TolC system is the neuroleptic pimozide (**254**, Figure 33). It was described that at 100 μM it was able to completely inhibit the system, but unable to decrease the MIC of the antibiotic series tested, being suggested that the activity of the efflux pump inhibitors could be highly selective for certain substrates [193]. Within this perspective of repositioning, a docking study was carried out with substances for cancer treatment where the substances paclitaxel and vinblastine were identified as potential inhibitors of AcrAB-TolC from *E. coli* [194].

Another substance capable of modulating this system is artesunate (**198**, Figure 25) which has no antibacterial activity but has been shown to decrease the synthesis of mRNA related to the production of AcrAB-TolC and to increase the sensitivity of *E. coli* to a series of *β*-lactam antibiotics [195]. Based on the structures of artesunate and dihydroartemisinin 7, 86 substances were designed via molecular docking and 3D-SAR. **255** (Figure 33) was identified as the best compound, potentiating the effect of ampicillin against standard strains and clinical isolates of *E. coli* [196]. Another antimalarial that demonstrated the ability to inhibit AcrAB-TolC was mefloquine (**256**, Figure 33), capable of restoring 95% sensitivity of *E. coli* to levofloxacin [197].

A substance without antimicrobial activity but which may be useful in modulating the resistance is EDTA (**4**, Figure 1), and it has been shown that it, at a dose of 10 mM was able to increase the sensitivity of *P. aeruginosa* to a series of antimicrobials, as well as to reduce in three times the expression of the genes *mexA* and *mexB* [198]. It was also reported that catharanthine, isolated from *Catharanthus roseous*, had a good binding affinity with the MexAB-OprM system *in silico* evaluation, later validated observation *in vitro*, demonstrating its ability to reduce MICs of tetracycline and streptomycin in 16 and 8 times, respectively, in addition to reducing the efflux of EtdBr, using as model *P. aeruginosa* P01 [199]. Another technique that can be used to reduce the efflux of the antimicrobial is the structural modification of the substance, as in the case of polymyxin, where the insertion of a second lipid chain caused it to be no longer recognized by the pump efflux, maintaining its activity against *P. aeruginosa* [200].

In vivo study of a *Galleria mellonella* larva infection model demonstrated the feasibility of trimethoprim, sertraline (**187**, Figure 24) and PA*β*N inhibitors in the control of *P. aeruginosa* infections expressing MexAB-OprM, MexCD-OprJ, and MexEF-OprN [201].

In the field of microbial growth control in the environment, the effect of ultraviolet (UV) radiation was studied as a sensitizer to the action of antimicrobials, and it was described that *P. aeruginosa* bacteria that survived the treatment by UV radiation had a decreased expression of the efflux pumps MexAB-OprM, MexCD-OprJ, and MexXY-OprM, making them more susceptible to antimicrobials such as norfloxacin, carbenicillin, rifampicin, vancomycin, and chloromycetin [202].

An alternative to avoiding the antimicrobial efflux by the MexAB-OprM system was the binding of the antimicrobial molecule to silver nanoparticles, and it has been described that a high charge of ofloxacin per particle is associated with a greater reduction in antimicrobial MIC [203].

ABC SUPERFAMILY

ABC superfamily is a family of efflux pumps that could also be found in other cell types different from bacterial ones. In a large-scale assay with microbial fermentation extracts, the fungal metabolites enniatins and beauvericin were identified as potent and ubiquitous inhibitors of transporters of this family [149]. Milbemycins, especially milbemycin $\alpha 9$ (**256**, Figure 34), were also potent inhibitors of the CDR1 protein of *Candida albicans*, increasing the sensitivity of the fungus to fluconazole in several clinical isolates [149]. Another substance capable of reducing the expression of this protein is plagiochin E, isolated from *Marchantia polymorpha*, capable of increasing the sensitivity of *C. albicans* to fluconazole four times [204].

A study with the diosmin and diosmetin flavonoids showed that they were able to inhibit the enzyme pyruvate kinase of *S. aureus* overexpressing MsrA pump, inducing an ATP deficiency necessary for the proper functioning and performing synergism indirectly with erythromycin [205]. Another substance that was able to modulate the activity of MsrA was vitamin K3 (**257**, Figure 34), decreasing the MIC of the EtdBr in 16 times in the dose of 46 μM [206]. Diterpenes 4, 2 and 3, isolated from *C. cervicornis*, were also able to block MsrA, increasing the sensitivity of S. aureus to erythromycin 16 fold [35].

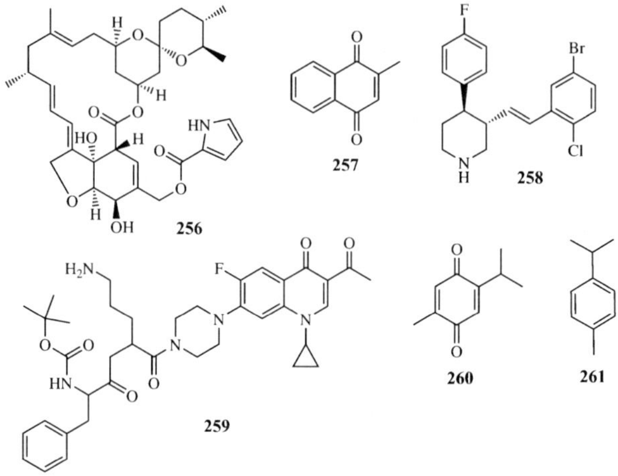

Figure 34. Resistance modulators.

MATE SUPERFAMILY

In a study with SSRI phenylpiperidines, substances synthesized on the basis of SSRI that had already been shown to inhibit NorA identified compound **258** (Figure 34) as capable of reducing the MIC of 32-fold EtdBr at the dose of 3 μM against MepA-producing *S. aureus* [119]. Another strategy used in the development of MepA inhibitors was to take as inspiration the structure of the substrate, in this case, fluoroquinolones. Compound **259** (Figure 34) was identified as capable of reducing in 84.1% the efflux of EtdBr at the dose of 10 μM [71].

In the area of natural products, the substances thymomoquine (**260**) and p-cymene (**261**, Figure 34), present in the essential oil of *Nigella sativa*, showed potential to modulate the expression of the *mepA* gene, where the first substance caused a 16-fold increase in tetracycline activity and in 8 times in the ciprofloxacin activity and the second caused a twofold increase in the activity of ciprofloxacin [207].

FINAL CONSIDERATIONS

Antimicrobial efflux inhibition is an interesting strategy to overcome the growing issues with microbial susceptibility. The administration of these substances could improve the activity of diverse antibiotic against resistant bacteria and fungi, even restoring completely the susceptibility of microorganisms like MRSA. It is also important that as they do not have intrinsic antimicrobial, it is more difficult to microorganisms to develop resistance to these substances.

The vast majority of studies published in this area only make a preliminary *in vitro* evaluation of the substances, being required greater investment in *in vivo* tests to detect problems related with possible toxicity of them and pave the way for the development of drugs that may serve as a therapeutic alternative to antibiotic-resistant bacteria. Finally, there is still place to the development of novel substances, especially against ABC, MATE, and SMR superfamilies, which are poorly studied.

REFERENCES

[1] Christena LR, Subramaniam S, Vidhyalakshmi M, Mahadevan V, Sivasubramanian A, Nagarajan S. Dual role of pinostrobin-a flavonoid nutraceutical as an efflux pump inhibitor and antibiofilm agent to mitigate food borne pathogens. *RSC Adv* 2015;5:61881–7. doi:10.1039/C5RA07165H.

[2] Machado D, Cannalire R, Santos Costa S, Manfroni G, Tabarrini O, Cecchetti V, et al. Boosting Effect of 2-Phenylquinoline Efflux Inhibitors in Combination with Macrolides against Mycobacterium smegmatis and Mycobacterium avium. *ACS Infect Dis* 2015;1:593–603. doi:10.1021/acsinfecdis.5b00052.

[3] Abriouel H, Lavilla Lerma L, Pérez Montoro B, Alonso E, Knapp CW, Caballero Gómez N, et al. Efficacy of "HLE"—a multidrug efflux-pump inhibitor—as a disinfectant against surface bacteria. *Environ Res* 2018;165:133–9. doi:10.1016/j.envres.2018.04.020.

[4] Munita JM, Arias CA. Mechanisms of Antibiotic Resistance. *Virulence Mech. Bact. Pathog.* Fifth Ed., vol. 4, American Society of Microbiology; 2016, p. 481–511. doi:10.1128/microbiolspec.VMBF-0016-2015.

[5] Reddy VS, Shlykov MA, Castillo R, Sun EI, Saier MH. The major facilitator superfamily (MFS) revisited. *FEBS J* 2012;279:2022–35. doi:10.1111/j.1742-4658.2012.08588.x.

[6] Riordan JT, Dupre JM, Cantore-Matyi SA, Kumar-Singh A, Song Y, Zaman S, et al. Alterations in the transcriptome and antibiotic susceptibility of Staphylococcus aureus grown in the presence of diclofenac. *Ann Clin Microbiol Antimicrob* 2011;10:30. doi:10.1186/1476-0711-10-30.

[7] Ospina Barrero MA, Pietralonga PAG, Schwarz DGG, Silva Junior A, Paula SO, Moreira MAS. Effect of the inhibitors phenylalanine arginyl ß-naphthylamide (PAßN) and 1-(1-naphthylmethyl)-piperazine (NMP) on expression of genes in multidrug efflux systems of Escherichia coli isolates from bovine mastitis. *Res Vet Sci* 2014;97:176–81. doi:10.1016/j.rvsc.2014.05.013.

[8] Bruns MM, Kakarla P, Floyd JT, Mukherjee MM, Ponce RC, Garcia JA, et al. Modulation of the multidrug efflux pump EmrD-3 from Vibrio cholerae by Allium sativum extract and the bioactive agent allyl sulfide plus synergistic enhancement of antimicrobial susceptibility by A. sativum extract. *Arch Microbiol* 2017;199:1103–12. doi:10.1007/s00203-017-1378-x.

[9] Kaatz GW, Seo SM, Ruble CA. Mechanisms of fluoroquinolone resistance in staphylococcus aureus. *J Infect Dis* 1991;163:1080–6. doi:10.1093/infdis/163.5.1080.

[10] Neyfakh AA, Bidnenko VE, Chen LB. Efflux-mediated multidrug resistance in Bacillus subtilis: similarities and dissimilarities with the mammalian system. *Proc Natl Acad Sci* 1991;88:4781–5. doi:10.2307/2357137.

[11] Neyfakh AA. The multidrug efflux transporter of Bacillus subtilis is a structural and functional homolog of the Staphylococcus norA

protein. *Antimicrob Agents Chemother* 1992;36:484–5. doi:10.1128/AAC.36.2.484.

[12] Neyfakh AA, Borsch CM, Kaatz GW. Fluoroquinolone resistance protein NorA of Staphylococcus aureus is a multidrug efflux transporter. *Antimicrob Agents Chemother* 1993;37:128–9. doi:10.1128/AAC.37.1.128.

[13] Markham PN, Neyfakh AA. Inhibition of the multidrug transporter NorA prevents emergence of norfloxacin resistance in Staphylococcus aureus. *Antimicrob Agents Chemother* 1996;40:2673–4. doi:10.1128/AAC.40.11.2673.

[14] Kaatz GW, Seo SM. Inducible NorA-mediated multidrug resistance in Staphylococcus aureus. *Antimicrob Agents Chemother* 1995;39:2650–5. doi:10.1128/AAC.39.12.2650.Updated.

[15] Schmitz FJ, Fluit AC, Luckefahr M, Engler B, Hofmann B, Verhoef J, et al. The effect of reserpine, an inhibitor of multidrug efflux pumps, on the in-vitro activities of ciprofloxacin, sparfloxacin and moxifloxacin against clinical isolates of Staphylococcus aureus. *J Antimicrob Chemother* 1998;42:807–10. doi:10.1093/jac/42.6.807.

[16] Stermitz FR, Lorenz P, Tawara JN, Zenewicz LA, Lewis K. Synergy in a medicinal plant: Antimicrobial action of berberine potentiated by 5'-methoxyhydnocarpin, a multidrug pump inhibitor. *Proc Natl Acad* Sci 2000;97:1433–7. doi:10.1073/pnas.030540597.

[17] Guz NR, Stermitz FR, Johnson JB, Beeson TD, Willen S, Hsiang J-FF, et al. Flavonolignan and Flavone Inhibitors of a Staphylococcus a ureus Multidrug Resistance Pump: Structure−Activity Relationships. *J Med Chem* 2001;44:261–8. doi:10.1021/jm0004190.

[18] Wang D, Xie K, Zou D, Meng M, Xie M. Inhibitory effects of silybin on the efflux pump of methicillin-resistant Staphylococcus aureus. *Mol Med Rep* 2018;18:827–33. doi:10.3892/mmr.2018.9021.

[19] Belofsky G, Percivill D, Lewis K, Tegos GP, Ekart J. Phenolic Metabolites of Dalea v ersicolor that Enhance Antibiotic Activity against Model Pathogenic Bacteria. *J Nat Prod* 2004;67:481–4. doi:10.1021/np030409c.

[20] Belofsky G, Carreno R, Lewis K, Ball A, Casadei G, Tegos GP. Metabolites of the "Smoke Tree", Dalea spinosa, Potentiate Antibiotic Activity against Multidrug-Resistant Staphylococcus aureus. *J Nat Prod* 2006;69:261–4. doi:10.1021/np058057s.

[21] Chan BCL, Ip M, Lau CBS, Lui SL, Jolivalt C, Ganem-Elbaz C, et al. Synergistic effects of baicalein with ciprofloxacin against NorA over-expressed methicillin-resistant Staphylococcus aureus (MRSA) and inhibition of MRSA pyruvate kinase. *J Ethnopharmacol* 2011;137:767–73. doi:10.1016/j.jep.2011.06.039.

[22] Oluwatuyi M, Kaatz G, Gibbons S. Antibacterial and resistance modifying activity of Rosmarinus officinalis. *Phytochemistry* 2004;65:3249–54. doi:10.1016/j.phytochem.2004.10.009.

[23] Stermitz FR, Tawara-matsuda J, Lorenz P, Mueller P, Zenewicz L, Lewis K. 5'-Methoxyhydnocarpin-D and pheophorbide A: Berberis species components that potentiate berberine growth inhibition of resistant Staphlococcus aureus. *J Nat Prod* 2000;63:1146–9. doi:10.1007/978-3-540-72816-0_1455.

[24] Junio HA, Sy-Cordero AA, Ettefagh KA, Burns JT, Micko KT, Graf TN, et al. Synergy-Directed Fractionation of Botanical Medicines: A Case Study with Goldenseal (Hydrastis canadensis). *J Nat Prod* 2011;74:1621–9. doi:10.1021/np200336g.

[25] Holler JG, Christensen SB, Slotved H-C, Rasmussen HB, Guzman A, Olsen C-E, et al. Novel inhibitory activity of the Staphylococcus aureus NorA efflux pump by a kaempferol rhamnoside isolated from Persea lingue Nees. *J Antimicrob Chemother* 2012;67:1138–44. doi:10.1093/jac/dks005.

[26] Qin R, Xiao K, Li B, Jiang W, Peng W, Zheng J, et al. The Combination of Catechin and Epicatechin Gallate from Fructus Crataegi Potentiates β-Lactam Antibiotics Against Methicillin-Resistant Staphylococcus aureus (MRSA) *in Vitro* and *in Vivo*. *Int J Mol Sci* 2013;14:1802–21. doi:10.3390/ijms14011802.

[27] Marquez B, Neuville L, Moreau NJ, Genet J-P, dos Santos AF, Caño de Andrade MC, et al. Multidrug resistance reversal agent from

Jatropha elliptica. *Phytochemistry* 2005;66:1804–11. doi:10.1016/j.phytochem.2005.06.008.

[28] Michalet S, Cartier G, David B, Mariotte A-M, Dijoux-franca M-G, Kaatz GW, et al. N-Caffeoylphenalkylamide derivatives as bacterial efflux pump inhibitors. *Bioorg Med Chem Lett* 2007;17:1755–8. doi:10.1016/j.bmcl.2006.12.059.

[29] Pereda-Miranda R, Kaatz GW, Gibbons S. Polyacylated Oligosaccharides from Medicinal Mexican Morning Glory Species as Antibacterials and Inhibitors of Multidrug Resistance in Staphylococcus aureus ⊥. *J Nat Prod* 2006;69:406–9. doi:10.1021/np050227d.

[30] Chérigo L, Pereda-Miranda R, Fragoso-Serrano M, Jacobo-Herrera N, Kaatz GW, Gibbons S. Inhibitors of Bacterial Multidrug Efflux Pumps from the Resin Glycosides of Ipomoea murucoides ⊥. *J Nat Prod* 2008;71:1037–45. doi:10.1021/np800148w.

[31] Chérigo L, Pereda-Miranda R, Gibbons S. Bacterial resistance modifying tetrasaccharide agents from Ipomoea murucoides. *Phytochemistry* 2009;70:222–7. doi:10.1016/j.phytochem.2008.12.005.

[32] Smith ECJ, Williamson EM, Wareham N, Kaatz GW, Gibbons S. Antibacterials and modulators of bacterial resistance from the immature cones of Chamaecyparis lawsoniana. *Phytochemistry* 2007;68:210–7. doi:10.1016/j.phytochem.2006.10.001.

[33] Smith ECJ, Kaatz GW, Seo SM, Wareham N, Williamson EM, Gibbons S. The Phenolic Diterpene Totarol Inhibits Multidrug Efflux Pump Activity in Staphylococcus aureus. Antimicrob Agents *Chemother* 2007;51:4480–3. doi:10.1128/AAC.00216-07.

[34] Gupta VK, Tiwari N, Gupta P, Verma S, Pal A, Srivastava SK, et al. A clerodane diterpene from Polyalthia longifolia as a modifying agent of the resistance of methicillin resistant Staphylococcus aureus. *Phytomedicine* 2016;23:654–61. doi:10.1016/j.phymed.2016.03.001.

[35] de Figueiredo CS, Menezes Silva SMP de, Abreu LS, da Silva EF, da Silva MS, Cavalcanti de Miranda GE, et al. Dolastane diterpenes

from Canistrocarpus cervicornis and their effects in modulation of drug resistance in Staphylococcus aureus. *Nat Prod Res* 2018:1–9. doi:10.1080/14786419.2018.1470512.

[36] Cégiéla-Carlioz P, Bessière J-M, David B, Mariotte A-M, Gibbons S, Dijoux-Franca MG. Modulation of multi-drug resistance (MDR) inStaphylococcus aureus by Osha (Ligusticum porteri L., Apiaceae) essential oil compounds. *Flavour Fragr J* 2005;20:671–5. doi:10.1002/ffj.1584.

[37] Urmila, Jandaik S, Mehta J, Mohan M. Synergism Between Natural Products and Fluoroquinolones Against Staphylococcus aureus STRAINS. *Int J Biol Pharm* Allied Sci 2016;5:1965–76.

[38] Nouran HA, Yasser MI, Ahmed MA, Magdy AA. The effect of clove extract on the minimum inhibitory concentration of ciprofloxacin in fluoroquinolone resistant clinical isolates of Staphylococcus aureus. A*frican J Microbiol Res* 2012;6:1306–11. doi:10.5897/AJMR11.1588.

[39] Oyedemi SO, Oyedemi BO, Coopoosamy RM, Prieto JM, Stapleton P, Gibbons S. Antibacterial and norfloxacin potentiation activities of Ocimum americanum L. against methicillin resistant Staphylococcus aureus. *South African J Bot* 2017;109:308–14. doi:10.1016/j.sajb.2016.12.025.

[40] Coutinho HDM, Matias EFF, Santos KKA, Tintino SR, Souza CES, Guedes GMM, et al. Enhancement of the Norfloxacin Antibiotic Activity by Gaseous Contact with the Essential Oil of Croton zehntneri. *J Young Pharm* 2010;2:362–4. doi:10.4103/0975-1483.71625.

[41] de Medeiros VM, do Nascimento YM, Souto AL, Madeiro SAL, Costa VC de O, Silva SMPM, et al. Chemical composition and modulation of bacterial drug resistance of the essential oil from leaves of Croton grewioides. *Microb Pathog* 2017;111:468–71. doi:10.1016/j.micpath.2017.09.034.

[42] Coêlho ML, Ferreira JHL, de Siqueira Júnior JP, Kaatz GW, Barreto HM, de Carvalho Melo Cavalcante AA. Inhibition of the NorA

multi-drug transporter by oxygenated monoterpenes. *Microb Pathog* 2016;99:173–7. doi:10.1016/j.micpath.2016.08.026.

[43] de Morais Oliveira-Tintino CD, Tintino SR, Limaverde PW, Figueredo FG, Campina FF, da Cunha FAB, et al. Inhibition of the essential oil from Chenopodium ambrosioides L. and α-terpinene on the NorA efflux-pump of Staphylococcus aureus. *Food Chem* 2018;262:72–7. doi:10.1016/j.foodchem.2018.04.040.

[44] Falcão-Silva VS, Silva DA, Souza M de F V., Siqueira-Junior JP. Modulation of drug resistance in staphylococcus aureus by a kaempferol glycoside from herissantia tiubae (malvaceae). *Phyther Res* 2009;23:1367–70. doi:10.1002/ptr.2695.

[45] Ponnusamy K, Ramasamy M, Savarimuthu I, Paulraj MG. Indirubin potentiates ciprofloxacin activity in the NorA efflux pump of Staphylococcus aureus. *Scand J Infect Dis* 2010;42:500–5. doi:10.3109/00365541003713630.

[46] Maia GL de A, Falcão-Silva V dos S, Aquino PGV, Araújo-Júnior JX de, Tavares JF, Silva MS da, et al. Flavonoids from Praxelis clematidea R. M. King and Robinson Modulate Bacterial Drug Resistance. *Molecules* 2011;16:4828–35. doi:10.3390/molecules16064828.

[47] Kalia NP, Mahajan P, Mehra R, Nargotra A, Sharma JP, Koul S, et al. Capsaicin, a novel inhibitor of the NorA efflux pump, reduces the intracellular invasion of Staphylococcus aureus. *J Antimicrob Chemother* 2012;67:2401–8. doi:10.1093/jac/dks232.

[48] Roy SK, Kumari N, Pahwa S, Agrahari UC, Bhutani KK, Jachak SM, et al. NorA efflux pump inhibitory activity of coumarins from Mesua ferrea. *Fitoterapia* 2013;90:140–50. doi:10.1016/j.fitote.2013.07.015.

[49] Shiu WKP, Malkinson JP, Rahman MM, Curry J, Stapleton P, Gunaratnam M, et al. A new plant-derived antibacterial is an inhibitor of efflux pumps in Staphylococcus aureus. *Int J Antimicrob Agents* 2013;42:513–8. doi:10.1016/j.ijantimicag.2013.08.007.

[50] Farooq S, Wahab A-T-, Fozing CDA, Rahman A-U-, Choudhary MI. Artonin I inhibits multidrug resistance in Staphylococcus aureus and

potentiates the action of inactive antibiotics *in vitro*. *J Appl Microbiol* 2014;117:996–1011. doi:10.1111/jam.12595.

[51] Diniz-Silva HT, Cirino IC da S, Falcão-Silva V dos S, Magnani M, de Souza EL, Siqueira-Júnior JP. Tannic Acid as a Potential Modulator of Norfloxacin Resistance in Staphylococcus Aureus Overexpressing norA. *Chemotherapy* 2016;61:319–22. doi:10.1159/000443495.

[52] Tintino SR, Oliveira-Tintino CDM, Campina FF, Silva RLP, Costa M do S, Menezes IRA, et al. Evaluation of the tannic acid inhibitory effect against the NorA efflux pump of Staphylococcus aureus. *Microb Pathog* 2016;97:9–13. doi:10.1016/j.micpath.2016.04.003.

[53] Randhawa HK, Hundal KK, Ahirrao PN, Jachak SM, Nandanwar HS. Efflux pump inhibitory activity of flavonoids isolated from Alpinia calcarata against methicillin-resistant Staphylococcus aureus. *Biologia* (Bratisl) 2016;71. doi:10.1515/biolog-2016-0073.

[54] Singh S, Kalia NP, Joshi P, Kumar A, Sharma PR, Kumar A, et al. Boeravinone B, A Novel Dual Inhibitor of NorA Bacterial Efflux Pump of Staphylococcus aureus and Human P-Glycoprotein, Reduces the Biofilm Formation and Intracellular Invasion of Bacteria. *Front Microbiol* 2017;8. doi:10.3389/fmicb.2017.01868.

[55] Salaheen S, Peng M, Joo J, Teramoto H, Biswas D. Eradication and Sensitization of Methicillin Resistant Staphylococcus aureus to Methicillin with Bioactive Extracts of Berry Pomace. *Front Microbiol* 2017;8:253. doi:10.3389/fmicb.2017.00253.

[56] dos Santos JFS, Tintino SR, de Freitas TS, Campina FF, de A. Menezes IR, Siqueira-Júnior JP, et al. *In vitro* e in silico evaluation of the inhibition of Staphylococcus aureus efflux pumps by caffeic and gallic acid. *Comp Immunol Microbiol Infect Dis* 2018;57:22–8. doi:10.1016/j.cimid.2018.03.001.

[57] Holler JG, Slotved H-C, Mølgaard P, Olsen CE, Christensen SB. Chalcone inhibitors of the NorA efflux pump in Staphylococcus aureus whole cells and enriched everted membrane vesicles. *Bioorg Med Chem* 2012;20:4514–21. doi:10.1016/j.bmc.2012.05.025.

[58] Thai K-M, Ngo T-D, Phan T-V, Tran T-D, Nguyen N-V, Nguyen T-H, et al. Virtual Screening for Novel Staphylococcus Aureus NorA Efflux Pump Inhibitors From Natural Products. *Med Chem* (Los Angeles) 2015;11:135–55. doi:10.2174/1573406410666140902110903.

[59] Joshi P, Singh S, Wani A, Sharma S, Jain SK, Singh B, et al. Osthol and curcumin as inhibitors of human Pgp and multidrug efflux pumps of Staphylococcus aureus: reversing the resistance against frontline antibacterial drugs. *Med Chem Commun* 2014;5:1540–7. doi:10.1039/C4MD00196F.

[60] Jaberi S, Fallah F, Hashemi A, Karimi AM, Azimi L. Inhibitory effects of curcumin on the expression of NorA efflux pump and reduce antibiotic resistance in staphylococcus aureus. *J Pure Appl Microbiol* 2018;12:95–102. doi:10.22207/JPAM.12.1.12.

[61] Sharma S, Kumar M, Sharma S, Nargotra A, Koul S, Khan IA. Piperine as an inhibitor of Rv1258c, a putative multidrug efflux pump of Mycobacterium tuberculosis. *J Antimicrob Chemother* 2010;65:1694–701. doi:10.1093/jac/dkq186.

[62] Tintino SR, Morais-Tintino CD, Campina FF, Pereira RL, Costa M do S, Braga MFBM, et al. Action of cholecalciferol and alpha-tocopherol on Staphylococcus aureus efflux pumps. *EXCLI J* 2016;15:315–22. doi:10.17179/excli2016-277.

[63] Markham PN, Westhaus E, Klyachko K, Johnson ME, Neyfakh AA. Multiple novel inhibitors of the NorA multidrug transporter of Staphylococcus aureus. *Antimicrob Agents Chemother* 1999;43: 2404–8.

[64] Ambrus JI, Kelso MJ, Bremner JB, Ball AR, Casadei G, Lewis K. Structure–activity relationships of 2-aryl-1H-indole inhibitors of the NorA efflux pump in Staphylococcus aureus. *Bioorg Med Chem Lett* 2008;18:4294–7. doi:10.1016/j.bmcl.2008.06.093.

[65] Dai Y, Zhang X, Zhang X, Wang H, Lu Z. DFT and GA Studies on the QSAR of 2-aryl-5-nitro-1H-indole derivatives as NorA Efflux Pump Inhibitors. *J Mol Model* 2008;14:807–12. doi:10.1007/s00894-008-0328-6.

[66] Hequet A, Burchak ON, Jeanty M, Guinchard X, Le Pihive E, Maigre L, et al. 1-(1 H -Indol-3-yl)ethanamine Derivatives as Potent Staphylococcus aureus NorA Efflux Pump Inhibitors. *Chem Med Chem* 2014;9:1534–45. doi:10.1002/cmdc.201400042.

[67] Lepri S, Buonerba F, Goracci L, Velilla I, Ruzziconi R, Schindler BD, et al. Indole Based Weapons to Fight Antibiotic Resistance: A Structure–Activity Relationship Study. *J Med Chem* 2016;59:867–91. doi:10.1021/acs.jmedchem.5b01219.

[68] Buonerba F, Lepri S, Goracci L, Schindler BD, Seo SM, Kaatz GW, et al. Improved Potency of Indole-Based NorA Efflux Pump Inhibitors: From Serendipity toward Rational Design and Development. *J Med Chem* 2017;60:517–23. doi:10.1021/acs.jmed chem.6b01281.

[69] Vidaillac C, Guillon J, Moreau S, Arpin C, Lagardère A, Larrouture S, et al. Synthesis of new 4-[2-(alkylamino)ethylthio]pyrrolo[1,2- a]quinoxaline and 5-[2-(alkylamino)ethylthio]pyrrolo[1,2- a]thieno[3,2- e]pyrazine derivatives, as potential bacterial multidrug resistance pump inhibitors. *J Enzyme Inhib Med Chem* 2007;22:620–31. doi:10.1080/14756360701485406.

[70] Samosorn S, Bremner JB, Ball A, Lewis K. Synthesis of functionalised 2-aryl-5-nitro-1H-indoles and their activity as bacterial NorA efflux pump inhibitors. *Bioorg Med Chem* 2006;14:857–65. doi:10.1016/j.bmc.2005.09.019.

[71] German N, Wei P, Kaatz GW, Kerns RJ. Synthesis and evaluation of fluoroquinolone derivatives as substrate-based inhibitors of bacterial efflux pumps. *Eur J Med Chem* 2008;43:2453–63. doi:10.1016/ j.ejmech.2008.01.042.

[72] Fournier dit Chabert J, Marquez B, Neville L, Joucla L, Broussous S, Bouhours P, et al. Synthesis and evaluation of new arylbenzo[b]thiophene and diarylthiophene derivatives as inhibitors of the NorA multidrug transporter of Staphylococcus aureus. *Bioorg Med Chem* 2007;15:4482–97. doi:10.1016/j.bmc.2007.04.023.

[73] Liger F, Bouhours P, Ganem-Elbaz C, Jolivalt C, Pellet-Rostaing S, Popowycz F, et al. C2 Arylated Benzo[b]thiophene Derivatives as

Staphylococcus aureus NorA Efflux Pump Inhibitors. *Chem Med Chem* 2016;11:320–30. doi:10.1002/cmdc.201500463.

[74] Dantas N, de Aquino TM, de Araújo-Júnior JX, da Silva-Júnior E, Gomes EA, Gomes AAS, et al. Aminoguanidine hydrazones (AGH's) as modulators of norfloxacin resistance in Staphylococcus aureus that overexpress NorA efflux pump. *Chem Biol Interact* 2018;280:8–14. doi:10.1016/j.cbi.2017.12.009.

[75] Brincat JP, Carosati E, Sabatini S, Manfroni G, Fravolini A, Raygada JL, et al. Discovery of Novel Inhibitors of the NorA Multidrug Transporter of Staphylococcus aureus. *J Med Chem* 2011;54:354–65. doi:10.1021/jm1011963.

[76] Samosorn S, Tanwirat B, Muhamad N, Casadei G, Tomkiewicz D, Lewis K, et al. Antibacterial activity of berberine-NorA pump inhibitor hybrids with a methylene ether linking group. *Bioorg Med Chem* 2009;17:3866–72. doi:10.1016/j.bmc.2009.04.028.

[77] Tomkiewicz D, Casadei G, Larkins-Ford J, Moy TI, Garner J, Bremner JB, et al. Berberine-INF55 (5-Nitro-2-Phenylindole) Hybrid Antimicrobials: Effects of Varying the Relative Orientation of the Berberine and INF55 Components. *Antimicrob Agents Chemother* 2010;54:3219–24. doi:10.1128/AAC.01715-09.

[78] Dolla NK, Chen C, Larkins-Ford J, Rajamuthiah R, Jagadeesan S, Conery AL, et al. On the Mechanism of Berberine–INF55 (5-Nitro-2-phenylindole) Hybrid Antibacterials. *Aust J Chem* 2014;67:1471. doi:10.1071/CH14426.

[79] Kapkoti DS, Gupta VK, Darokar MP, Bhakuni RS. Glabridin-chalcone hybrid molecules: drug resistance reversal agent against clinical isolates of methicillin-resistant Staphylococcus aureus. *Medchemcomm* 2016;7:693–705. doi:10.1039/C5MD00527B.

[80] Rineh A, Dolla NK, Ball AR, Magana M, Bremner JB, Hamblin MR, et al. Attaching the NorA Efflux Pump Inhibitor INF55 to Methylene Blue Enhances Antimicrobial Photodynamic Inactivation of Methicillin-Resistant Staphylococcus aureus *in Vitro* and *in Vivo*. *ACS Infect Dis* 2017;3:756–66. doi:10.1021/acsinfecdis.7b00095.

[81] Rineh A, Bremner JB, Hamblin MR, Ball AR, Tegos GP, Kelso MJ. Attaching NorA efflux pump inhibitors to methylene blue enhances antimicrobial photodynamic inactivation of Escherichia coli and Acinetobacter baumannii *in vitro* and *in vivo*. *Bioorg Med Chem Lett* 2018;28:2736–40. doi:10.1016/j.bmcl.2018.02.041.

[82] Pieroni M, Dimovska M, Brincat JP, Sabatini S, Carosati E, Massari S, et al. From 6-Aminoquinolone Antibacterials to 6-Amino-7-thiopyranopyridinylquinolone Ethyl Esters as Inhibitors of Staphylococcus aureus Multidrug Efflux Pumps. *J Med Chem* 2010;53:4466–80. doi:10.1021/jm1003304.

[83] Sabatini S, Gosetto F, Manfroni G, Tabarrini O, Kaatz GW, Patel D, et al. Evolution from a Natural Flavones Nucleus to Obtain 2-(4-Propoxyphenyl)quinoline Derivatives As Potent Inhibitors of the S. aureus NorA Efflux Pump. *J Med Chem* 2011;54:5722–36. doi:10.1021/jm200370y.

[84] Sabatini S, Gosetto F, Iraci N, Barreca ML, Massari S, Sancineto L, et al. Re-evolution of the 2-Phenylquinolines: Ligand-Based Design, Synthesis, and Biological Evaluation of a Potent New Class of Staphylococcus aureus NorA Efflux Pump Inhibitors to Combat Antimicrobial Resistance. *J Med Chem* 2013;56:4975–89. doi:10.1021/jm400262a.

[85] Carotti A, Ianni F, Sabatini S, Di Michele A, Sardella R, Kaatz GW, et al. The "racemic approach" in the evaluation of the enantiomeric NorA efflux pump inhibition activity of 2-phenylquinoline derivatives. *J Pharm Biomed Ana*l 2016;129:182–9. doi:10.1016/j.jpba.2016.07.003.

[86] Felicetti T, Cannalire R, Nizi MG, Tabarrini O, Massari S, Barreca ML, et al. Studies on 2-phenylquinoline Staphylococcus aureus NorA efflux pump inhibitors: New insights on the C-6 position. *Eur J Med Chem* 2018;155:428–33. doi:10.1016/j.ejmech.2018.06.013.

[87] Felicetti T, Cannalire R, Pietrella D, Latacz G, Lubelska A, Manfroni G, et al. 2-Phenylquinoline S. aureus NorA Efflux Pump Inhibitors: Evaluation of the Importance of Methoxy Group

Introduction. *J Med Chem* 2018;61:7827–48. doi:10.1021/acs.jmedchem.8b00791.

[88] Pieroni M, Sabatini S, Massari S, Kaatz GW, Cecchetti V, Tabarrini O. Searching for innovative quinolone-like scaffolds: synthesis and biological evaluation of 2,1-benzothiazine 2,2-dioxide derivatives. *Medchemcomm* 2012;3:1092. doi:10.1039/c2md20101a.

[89] Doléans-Jordheim A, Veron JB, Fendrich O, Bergeron E, Montagut-Romans A, Wong YS, et al. 3-Aryl-4-methyl-2-quinolones Targeting Multiresistant Staphylococcus aureus Bacteria. *ChemMedChem* 2013;8:652–7. doi:10.1002/cmdc.201200551.

[90] Fontaine F, Hequet A, Voisin-Chiret A-S, Bouillon A, Lesnard A, Cresteil T, et al. First Identification of Boronic Species as Novel Potential Inhibitors of the Staphylococcus aureus NorA Efflux Pump. *J Med Chem* 2014;57:2536–48. doi:10.1021/jm401808n.

[91] Fontaine F, Héquet A, Voisin-Chiret AS, Bouillon A, Lesnard A, Cresteil T, et al. Boronic species as promising inhibitors of the Staphylococcus aureus NorA efflux pump: Study of 6-substituted pyridine-3-boronic acid derivatives. *Eur J Med Chem* 2015;95:185–98. doi:10.1016/j.ejmech.2015.02.056.

[92] Lowrence RC, Raman T, Makala H V., Ulaganathan V, Subramaniapillai SG, Kuppuswamy AA, et al. Dithiazole thione derivative as competitive NorA efflux pump inhibitor to curtail multi drug resistant clinical isolate of MRSA in a zebrafish infection model. *Appl Microbiol Biotechnol* 2016;100:9265–81. doi:10.1007/s00253-016-7759-2.

[93] Cannalire R, Machado D, Felicetti T, Santos Costa S, Massari S, Manfroni G, et al. Natural isoflavone biochanin A as a template for the design of new and potent 3-phenylquinolone efflux inhibitors against Mycobacterium avium. *Eur J Med Chem* 2017;140:321–30. doi:10.1016/j.ejmech.2017.09.014.

[94] Radix S, Jordheim AD, Rocheblave L, N'Digo S, Prignon A-LL, Commun C, et al. N,N′ -disubstituted cinnamamide derivatives potentiate ciprofloxacin activity against overexpressing NorA efflux

pump Staphylococcus aureus 1199B strains. *Eur J Med Chem* 2018;150:900–7. doi:10.1016/j.ejmech.2018.03.028.

[95] Rath SK, Singh S, Kumar S, Wani NA, Rai R, Koul S, et al. Synthesis of amides from (E)-3-(1-chloro-3,4-dihydronaphthalen-2-yl)acrylic acid and substituted amino acid esters as NorA efflux pump inhibitors of Staphylococcus aureus. *Bioorg Med Chem* 2019;27:343–53. doi:10.1016/j.bmc.2018.12.008.

[96] Chaudhary M, Patnaik SK, Payasi A. Evaluation of Different Drugs in Down-Regulation of Efflux Pump Genes expression in Methicillin-Resistant Staphylococcus aureus Strains. *Am J Infect Dis* 2014;10:184–9. doi:10.3844/ajidsp.2014.184.189.

[97] Shokoofeh N, Moradi-Shoeili Z, Naeemi AS, Jalali A, Hedayati M, Salehzadeh A. Biosynthesis of Fe3O4@Ag Nanocomposite and Evaluation of Its Performance on Expression of norA and norB Efflux Pump Genes in Ciprofloxacin-Resistant Staphylococcus aureus. *Biol Trace Elem Res* 2019. doi:10.1007/s12011-019-1632-y.

[98] Zloh M, Gibbons S. The Role of Small Molecule–small Molecule Interactions in Overcoming Biological Barriers for Antibacterial Drug Action. *Theor Chem Acc* 2007;117:231–8. doi:10.1007/s00214-006-0149-6.

[99] Bhaskar BV, Babu TMC, Reddy NV, Rajendra W. Homology modeling, molecular dynamics, and virtual screening of NorA efflux pump inhibitors of Staphylococcus aureus. *Drug Des Devel Ther* 2016;Volume 10:3237–52. doi:10.2147/DDDT.S113556.

[100] Thota N, Koul S, Reddy M V., Sangwan PL, Khan IA, Kumar A, et al. Citral derived amides as potent bacterial NorA efflux pump inhibitors. *Bioorg Med Chem* 2008;16:6535–43. doi:10.1016/j.bmc.2008.05.030.

[101] Mohammed NH, Mostafa MI, Al-Taher AY. Augmentation Effects of Novel Narigenin Analogues and Ciprofloxacin as Inhibitors for NorA Efflux Pump (EPIs) and Pyruvate Kinase (PK) Against MRSA. *J Anim Vet Adv* 2015;14:386–92. doi:10.3923/javaa.2015.386.392.

[102] Sharma P, Kumar S, Ali F, Anthal S, Gupta VK, Khan IA, et al. Synthesis and biologic activities of some novel heterocyclic chalcone derivatives. *Med Chem Res* 2013;22:3969–83. doi:10.1007/s00044-012-0401-7.

[103] Costa LM, de Macedo EV, Oliveira FAA, Ferreira JHL, Gutierrez SJC, Peláez WJ, et al. Inhibition of the NorA efflux pump of Staphylococcus aureus by synthetic riparins. *J Appl Microbiol* 2016;121:1312–22. doi:10.1111/jam.13258.

[104] Ganesan A, Christena LR, Venkata Subbarao HM, Venkatasubramanian U, Thiagarajan R, Sivaramakrishnan V, et al. Identification of benzochromene derivatives as a highly specific NorA efflux pump inhibitor to mitigate the drug resistant strains of S. aureus. *RSC Adv* 2016;6:30258–67. doi:10.1039/C6RA01981A.

[105] Bharate JB, Singh S, Wani A, Sharma S, Joshi P, Khan IA, et al. Discovery of 4-acetyl-3-(4-fluorophenyl)-1-(p-tolyl)-5-methylpyrrole as a dual inhibitor of human P-glycoprotein and Staphylococcus aureus Nor A efflux pump. *Org Biomol Chem* 2015;13:5424–31. doi:10.1039/C5OB00246J.

[106] Sundaramoorthy NS, Mitra K, Ganesh JS, Makala H, Lotha R, Bhanuvalli SR, et al. Ferulic acid derivative inhibits NorA efflux and in combination with ciprofloxacin curtails growth of MRSA *in vitro* and *in vivo*. *Microb Pathog* 2018;124:54–62. doi:10.1016/j.micpath.2018.08.022.

[107] Gupta VK, Gaur R, Sharma A, Akther J, Saini M, Bhakuni RS, et al. A novel bi-functional chalcone inhibits multi-drug resistant Staphylococcus aureus and potentiates the activity of fluoroquinolones. *Bioorg Chem* 2019;83:214–25. doi:10.1016/j.bioorg.2018.10.024.

[108] Kumar A, Khan IA, Koul S, Koul JL, Taneja SC, Ali I, et al. Novel structural analogues of piperine as inhibitors of the NorA efflux pump of Staphylococcus aureus. *J Antimicrob Chemother* 2008;61:1270–6. doi:10.1093/jac/dkn088.

[109] Sangwan PL, Koul JL, Koul S, Reddy M V., Thota N, Khan IA, et al. Piperine analogs as potent Staphylococcus aureus NorA efflux

pump inhibitors. *Bioorg Med Chem* 2008;16:9847–57. doi:10.1016/j.bmc.2008.09.042.

[110] Nargotra A, Sharma S, Koul JL, Sangwan PL, Khan IA, Kumar A, et al. Quantitative structure activity relationship (QSAR) of piperine analogsfor bacterial NorA efflux pump inhibitors. *Eur J Med Chem* 2009;44:4128–35. doi:10.1016/j.ejmech.2009.05.004.

[111] Wani NA, Singh S, Farooq S, Shankar S, Koul S, Khan IA, et al. Amino acid amides of piperic acid (PA) and 4-ethylpiperic acid (EPA) as NorA efflux pump inhibitors of Staphylococcus aureus. *Bioorg Med Chem Lett* 2016;26:4174–8. doi:10.1016/j.bmcl.2016.07.062.

[112] Aeschlimann JR, Dresser LD, Kaatz GW, Rybak MJ. Effects of NorA Inhibitors on *In Vitro* Antibacterial Activities and Postantibiotic Effects of Levofloxacin, Ciprofloxacin, and Norfloxacin in Genetically Related Strains of Staphylococcus aureus. *Antimicrob Agents Chemother* 1999;43:335–40. doi:10.1128/AAC.43.2.335.

[113] Aeschlimann JR, Kaatz GW, Rybak MJ. The effects of NorA inhibition on the activities of levofloxacin, ciprofloxacin and norfloxacin against two genetically related strains of Staphylococcus aureus in an in-vitro infection model. *J Antimicrob Chemother* 1999;44:343–9. doi:10.1093/jac/44.3.343.

[114] Vidaillac C, Guillon J, Arpin C, Forfar-Bares I, Ba BB, Grellet J, et al. Synthesis of Omeprazole Analogues and Evaluation of These as Potential Inhibitors of the Multidrug Efflux Pump NorA of Staphylococcus aureus. *Antimicrob Agents Chemother* 2007;51:831–8. doi:10.1128/AAC.01306-05.

[115] Gibbons S. A novel inhibitor of multidrug efflux pumps in Staphylococcus aureus. *J Antimicrob Chemother* 2003;51:13–7. doi:10.1093/jac/dkg044.

[116] Brincat JP, Broccatelli F, Sabatini S, Frosini M, Neri A, Kaatz GW, et al. Ligand Promiscuity between the Efflux Pumps Human P-Glycoprotein and S. aureus NorA. *ACS Med Chem Lett* 2012;3:248–51. doi:10.1021/ml200293c.

[117] Ngo TD, Tran TD, Le MT, Thai KM. Machine learning-, rule- and pharmacophore-based classification on the inhibition of P-glycoprotein and NorA. *SAR QSAR Environ Res* 2016;27:747–80. doi:10.1080/1062936X.2016.1233137.

[118] Kaatz G, Moudgal V V., Seo SM, Hansen JB, Kristiansen JE. Phenylpiperidine selective serotonin reuptake inhibitors interfere with multidrug efflux pump activity in Staphylococcus aureus. *Int J Antimicrob Agents* 2003;22:254–61. doi:10.1016/S0924-8579(03) 00220-6.

[119] German N, Kaatz GW, Kerns RJ. Synthesis and evaluation of PSSRI-based inhibitors of Staphylococcus aureus multidrug efflux pumps. *Bioorg Med Chem Lett* 2008;18:1368–73. doi:10.1016/ j.bmcl.2008.01.014.

[120] Ayaz M, Subhan F, Ahmed J, Khan A, Ullah F, Ullah I, et al. Sertraline enhances the activity of antimicrobial agents against pathogens of clinical relevance. *J Biol Re*s 2015;22:4. doi:10.1186/ s40709-015-0028-1.

[121] Kaatz GW, Moudgal VV., Seo SM, Kristiansen JE. Phenothiazines and Thioxanthenes Inhibit Multidrug Efflux Pump Activity in Staphylococcus aureus. *Antimicrob Agents Chemother* 2003;47:719–26. doi:10.1128/AAC.47.2.719-726.2003.

[122] Sabatini S, Kaatz GW, Rossolini GM, Brandini D, Fravolini A. From Phenothiazine to 3-Phenyl-1,4-benzothiazine Derivatives as Inhibitors of the Staphylococcus aureus NorA Multidrug Efflux Pump. *J Med Chem* 2008;51:4321–30. doi:10.1021/jm701623q.

[123] Felicetti T, Cannalire R, Burali MS, Massari S, Manfroni G, Barreca ML, et al. Searching for Novel Inhibitors of the S. aureus NorA Efflux Pump: Synthesis and Biological Evaluation of the 3-Phenyl-1,4-benzothiazine Analogues. *ChemMedChem* 2017;12:1293–302. doi:10.1002/cmdc.201700286.

[124] Astolfi A, Felicetti T, Iraci N, Manfroni G, Massari S, Pietrella D, et al. Pharmacophore-Based Repositioning of Approved Drugs as Novel Staphylococcus aureus NorA Efflux Pump Inhibitors. *J Med Chem* 2017;60:1598–604. doi:10.1021/acs.jmedchem.6b01439.

[125] Sabatini S, Gosetto F, Serritella S, Manfroni G, Tabarrini O, Iraci N, et al. Pyrazolo[4,3- c][1,2]benzothiazines 5,5-Dioxide: A Promising New Class of Staphylococcus aureus NorA Efflux Pump Inhibitors. *J Med Chem* 2012;55:3568–72. doi:10.1021/jm201446h.

[126] Jiang W, Li B, Zheng X, Liu X, Pan X, Qing R, et al. Artesunate has its enhancement on antibacterial activity of β-lactams via increasing the antibiotic accumulation within methicillin-resistant Staphylococcus aureus (MRSA). *J Antibiot* (Tokyo) 2013;66:339–45. doi:10.1038/ja.2013.22.

[127] Amaral L, Fanning S, Pagès J-M. Efflux Pumps of Gram-Negative Bacteria: Genetic Responses to Stress and the Modulation of their Activity by pH, Inhibitors, and Phenothiazines. In: Toone EJ, editor. Adv. Enzymol. Relat. *Areas Mol. Biol.*, 2011, p. 61–108. doi:10.1002/9780470920541.ch2.

[128] Fernando D, Kumar A. *Resistance-Nodulation-Division Multidrug Efflux Pumps in Gram-Negative Bacteria: Role in Virulence. Antibiotics* 2013;2:163–81. doi:10.3390/antibiotics2010163.

[129] Venter H, Mowla R, Ohene-Agyei T, Ma S. RND-type drug efflux pumps from Gram-negative bacteria: molecular mechanism and inhibition. *Front Microbiol* 2015;06. doi:10.3389/fmicb.2015.00377.

[130] Poole K. Outer Membranes and Efflux: The Path to Multidrug Resistance in Gram- Negative Bacteria. *Curr Pharm Biotechnol* 2002;3:77–98. doi:10.2174/1389201023378454.

[131] Baugh S, Ekanayaka AS, Piddock LJ V., Webber MA. Loss of or inhibition of all multidrug resistance efflux pumps of Salmonella enterica serovar Typhimurium results in impaired ability to form a biofilm. *J Antimicrob Chemother* 2012;67:2409–17. doi:10.1093/jac/dks228.

[132] Baugh S, Phillips CR, Ekanayaka AS, Piddock LJ V., Webber MA. Inhibition of multidrug efflux as a strategy to prevent biofilm formation. *J Antimicrob Chemother* 2014;69:673–81. doi:10.1093/jac/dkt420.

[133] Mavri A, Smole Mozina S. Involvement of efflux mechanisms in biocide resistance of Campylobacter jejuni and Campylobacter coli. *J Med Microbiol* 2012;61:800–8. doi:10.1099/jmm.0.041467-0.

[134] Mao W, Warren MS, Lee A, Mistry A, Lomovskaya O. MexXY-OprM Efflux Pump Is Required for Antagonism of Aminoglycosides by Divalent Cations in Pseudomonas aeruginosa. *Antimicrob Agents Chemother* 2001;45:2001–7. doi:10.1128/AAC.45.7.2001-2007.2001.

[135] Ruiz J, Ribera A, Jurado A, Marco F, Vila J. Evidence for a reserpine-affected mechanism of resistance to tetracycline in Neisseria gonorrhoeae. *APMIS* 2005;113:670–4. doi:10.1111/j.1600-0463.2005.apm_303.x.

[136] Tamburrino G, Llabrés S, Vickery ON, Pitt SJ, Zachariae U. Modulation of the Neisseria gonorrhoeae drug efflux conduit MtrE. *Sci Rep* 2017;7:17091. doi:10.1038/s41598-017-16995-x.

[137] Zhang L, Li XZ, Poole K. SmeDEF Multidrug Efflux Pump Contributes to Intrinsic Multidrug Resistance in Stenotrophomonas maltophilia. *Antimicrob Agents Chemother* 2001;45:3497–503. doi:10.1128/AAC.45.12.3497-3503.2001.

[138] Kim HR, Lee D, Eom YB. Anti-biofilm and Anti-Virulence Efficacy of Celastrol Against Stenotrophomonas maltophilia. *Int J Med Sci* 2018;15:617–27. doi:10.7150/ijms.23924.

[139] Aparna V, Dineshkumar K, Mohanalakshmi N, Velmurugan D, Hopper W. Identification of Natural Compound Inhibitors for Multidrug Efflux Pumps of Escherichia coli and Pseudomonas aeruginosa Using In Silico High-Throughput Virtual Screening and *In Vitro* Validation. *PLoS One* 2014;9:e101840. doi:10.1371/journal.pone.0101840.

[140] Whalen KE, Poulson-Ellestad KL, Deering RW, Rowley DC, Mincer TJ. Enhancement of antibiotic activity against multidrug-resistant bacteria by the efflux pump inhibitor 3,4-dibromopyrrole-2,5-dione isolated from a pseudoalteromonas sp. *J Nat Prod* 2015;78:402–12. doi:10.1021/np500775e.

[141] Seukep JA, Sandjo LP, Ngadjui BT, Kuete V. Antibacterial and antibiotic-resistance modifying activity of the extracts and compounds from Nauclea pobeguinii against Gram-negative multi-drug resistant phenotypes. *BMC Complement Altern Med* 2016;16:193. doi:10.1186/s12906-016-1173-2.

[142] Zhou X, Jia F, Liu X, Wang Y. Total Alkaloids of Sophorea alopecuroides -induced Down-regulation of AcrAB-ToLC Efflux Pump Reverses Susceptibility to Ciprofloxacin in Clinical Multidrug Resistant Escherichia coli isolates. *Phyther Res* 2012;26:1637–43. doi:10.1002/ptr.4623.

[143] Lieutaud A, Guinoiseau E, Lorenzi V, Giuliani MC, Lome V, Brunel J-M, et al. Inhibitors of Antibiotic Efflux by AcrAB-TolC in Enterobacter aerogenes. *Anti-Infective Agents* 2013;11:168–78. doi:10.2174/22113525113110200011.

[144] Ohene-Agyei T, Mowla R, Rahman T, Venter H. Phytochemicals increase the antibacterial activity of antibiotics by acting on a drug efflux pump. *Microbiologyopen* 2014;3:885–96. doi:10.1002/mbo3.212.

[145] Mehta J, Jandaik S, U. Evaluation of Phytochemicals And Synergistic Interaction between Plant Extracts and Antibiotics for Efflux Pump Inhibitory Activity Against Salmonella Enterica Serovar Typhimurium Strains. *Int J Pharm Pharm Sci* 2016;8:217. doi:10.22159/ijpps.2016v8i10.14062.

[146] Rafiq Z, Sivaraj S, Vaidyanathan R. Computational Docking and in Silico Analysis of Potential Efflux Pump Inhibitor Punigratane. *Int J Pharm Pharm Sci* 2018;10:27. doi:10.22159/ijpps.2018v10i3.21629.

[147] Hwang D, Lim YH. Resveratrol controls Escherichia coli growth by inhibiting the AcrAB-TolC efflux pump. *FEMS Microbiol Lett* 2019;366. doi:10.1093/femsle/fnz030.

[148] Chusri S, Villanueva I, Voravuthikunchai SP, Davies J. Enhancing antibiotic activity: a strategy to control Acinetobacter infections. *J Antimicrob Chemother* 2009;64:1203–11. doi:10.1093/jac/dkp381.

[149] Lee MD, Galazzo JL, Staley AL, Lee JC, Warren MS, Fuernkranz H, et al. Microbial fermentation-derived inhibitors of efflux-pump-

mediated drug resistance. *Farm* 2001;56:81–5. doi:10.1016/S0014-827X(01)01002-3.

[150] Morita Y, Nakashima K, Nishino K, Kotani K, Tomida J, Inoue M, et al. Berberine Is a Novel Type Efflux Inhibitor Which Attenuates the MexXY-Mediated Aminoglycoside Resistance in Pseudomonas aeruginosa. *Front Microbiol* 2016;7. doi:10.3389/fmicb.2016.01223.

[151] Rahbar Takrami S, Ranji N, Sadeghizadeh M. Antibacterial effects of curcumin encapsulated in nanoparticles on clinical isolates of Pseudomonas aeruginosa through downregulation of efflux pumps. *Mol Biol Rep* 2019;0:0. doi:10.1007/s11033-019-04700-2.

[152] Lomovskaya O, Warren MS, Lee A, Galazzo J, Fronko R, Lee M, et al. Identification and Characterization of Inhibitors of Multidrug Resistance Efflux Pumps in Pseudomonas aeruginosa: Novel Agents for Combination Therapy. *Antimicrob Agents Chemother* 2001;45:105–16. doi:10.1128/AAC.45.1.105-116.2001.

[153] Coban AY, Ekinci B, Durupinar B. A Multidrug Efflux Pump Inhibitor Reduces Fluoroquinolone Resistance in Pseudomonas aeruginosa Isolates. *Chemotherapy* 2004;50:22–6. doi:10.1159/000077280.

[154] Avakh A, Rezaei K. Inhibition of the Mex Pumps of Pseudomonas aeruginosa with a Newly Characterized Member of Peptidomimetic Family. *Br Microbiol Res J* 2016;16:1–13. doi:10.9734/BMRJ/2016/27374.

[155] Bohnert JA, Kern WV. Selected Arylpiperazines Are Capable of Reversing Multidrug Resistance in Escherichia coli Overexpressing RND Efflux Pumps. *Antimicrob Agents Chemother* 2005;49:849–52. doi:10.1128/AAC.49.2.849-852.2005.

[156] Kern WV., Steinke P, Schumacher A, Schuster S, von Baum H, Bohnert JA. Effect of 1-(1-naphthylmethyl)-piperazine, a novel putative efflux pump inhibitor, on antimicrobial drug susceptibility in clinical isolates of Escherichia coli. *J Antimicrob Chemother* 2006;57:339–43. doi:10.1093/jac/dki445.

[157] Chevalier J, Atifi S, Eyraud A, Mahamoud A, Barbe J, Pagès JM. New pyridoquinoline derivatives as potential inhibitors of the

fluoroquinolone efflux pump in resistant enterobacter aerogenes strains. *J Med Chem* 2001;44:4023–6. doi:10.1021/jm010911z.

[158] Malléa M, Mahamoud A, Chevalier J, Alibert-Franco S, Brouant P, Barbe J, et al. Alkylaminoquinolines inhibit the bacterial antibiotic efflux pump in multidrug-resistant clinical isolates. *Biochem J* 2003;376:801–5. doi:10.1042/bj20030963.

[159] Chevalier J, Bredin J, Mahamoud A, Mallea M, Barbe J, Pages J-M. Inhibitors of Antibiotic Efflux in Resistant Enterobacter aerogenes and Klebsiella pneumoniae Strains. *Antimicrob Agents Chemother* 2004;48:1043–6. doi:10.1128/AAC.48.3.1043-1046.2004.

[160] Tang J, Wang H. Indole derivatives as efflux pump inhibitors for TolC protein in a clinical drug-resistant Escherichia coli isolated from a pig farm. *Int J Antimicrob Agents* 2008;31:497–8. doi:10.1016/j.ijantimicag.2008.01.007.

[161] Zeng B, Wang H, Zou L, Zhang A, Yang X, Guan Z. Evaluation and Target Validation of Indole Derivatives as Inhibitors of the AcrAB-TolC Efflux Pump. *Biosci Biotechnol Biochem* 2010;74:2237–41. doi:10.1271/bbb.100433.

[162] Handzlik J, Szymańska E, Chevalier J, Otrębska E, Kieć-Kononowicz K, Pagès JM, et al. Amine–alkyl derivatives of hydantoin: New tool to combat resistant bacteria. *Eur J Med Chem* 2011;46:5807–16. doi:10.1016/j.ejmech.2011.09.032.

[163] Handzlik J, Szymańska E, Alibert S, Chevalier J, Otrębska E, Pękala E, et al. Search for new tools to combat Gram-negative resistant bacteria among amine derivatives of 5-arylidenehydantoin. *Bioorg Med Chem* 2013;21:135–45. doi:10.1016/j.bmc.2012.10.053.

[164] Opperman TJ, Kwasny SM, Kim HS, Nguyen ST, Houseweart C, D'Souza S, et al. Characterization of a Novel Pyranopyridine Inhibitor of the AcrAB Efflux Pump of Escherichia coli. *Antimicrob Agents Chemother* 2014;58:722–33. doi:10.1128/AAC.01866-13.

[165] Vargiu A V., Ruggerone P, Opperman TJ, Nguyen ST, Nikaido H. Molecular Mechanism of MBX2319 Inhibition of Escherichia coli AcrB Multidrug Efflux Pump and Comparison with Other Inhibitors.

Antimicrob Agents Chemother 2014;58:6224–34. doi:10.1128/AAC. 03283-14.

[166] Sjuts H, Vargiu AV., Kwasny SM, Nguyen ST, Kim HS, Ding X, et al. Molecular basis for inhibition of AcrB multidrug efflux pump by novel and powerful pyranopyridine derivatives. *Proc Natl Acad Sci* 2016;113:3509–14. doi:10.1073/pnas.1602472113.

[167] Silva L, Carrion LL, von Groll A, Costa SS, Junqueira E, Ramos DF, et al. *In vitro* and *in silico* analysis of the efficiency of tetrahydropyridines as drug efflux inhibitors in Escherichia coli. *Int J Antimicrob Agents* 2017;49:308–14. doi:10.1016/j.ijantimicag.2016. 11.024.

[168] Yilmaz S, Altinkanat-Gelmez G, Bolelli K, Guneser-Merdan D, Ufuk Over-Hasdemir M, Aki-Yalcin E, et al. Binding site feature description of 2-substituted benzothiazoles as potential AcrAB-TolC efflux pump inhibitors in E. coli. SAR QSAR *Environ Res* 2015;26:853–71. doi:10.1080/1062936X.2015.1106581.

[169] Bohnert JA, Schuster S, Kern WV., Karcz T, Olejarz A, Kaczor A, et al. Novel Piperazine Arylideneimidazolones Inhibit the AcrAB-TolC Pump in Escherichia coli and Simultaneously Act as Fluorescent Membrane Probes in a Combined Real-Time Influx and Efflux Assay. *Antimicrob Agents Chemother* 2016;60:1974–83. doi:10. 1128/AAC.01995-15.

[170] Szabó Ám, Kincses A, Watanabe G, Molnár J, Saijo R, Spengler G, et al. Fluorinated Beta-diketo Phosphorus Ylides Are Novel Efflux Pump Inhibitors in Bacteria. *In Vivo* (Brooklyn) 2016;30:813–8. doi:10.21873/invivo.10999.

[171] Kaczor A, Witek K, Podlewska S, Czekajewska J, Lubelska A, Żesławska E, et al. 5-Arylideneimidazolones with Amine at Position 3 as Potential Antibiotic Adjuvants against Multidrug Resistant Bacteria. *Molecules* 2019;24:438. doi:10.3390/molecules24030438.

[172] Wang Y, Mowla R, Guo L, Ogunniyi AD, Rahman T, De Barros Lopes MA, et al. Evaluation of a series of 2-napthamide derivatives as inhibitors of the drug efflux pump AcrB for the reversal of

antimicrobial resistance. *Bioorg Med Chem Lett* 2017;27:733–9. doi:10.1016/j.bmcl.2017.01.042.

[173] Jin C, Venter H, Guo L, Ji S, Ma S, De Barros Lopes MA, et al. Design, synthesis and biological activity evaluation of novel 4-subtituted 2-naphthamide derivatives as AcrB inhibitors. *Eur J Med Chem* 2017;143:699–709. doi:10.1016/j.ejmech.2017.11.102.

[174] Wang Y, Mowla R, Ji S, Guo L, De Barros Lopes MA, Jin C, et al. Design, synthesis and biological activity evaluation of novel 4-subtituted 2-naphthamide derivatives as AcrB inhibitors. *Eur J Med Chem* 2018;143:699–709. doi:10.1016/j.ejmech.2017.11.102.

[175] Jin C, Alenazy R, Wang Y, Mowla R, Qin Y, Tan JQE, et al. Design, synthesis and evaluation of a series of 5-methoxy-2,3-naphthalimide derivatives as AcrB inhibitors for the reversal of bacterial resistance. Bi*oorg Med Chem Lett* 2019;29:882–9. doi:10.1016/j.bmcl.2019.02.003.

[176] Meng J, Bai H, Jia M, Ma X, Hou Z, Xue X, et al. Restoration of antibiotic susceptibility in fluoroquinolone-resistant Escherichia coli by targeting acrB with antisense phosphorothioate oligonucleotide encapsulated in novel anion liposome. *J Antibiot* (Tokyo) 2012;65:129–34. doi:10.1038/ja.2011.125.

[177] Kinana AD, Vargiu A V., May T, Nikaido H. Aminoacyl β-naphthylamides as substrates and modulators of AcrB multidrug efflux pump. *Proc Natl Acad Sci* 2016;113:1405–10. doi:10.1073/pnas.1525143113.

[178] Nakayama K, Ishida Y, Ohtsuka M, Kawato H, Yoshida K, Yokomizo Y, et al. MexAB-OprM-Specific efflux pump inhibitors in Pseudomonas aeruginosa. Part 1: Discovery and early strategies for lead optimization. *Bioorg Med Chem Lett* 2003;13:4201–4. doi:10.1016/j.bmcl.2003.07.024.

[179] Nakayama K, Ishida Y, Ohtsuka M, Kawato H, Yoshida K, Yokomizo Y, et al. MexAB-OprM specific efflux pump inhibitors in Pseudomonas aeruginosa. Part 2: achieving activity *in vivo* through the use of alternative scaffolds. *Bioorg Med Chem Lett* 2003;13:4205–8. doi:10.1016/j.bmcl.2003.07.027.

[180] Nakayama K, Kawato H, Watanabe J, Ohtsuka M, Yoshida K, Yokomizo Y, et al. MexAB-OprM specific efflux pump inhibitors in Pseudomonas aeruginosa. Part 3: Optimization of potency in the pyridopyrimidine series through the application of a pharmacophore model. *Bioorg Med Chem Lett* 2004;14:475–9. doi:10.1016/j.bmcl.2003.10.060.

[181] Nakayama K, Kuru N, Ohtsuka M, Yokomizo Y, Sakamoto A, Kawato H, et al. MexAB-OprM specific efflux pump inhibitors in Pseudomonas aeruginosa. Part 4: Addressing the problem of poor stability due to photoisomerization of an acrylic acid moiety. *Bioorg Med Chem Lett* 2004;14:2493–7. doi:10.1016/j.bmcl.2004.03.007.

[182] Yoshida K, Nakayama K, Kuru N, Kobayashi S, Ohtsuka M, Takemura M, et al. MexAB-OprM specific efflux pump inhibitors in Pseudomonas aeruginosa. Part 5: Carbon-substituted analogues at the C-2 position. *Bioorg Med Chem* 2006;14:1993–2004. doi:10.1016/j.bmc.2005.10.043.

[183] Yoshida K, Nakayama K, Yokomizo Y, Ohtsuka M, Takemura M, Hoshino K, et al. MexAB-OprM specific efflux pump inhibitors in Pseudomonas aeruginosa. Part 6: Exploration of aromatic substituents. *Bioorg Med Chem* 2006;14:8506–18. doi:10.1016/j.bmc.2006.08.037.

[184] Yoshida K, Nakayama K, Ohtsuka M, Kuru N, Yokomizo Y, Sakamoto A, et al. MexAB-OprM specific efflux pump inhibitors in Pseudomonas aeruginosa. Part 7: Highly soluble and *in vivo* active quaternary ammonium analogue D13-9001, a potential preclinical candidate. *Bioorg Med Chem* 2007;15:7087–97. doi:10.1016/j.bmc.2007.07.039.

[185] Hirakata Y, Kondo A, Hoshino K, Yano H, Arai K, Hirotani A, et al. Efflux pump inhibitors reduce the invasiveness of Pseudomonas aeruginosa. *Int J Antimicrob Agents* 2009;34:343–6. doi:10.1016/j.ijantimicag.2009.06.007.

[186] Zuo Z, Weng J, Wang W. Insights into the Inhibitory Mechanism of D13-9001 to the Multidrug Transporter AcrB through Molecular

Dynamics Simulations. *J Phys Chem B* 2016;120:2145–54. doi:10.1021/acs.jpcb.5b11942.

[187] Wang H, Meng J, Jia M, Ma X, He G, Yu J, et al. oprM as a new target for reversion of multidrug resistance in Pseudomonas aeruginosa by antisense phosphorothioate oligodeoxynucleotides. *FEMS Immunol Med Microbiol* 2010;60:275–82. doi:10.1111/j.1574-695X.2010.00742.x.

[188] Dwivedi GR, Tiwari N, Singh A, Kumar A, Roy S, Negi AS, et al. Gallic acid-based indanone derivative interacts synergistically with tetracycline by inhibiting efflux pump in multidrug resistant E. coli. *Appl Microbiol Biotechnol* 2016;100:2311–25. doi:10.1007/s00253-015-7152-6.

[189] Bohnert JA, Szymaniak-Vits M, Schuster S, Kern W V. Efflux inhibition by selective serotonin reuptake inhibitors in Escherichia coli. *J Antimicrob Chemother* 2011;66:2057–60. doi:10.1093/jac/dkr258.

[190] Bailey AM, Paulsen IT, Piddock LJ V. RamA Confers Multidrug Resistance in Salmonella enterica via Increased Expression of acrB, Which Is Inhibited by Chlorpromazine. *Antimicrob Agents Chemother* 2008;52:3604–11. doi:10.1128/AAC.00661-08.

[191] Rahbar M, Mehrgan H, Hadji-nejad S. Enhancement of Vancomycin Activity by Phenothiazines against Vancomycin-Resistant Enterococcus Faecium *in vitro. Basic Clin Pharmacol Toxicol* 2010;107:676–9. doi:10.1111/j.1742-7843.2010.00558.x.

[192] Spengler G, Takács D, Horváth A, Szabó AM, Riedl Z, Hajós G, et al. Efflux pump inhibiting properties of racemic phenothiazine derivatives and their enantiomers on the bacterial AcrAB-TolC system. *In Vivo* (Brooklyn) 2014;28:1071–5.

[193] Bohnert JA. Pimozide Inhibits the AcrAB-TolC Efflux Pump in Escherichia coli. *Open Microbiol J* 2013;7:83–6. doi:10.2174/1874285801307010083.

[194] Gupta P, Rai N, Gautam P. Anticancer drugs as potential inhibitors of AcrAB-TolC of multidrug resistant Escherichia coli: an in silico

molecular modeling and docking study. *Asian J Pharm Clin Res* 2015;8:351–8.

[195] Li B, Yao Q, Pan XC, Wang N, Zhang R, Li J, et al. Artesunate enhances the antibacterial effect of β-lactam antibiotics against Escherichia coli by increasing antibiotic accumulation via inhibition of the multidrug efflux pump system AcrAB-TolC. *J Antimicrob Chemother* 2011;66:769–77. doi:10.1093/jac/dkr017.

[196] Song Y, Qin R, Pan X, Ouyang Q, Liu T, Zhai Z, et al. Design of New Antibacterial Enhancers Based on AcrB's Structure and the Evaluation of Their Antibacterial Enhancement Activity. *Int J Mol Sci* 2016;17:1934. doi:10.3390/ijms17111934.

[197] Helaly GF, Shawky S, Amer R, El-sawaf G, Kholy MA El. Expression of AcrAB Efflux Pump and Role of Mefloquine as Efflux Pump Inhibitor in MDR E . coli 2016;4:6–13. doi:10.12691/ajidm-4-1-2.

[198] Chaudhary M, Payasi A. Ethylenediaminetetraacetic acid: A non antibiotic adjuvant enhancing Pseudomonas aeruginosa susceptibility. *African J Microbiol Res* 2012;6:6799–804. doi:10.5897/AJMR12.1407.

[199] Dwivedi GR, Tyagi R, Sanchita, Tripathi S, Pati S, Srivastava SK, et al. Antibiotics potentiating potential of catharanthine against superbug Pseudomonas aeruginosa. *J Biomol Struct Dyn* 2018;36:4270–84. doi:10.1080/07391102.2017.1413424.

[200] Domalaon R, Berry L, Tays Q, Zhanel GG, Schweizer F. Development of dilipid polymyxins: Investigation on the effect of hydrophobicity through its fatty acyl component. *Bioorg Chem* 2018;80:639–48. doi:10.1016/j.bioorg.2018.07.018.

[201] Adamson DH, Krikstopaityte V, Coote PJ. Enhanced efficacy of putative efflux pump inhibitor/antibiotic combination treatments versus MDR strains of Pseudomonas aeruginosa in a Galleria mellonella *in vivo* infection model. *J Antimicrob Chemother* 2015;70:2271–8. doi:10.1093/jac/dkv111.

[202] Zhao F, Hu Q, Ren H, Zhang X-X. Ultraviolet irradiation sensitizes: Pseudomonas aeruginosa PAO1 to multiple antibiotics. *Environ Sci Water Res Technol* 2018;4:2051–7. doi:10.1039/c8ew00293b.

[203] Ding F, Songkiatisak P, Cherukuri PK, Huang T, Xu XHN. Size-Dependent Inhibitory Effects of Antibiotic Drug Nanocarriers against Pseudomonas aeruginosa. *ACS Omega* 2018;3:1231–43. doi:10.1021/acsomega.7b01956.

[204] Guo XL, Leng P, Yang Y, Yu LG, Lou HX. Plagiochin E, a botanic-derived phenolic compound, reverses fungal resistance to fluconazole relating to the efflux pump. *J Appl Microbiol* 2008;104:831–8. doi:10.1111/j.1365-2672.2007.03617.x.

[205] Chan BCL, Ip M, Gong H, Lui SL, See RH, Jolivalt C, et al. Synergistic effects of diosmetin with erythromycin against ABC transporter over-expressed methicillin-resistant Staphylococcus aureus (MRSA) RN4220/pUL5054 and inhibition of MRSA pyruvate kinase. *Phytomedicine* 2013;20:611–4. doi:10.1016/j.phymed.2013.02.007.

[206] Tintino SR, Leal-Balbino TC, Quintans-Júnior LJ, Pereira PS, Oliveira-Tintino CDM, da Silva TG, et al. Vitamin K enhances the effect of antibiotics inhibiting the efflux pumps of Staphylococcus aureus strains. *Med Chem Res* 2017;27:261–7. doi:10.1007/s00044-017-2063-y.

[207] Mouwakeh A, Kincses A, Nové M, Mosolygó T, Mohácsi-Farkas C, Kiskó G, et al. Nigella sativa essential oil and its bioactive compounds as resistance modifiers against Staphylococcus aureus. *Phyther Res* 2019. doi:10.1002/ptr.6294.

In: Advances in Medicinal Chemistry … ISBN: 978-1-53616-368-1
Editor: E. Ferreira da Silva-Júnior

Chapter 3

THE ROLE OF THIOPHENE CORE IN MEDICINAL CHEMISTRY OF NEGLECTED TROPICAL DISEASES

Igor José dos Santos Nascimento[1], PhD, Paulo Fernando da Silva Santos Júnior[1], PhD, Rodrigo Santos Aquino de Araújo[2], PhD, Francisco Jaime B. Mendonça-Junior[2], PhD and Thiago Mendonça de Aquino[1,*], PhD

[1]Laboratory of Medicinal Chemistry,
Federal University of Alagoas, Maceió, Brazil
[2]Laboratory of Synthesis and Drug Delivery,
State University of Paraiba, João Pessoa, Brazil

ABSTRACT

Thiophene scaffold and its derivatives 2-aminothiophene, nitrothiophene, and benzo[*b*]thiophene are privileged structures in drug

* Corresponding Author's E-mail: thiago.aquino@iqb.ufal.br.

discovery and development. Their wide range of biological activities culminated in the discovery of several hit and lead compounds, in special against neglected tropical diseases (NTD). The present chapter reports the attractive feature of these derivatives as building blocks in the synthesis of biological active thiophene-based and hybrid compounds, and highlight new progress of rational design of thiophene-containing medicinal drugs against tuberculosis, leishmaniasis, and hepatitis. Various mechanisms of action as selective inhibitors and structure-activiety relationship (SAR) studies have been highlighted, comparing chemical groups or scaffolds which evoke the biological potential of thiophene derivatives. In addition, we focused different computational techniques, as molecular docking and virtual screening, which were reported in the past years, in order to identify correlations between activity against biological activity and thiophene-based structures.

Keywords: thiophene derivatives, neglected tropical diseases, tuberculosis, leishmaniasis and hepatitis

INTRODUCTION

The thiophene scaffold and its derivatives, 2-aminothiophene, nitrothiophene, and benzo[*b*]thiophene are privileged structures in drug discovery and development. Their wide range of biological activities led to the discovery of several hit and lead compounds, as well as FDA approved drugs (Figure 1). The present chapter will describe some of the compelling features of these derivatives as building blocks in the synthesis of biologically active thiophene-based and hybrid compounds. We also highlight progress in rational design of thiophene-containing medicinal drugs against neglected tropical diseases (NTD), including tuberculosis, leishmaniasis, and hepatitis. We highlight various mechanisms of action of selective inhibitors and structure–activity relationship (SAR) studies, comparing chemical groups or scaffolds reveal the biological potential of thiophene derivatives. In addition, we focus on various recently reported computational techniques to identify correlations between activity against biological activity and thiophene- or benzo[*b*]thiophene-based structures.

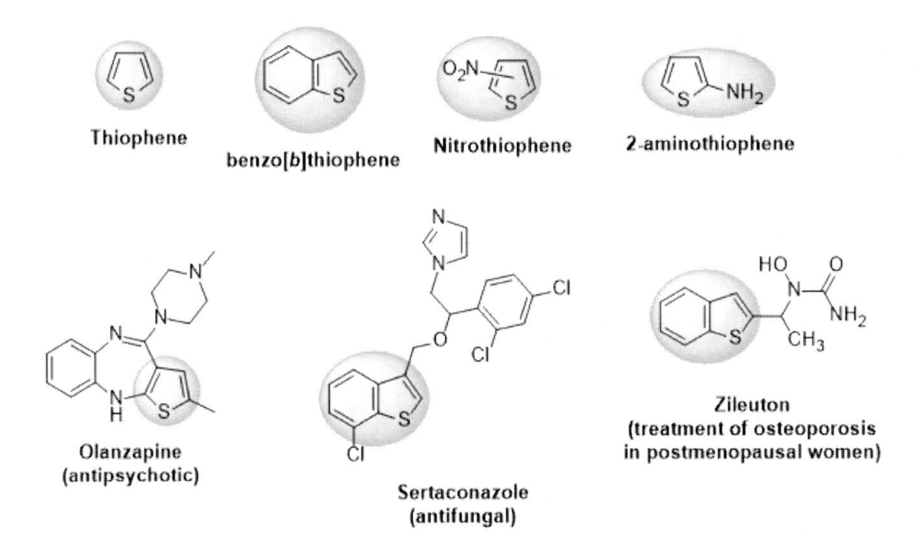

Figure 1. Chemical structures of thiophene, nitrothiophene, 2-aminothiophene and benzo[*b*]thiophene, and some drugs containing these important scaffolds in medicinal chemistry.

TUBERCULOSIS

Tuberculosis (TB) is caused by *Mycobacterium tuberculosis* (MTB). TB presents in two stages of development: asymptomatic and symptomatic. After infection, the asymptomatic (or latent) phase can last for years. In the symptomatic phase (when bacterial replication occurs with consequent multiplication of microorganisms in the body), the disease manifests itself as fever, chronic cough, asthenia, loss of appetite, and weight loss [1].

The standard therapeutic regimen for the treatment of TB initially consists of a cocktail of four first-line drugs: isoniazid (INH), rifampicin (RIF), pyrazinamide (PZA), and ethambutol (ETB), for two months (Figure 2). Subsequently, the treatment continues with INH and RIF for four months [2].

The duration of treatment, adverse effects, and socioeconomic factors are the main reasons for non-adherence on the part of patients to this protocol that was promulgated by the World Health Organization (WHO).

The result has been generation of multidrug-resistant tuberculosis (MDR-TB). For MDR-TB, the WHO recommends treatment with second-line drugs: amikacin (AMK), ethionamide (ETH), moxifloxacin (MOX), and PZA, as well as other drugs. In recent years, extensively drug-resistant tuberculosis (XDR-TB) has emerged. These organisms are resistant to at least three of the six classes of second-line drugs (aminoglycosides, polypeptides, fluoroquinolones, thioamides, cycloserine, and *p*-aminosalicylic acid). In these cases, injectable drugs as capreomycin (CPM) and kanamycin (KM) are used (Figure 2) [3].

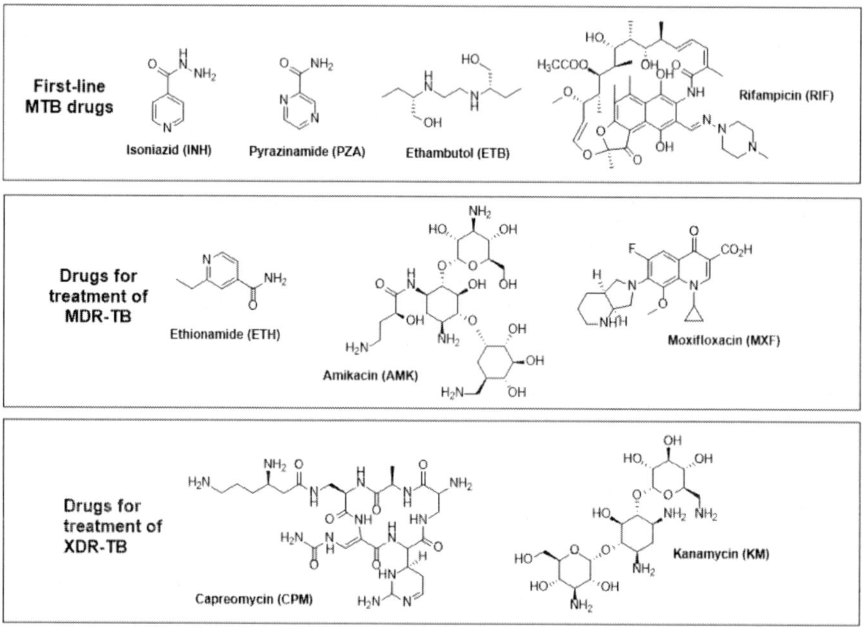

Figure 2. Drugs for treatment of TB, MDR-TB, and XDR-TB.

For the drugs included in the standard therapeutic regimen, as well as for treatment of MDR-TB or XDR-TB, there are three main mechanisms of action: (*i*) disruption of the cell wall; (*ii*) inhibition of RNA translation; and (*iii*) inhibition of DNA transcription. Several MTB enzymes are involved in the process of infection and drug resistance and may be considered promising targets. Enzymes involved in cell wall biosynthesis, metabolism, persistence, virulence, and signal transduction are thought to be the most attractive for rational development of new drugs. Some of these targets will be covered later [3].

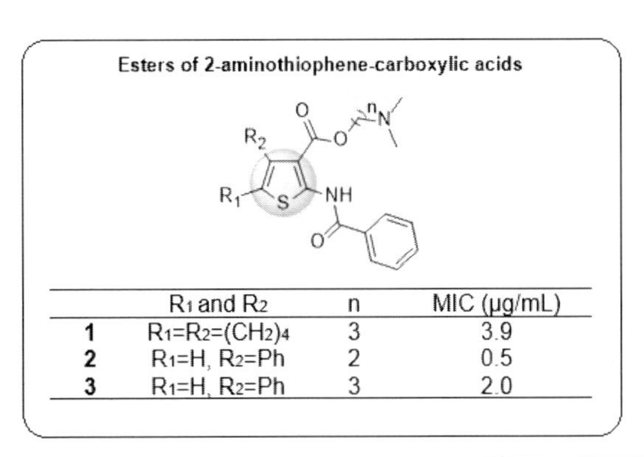

	R1 and R2	n	MIC (µg/mL)
1	$R_1=R_2=(CH_2)_4$	3	3.9
2	R_1=H, R_2=Ph	2	0.5
3	R_1=H, R_2=Ph	3	2.0

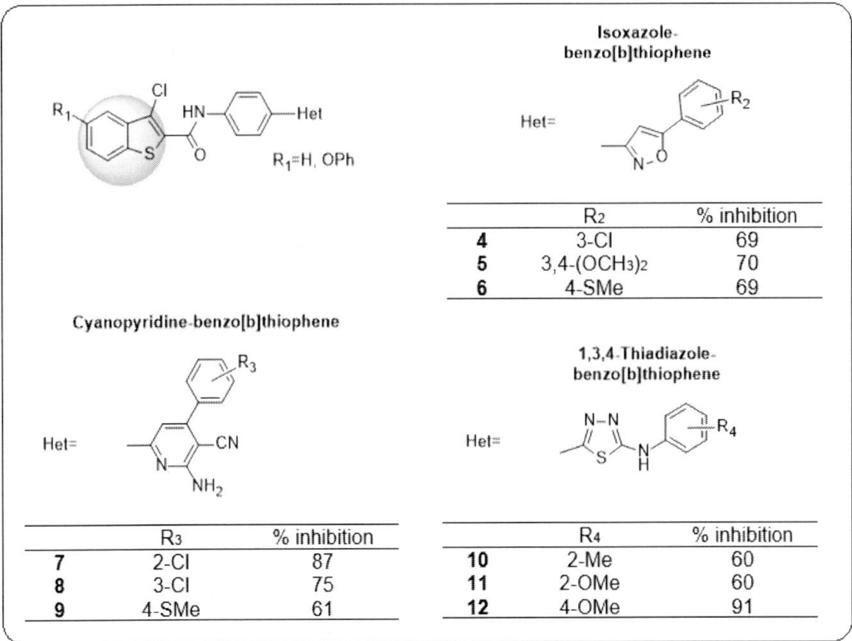

	R2	% inhibition
4	3-Cl	69
5	3,4-$(OCH_3)_2$	70
6	4-SMe	69

	R3	% inhibition
7	2-Cl	87
8	3-Cl	75
9	4-SMe	61

	R4	% inhibition
10	2-Me	60
11	2-OMe	60
12	4-OMe	91

Figure 3. First thiophene **(1-3)** and benzo[*b*]thiophene **(4-12)** derivatives synthesized and evaluated against MTB described in the literature.

Since 1981, thiophene derivatives has been synthesized and evaluated against MTB. Among the first studied compounds, three esters of 2-

aminothiophene-carboxylic acids **(1-3)** were reported by Grinev and collaborators [4], showing high anti-tuberculostatic activity, with MIC (minimal inhibitory cncentration) values ranged from 0.5 to 3.9 μg/mL. Only twenty-three years later, isoxazole- and cyanopyridine-benzo[*b*]thiophene derivatives were screened for their antitubercular activity against MTB $H_{37}Rv$ (considered the primary screening method for anti-MTB drugs). Variations of substituents at all positions of the phenyl group were planned, and among all evaluated compounds, eight **(4-9)** showed %inhibition ranging from 61 to 87% at 6.25 μg/mL. In general, derivatives substituted with chlorine and thiomethyl groups presented the best activities. In subsequent studies, these authors synthesized and evaluated azetidinone, acetyl oxadiazole, and 1,3,4-thiadiazole analogs bearing the benzo[*b*]thiophene nucleus. Only the benzo[*b*]thiophene-1,3,4-thiadiazole derivatives **(10-12)** were active, and in this last series, the compound substituted with methoxy group at 4 position in the aniline moiety **(12)** was the most potent compound, with %inhibition of 91%, and MIC > 6.25 μg/mL (Figure 3).

As the first nitrothiophene derivatives studied as anti-MTB agents, Foroumadi and coauthors [5]. reported the activity of 2-[5-nitrothiophene-1,3,4-thiadiazol-2-ylthio, sulfinyl and sulfonyl] propionic acid alkyl esters **(13-17)**. Among all evaluated compounds, the active anti-MTB agents were one α-propionate **(13)**, their analog synthesized by oxidation of S to SO_2 **(14)**, and other derivatives containing a methylene group on the linker attaching carbonyl and S moieties **(15-17)** in a special compound **16** substituted with propyl group, showing MIC of 1.56 μg/mL and inhibition of 98%. Another series of 5-nitrothiophene-based derivatives **(18-21)**, coupled with benzothiadiazine moiety, was synthesized and evaluated against MTB $H_{37}Rv$, exploring the antibacterial potential of benzothiadiazine as the cyclic sulfonamide class. Compounds substituted with phenyl and isopropyl groups showed the best MIC values (4 μg/mL); however, replacement with methyl and ethyl groups led to reduction of activity by two- and four-fold, respectively (Figure 4) [5].

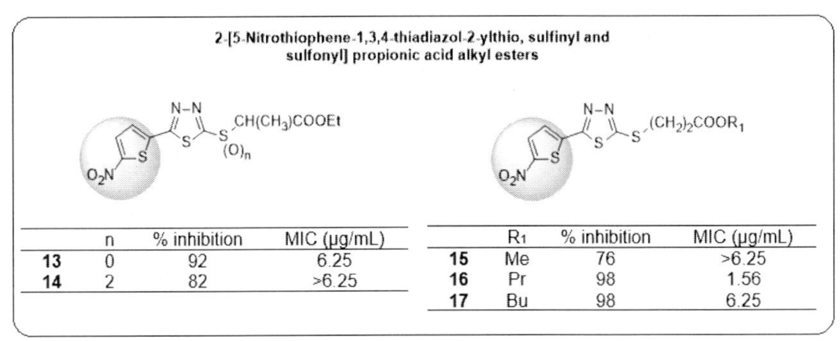

	n	% inhibition	MIC (µg/mL)
13	0	92	6.25
14	2	82	>6.25

	R1	% inhibition	MIC (µg/mL)
15	Me	76	>6.25
16	Pr	98	1.56
17	Bu	98	6.25

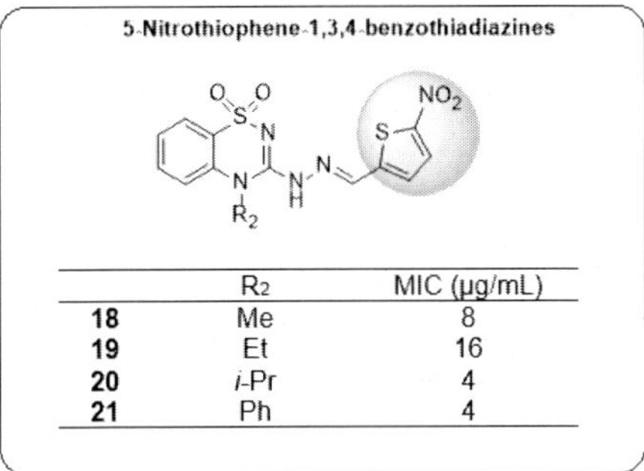

	R2	MIC (µg/mL)
18	Me	8
19	Et	16
20	*i*-Pr	4
21	Ph	4

Figure 4. First 5-nitrothiophene **(13-21)** derivatives synthesized and evaluated against MTB described in the literature.

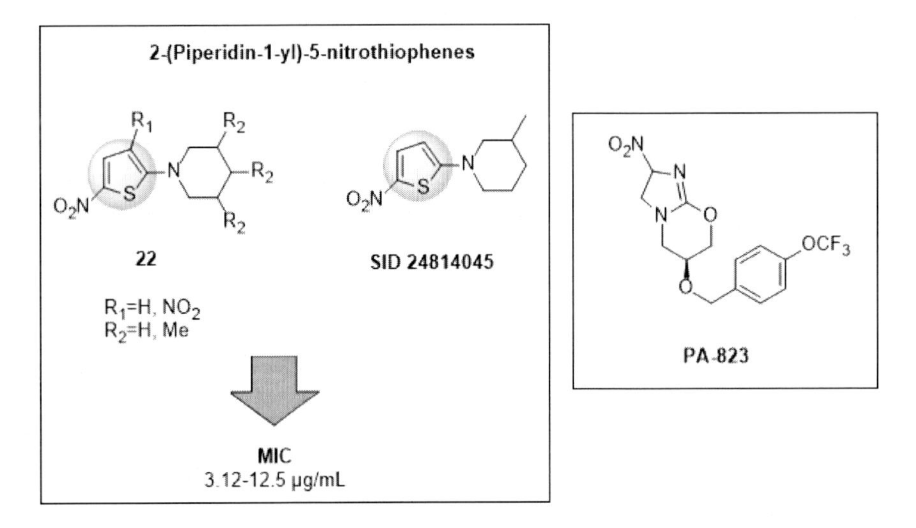

Figure 5. A – 5-Nitrothiophene derivatives active against MTB $H_{37}Rv$ and non-replicating (ss18b) strains through reduction by a nitroreductase and release of nitric oxide. B – Chemical structure of **PA-824**.

More recently, Hartkoorn and collaborators [6] demonstrated that the derivative 2-(3-methylpiperidin-1-yl)-5-nitrothiophene **(SID 24814045)** was equipotent against replicating $H_{37}Rv$ and nonreplicating ss18b, and showed no cytotoxicity against HepG2 and A549 cells at 20 μg/mL. Therefore, others analogs were synthesized **(22)**, exchanging position and number of methyl groups at the piperidine moiety and introducing a second nitro group at the C-3 position of the thiophene ring. This series of 5-nitrothiophene derivatives was investigated to determine the killing mechanisms against actively growing $H_{37}Rv$ and non-replicating (ss18b) strains. The authors found that all 5-nitrothiophenes were active against MTB $H_{37}Rv$, and the mechanism of action was similar to that of **PA-824** (nitroimidazole), involving the reduction of the nitrothiophene by a nitroreductase to release nitric oxide that kills MTB (Figure 5) [6].

The chemical structure of the first-line drug ETB and other amino alcohol derivatives were used as starting points for the design of new thiophene compounds containing diarylmethane and 2-hydroxy-amino functionalities **(23)** with antimycobacterial activity. The biological assays revealed that some compounds showed MIC values of 1.56 and 6.25 μg/mL against MTB $H_{37}Rv$; unfortunately, the introduction of 2-hydroxy-amino group led to inactive compounds (Figure 6). The same research group decided to continue the work by changing possible substitutions at the amino alkyl chain R2 and retain both electron-donating (EDG) and electron-withdrawing (EWG) groups at the benzene ring (R1) on **23**. Among thirty-one synthesized and screened compounds, nineteen showed MIC values ranging from 0.78 to 9.19 μg/mL, and four compounds **(24-27)** that were non-toxic against *Vero C-1008* cell line with SI > 10, were selected for *ex vivo* assays in mouse and human macrophage models of tuberculosis (Figure 6). In the mouse model, the efficacy of the four derivatives were comparable to those of the three first-line MTB drugs INH, RFM and PZA, and the compounds **26** and **27** displayed *ex vivo* activity comparable with that of INH. Finally, the thiophene derivative **27** **(S006-830)** induced approximately 15-fold reduction in viable MTB in the lungs of infected mice, and showed activity against MDR-TB (resistant to RFM and INH), with MIC value of 6.25 μg/mL [7].

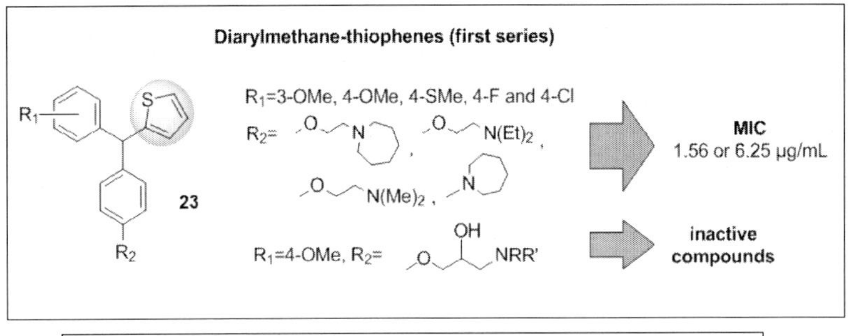

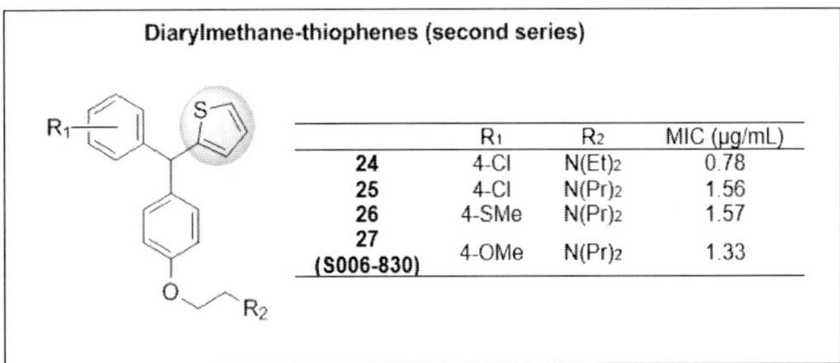

	R1	R2	MIC (µg/mL)
24	4-Cl	N(Et)2	0.78
25	4-Cl	N(Pr)2	1.56
26	4-SMe	N(Pr)2	1.57
27 (S006-830)	4-OMe	N(Pr)2	1.33

Figure 6. Two series of diarylmethane-thiophenes, including derivative **27 (S006-830)** that displayed *ex vivo* activity comparable with that of INH and showed activity against MDR-TB (resistant to RFM and INH).

The identified lead compound **27 (S006-830)** showed good pharmacokinetic (PK) properties including: (*i*) peak plasma concentration <1 h post-oral dose; (*ii*) elimination half-time ~9 h; (*iii*) mean residence time ~11 h; (*iv*) plasma protein binding ~60%; and (*v*) bioavailability ranging from 45% to 50%. These PK parameters characterize **S006-830** as a drug-like compound. However, **S006-830** is a racemic mixture and the two enantiomers forms *R* and *S* were separated on a chiral column and their anti-MTB activity were compared with the racemate [8].

Figure 7 summarizes the results: (*i*) the *S*-enantiomer (MIC of 3.12 µg/mL) was over 2- and 4-fold more potent against MTB $H_{37}Rv$ than the racemate (MIC ≥ 6.25 µg/mL) and the *R*-enantiomer (MIC of 12.5 µg/mL), respectively; (*ii*) only the *S*-enantiomer killed the bacilli *in inocula*, showing bactericidal activity, whereas the *R*-enantiomer and racemate appeared to be bacteriostatic; (*iii*) when infected macrophages were exposed to 2x MICs values, after 5 days of exposure, the *S*-enantiomer killed 91% of the bacilli, and was more active *ex vivo* than was

INH (80% killing) and the *R*-enantiomer (62% killing); and (*iv*) the combination of the *S*-enantiomer and RIF (2x MICs each) produced 100% killing of the bacilli, and this combination was more efficacious than combinations with INH or EMB. As for the mechanism of action, the *S*-enantiomer inhibited mycolic acid biosynthesis by inhibition of *β*-ketoacyl-CoA reductase, which utilizes NADH to reduce *β*-ketoacetyl-CoA to *β*-hydroxyacetyl-CoA, involved in type-II fatty acid synthase (FAS II) pathway. Mycolic acids are vital components of the MTB cell envelope [8].

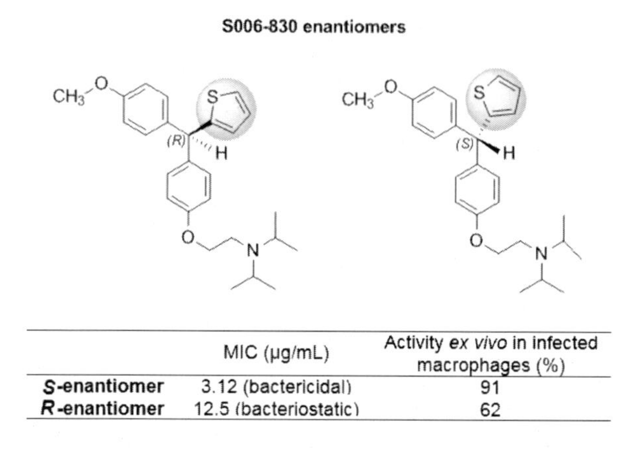

	MIC (μg/mL)	Activity *ex vivo* in infected macrophages (%)
***S*-enantiomer**	3.12 (bactericidal)	91
***R*-enantiomer**	12.5 (bacteriostatic)	62

Figure 7. *In vitro* and *ex vivo* activities of *S* and *R* enantiomers of **S006-830.**

There are other examples of prototypes used for design of new anti-MTB drugs. Thiolactomycin (TLM), a natural thiolactone, and other TLM chemically-modified analogs, target *β*-ketoacyl-ACP synthase (*mt*FabH and *mt*FabB), leading to inhibition of mycolic acid biosynthesis and, subsequent MTB death. Lu and coauthor [9] reported synthesis and anti-MTB activity of two series of 2-amino-5-(4-(2,6-dichlorobenzyloxy) phenyl)thiophene-3-carboxylic acid derivatives **(28-34)**, where the thiophene ring was thought to be an alternative and easily-accessible five-membered heterocyclic isostere of the thiolactone moiety (Figure 8). These compounds were the first thiophene derivatives described in the literature to be tested against MDR-TB and XDR-TB. In the first series evaluated against MTB *H37Rv*, substitutions at the phenyl ring, amide, or amine groups led to the most potent compounds **28-34**. Four of these that displayed lower MICs and better selectivity index (SI) values, were evaluated against MDR-TB and XDR-TB clinical strains. They found that compounds **29** and **34** were five- and four-times more active than INH [9].

2-Amino-5-(4-(2,6-dichlorobenzyloxy)phenyl) thiophene-3-carboxylic acids

NHR$_1$
COOH
for 34 COOEt
Cl
O
Cl

	R_1	SI	MIC (μM)		
			MTB	MDR-TB	XDR-TB
28	—CO–3-ClPh	13.6	6.6	-	-
29	—CO–4-ClPh	14.5	3.7	48	12
30	*n*-propyl	2.8	7.5	-	-
31	Isobutyl	4.4	7.8	-	-
32	*n*-butyl	4.6	7.7	128	128
33	CH_2COOH	5.8	3.7	32	32
34	O OH O	16.5	1.9	16	16

Figure 8. 2-Aminothiophene derivatives based on the chemical structures of thiolactomycin (TLM) evaluated against MTB, MDR-TB and XDR-TB.

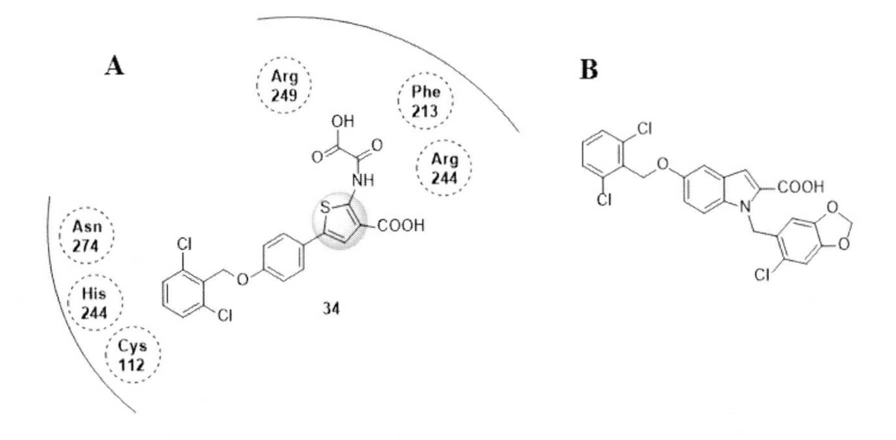

Figure 9. A - Interactions of 2-aminothiophene derivative **34** in the active site of *mt*FabH of MTB. B – Chemical structure of indole analogue potent inhibitor of *mt*FabH.

For the most active compound of this series **(34)**, a docking study in *mt*FabH revealed that 2,6-dichlorobenzyloxy moiety fit into the catalytic triad Cys[112], His[244], and Asn[274]. A hydrogen bond is formed between the carbonyl oxygen (carboethoxyl group) and the amino group of Arg[249], and the acylated group interacted with Arg[249], Phe[213], and Arg[244] residing at the top-right of the active site (Figure 9A). These molecular modeling studies revealed that the thiophene derivative share similar binding modes of the indole analogue (Figure 9B) as potent inhibitors of the enzyme [9].

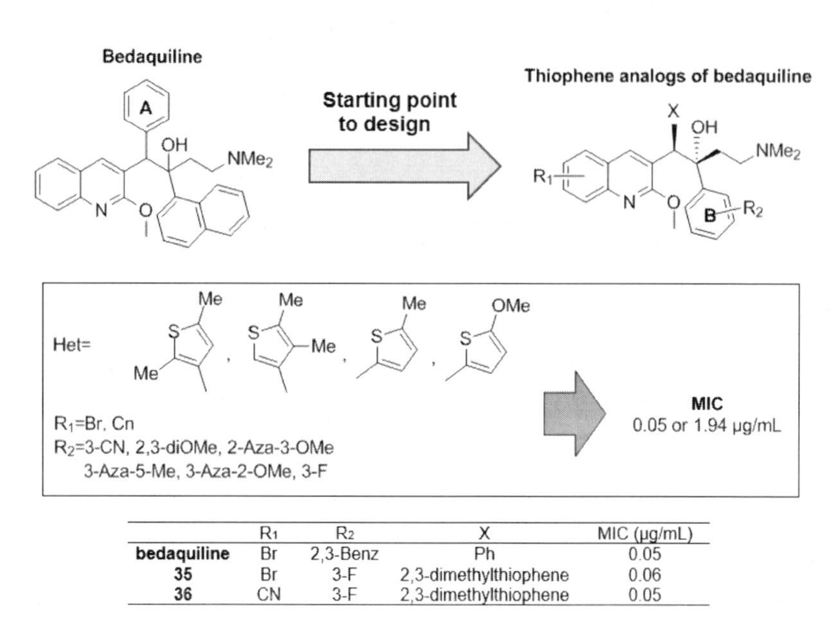

	R_1	R_2	X	MIC (μg/mL)
bedaquiline	Br	2,3-Benz	Ph	0.05
35	Br	3-F	2,3-dimethylthiophene	0.06
36	CN	3-F	2,3-dimethylthiophene	0.05

Figure 10. Thiophene derivatives analogs of bedaquiline active against MTB.

Bedaquiline (Figure 10) is an anti-MTB drug approved by the US FDA in 2012 for treatment of MDR-TB. However, its lipophilicity (logP 7.25) results in accumulation in tissue and contributes to its long terminal half-life of 5–6 months [10]. In order to develop less lipophilic compounds, Choi and coauthors [10] explored the synthesis and screening anti-MTB of new analogues of bedaquiline, replacing the phenyl A-ring with heterocycles, including a thiophene ring, and exploring some substituents at the B-ring (Figure 10). They reported that all thiophene analogs had clogP values lower than bedaquiline, and MICs values ranged from 0.05 to

1.94 μg/mL against replicating MTB $H_{37}Rv$. Compounds **35** and **36**, both presenting dimethyl substituted thiophene had comparable activity to that of bedaquiline, and derivative **36** presented a higher rate of clearance and a half-life approximately 4-fold lower on exposure to human liver microsomes [10].

In 2009, Balamurugan and colleagues [11] employed domino Gewald reaction–dehydrogenation sequence for the synthesis of 2-amino-5-arylthieno[2,3-*b*]thiophene derivatives, comprising two coupled thiophene rings (Figure 11). In their biological results, compounds substituted with ethyl ester group were generally more activity against MTB and MDR-TB. Among all synthesized and tested compounds, derivatives **37**, **38**, and **40** showed MIC values ranging from 1.1 to 8.4 μM, and they were more potent than the first-line drug ETB against MTB, whereas compounds **37-41** were 80–370 times more active against MDR-TB than the second line drug PZA [11].

2-Amino-5-arylthieno[2,3-b]thiophenes

	Ar	R_1	MIC (μg/mL)	
			MTB	MDR-TB
37	COOEt	4-MePh	4.9	4.9
38	COOEt	4-PrPh	2.3	2.3
39	COOEt	3,4-Cl_2Ph	8.4	4.2
40	COOEt	1-Naphthyl	1.1	1.1
41	CN	1-Naphthyl	10.2	5.1

Figure 11. Some derivatives containing two coupled thiophene rings were 80-370 times more active against MDR-TB than the second line drug pyrazinamide.

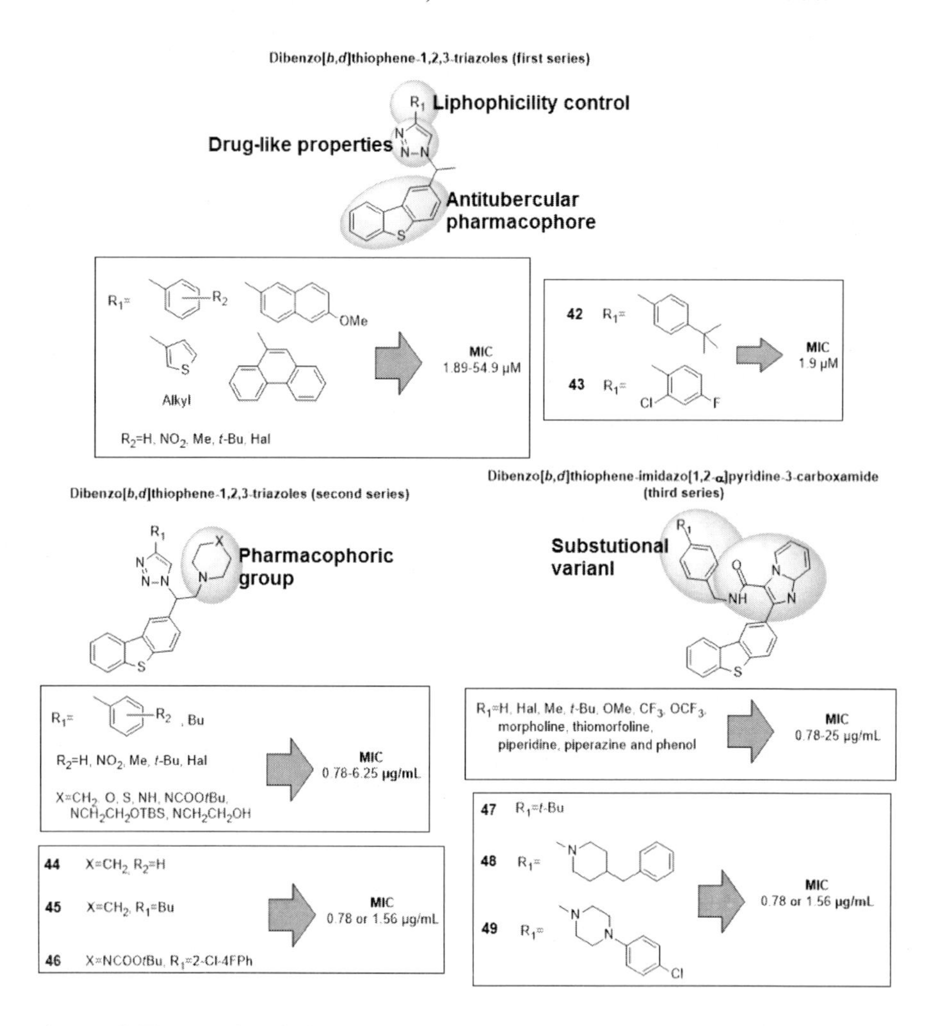

Figure 12. Three series of dibenzo[*b,d*]thiophene coupled with others pharmacophore fragments with antitubercular activity.

The heterotricyclic dibenzo[*b,d*]thiophene, consisting of a benzothiophene *ortho*-fused with another benzene ring across the 2,3- and 4,5-positions, was explored by Patip and collaborators [12] for the design of new anti-MTB compounds. In their study, a molecular hybridization approach was adopted, where the 1,2,3-triazole moiety was considered the central backbone, on which was attached dibenzo[*b,d*]thiophene as the antitubercular pharmacophore fragment, as well as aliphatic or aromatic groups for control of lipophilicity (Figure 12). Among twenty screened compounds, eleven were more active or equipotent to ethambutol (MIC of 1.9 μM), and special derivatives substituted with 4-*t*-BuPh **(42)** and 2-Cl-4-FPh **(43)** groups were 26 times more active than PZA and 4 times more active than ethambutol, while also exhibiting lower toxicity (SI ranging from 55 to 255) in four cell lines [12].

The same research group synthesized two other series of dibenzo[*b,d*]thiophene-1,2,3-triazoles. In the second series, the thiophene-triazole hit scaffold was coupled with morpholine, thiomorpholine, piperidine, and piperazine units; these were thought to be other pharmacophore fragments that inhibit MTB. The screening against MTB $H_{37}Rv$ revealed that, among twenty screened compounds, ten exhibited MIC values ranging from 0.78 to 6.25 μg/ml. Special derivatives **44-46** exhibited greater *in vitro* potency (MIC values of 0.78 and 1.56 μg/mL) than ETB and PZA, and were less toxic against Human Embryonic Kidney (HEK-293T) cells (Figure 12) [13].

In the third series, the dibenzo[*b,d*]thiophene hit scaffold, the 1,2,3-triazole moiety was replaced by the antitubercular fragment imidazo[1,2-*α*]pyridine-3-carboxamide appended with substituted benzyl amines. Similar screening with nineteen derivatives of this last series rendered compounds with MIC values ranging from 0.78 to 25 μg/mL, and compounds **47-49** were equipotent to previously reported thiophene-triazole derivatives **44-46** and were considered non-toxic (Figure 12) [14].

The pharmacophore fragments morpholine, thiomorpholine, and *N*-substituted piperazine were used as building blocks by Marvadi and coworkers [15] and were hybridized with a dihydroquinoline moiety substituted with a thiophene ring at the 2-position to generate a new scaffold for anti-MTB evaluation. Among sixteen synthesized and evaluated compounds, six **(50-55)** presented MIC values ranging from 1.56 to 6.25 μg/mL against MTB $H_{37}Rv$ (Figure 13). The most active compounds presented a piperazine unit *N*-substituted with benzyl **(52)** and pyridine groups **(55)** that were more potent than ETB, with lower cytotoxicity profiles. According to the authors, the clogP values (6.32 and 5.08, respectively), when compared with the previously reported bedaquiline (anti-MTB drug; clogP of 7.3) characterize these two compounds as hits for *in vivo* assays and mechanistic evaluations [15].

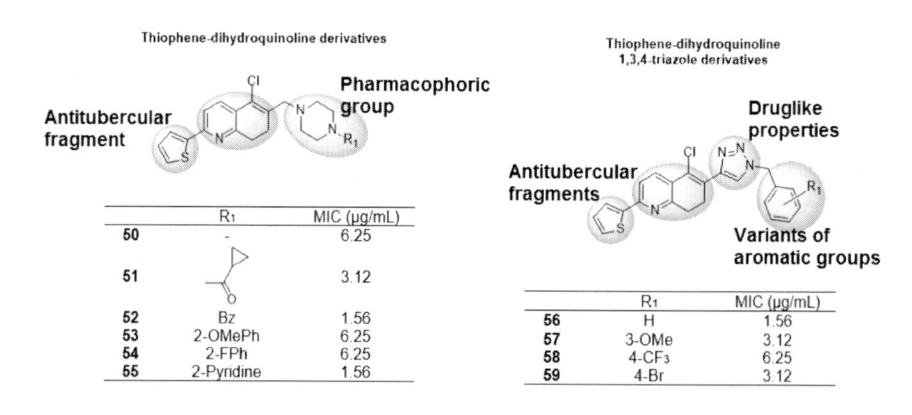

	R1	MIC (μg/mL)
50	-	6.25
51		3.12
52	Bz	1.56
53	2-OMePh	6.25
54	2-FPh	6.25
55	2-Pyridine	1.56

	R1	MIC (μg/mL)
56	H	1.56
57	3-OMe	3.12
58	4-CF_3	6.25
59	4-Br	3.12

Figure 13. Thiophene-dihydroquinoline derivatives hybridized with morpholine, *N*-substituted piperazine and 1,3,4-triazole moieties with antitubercular activity.

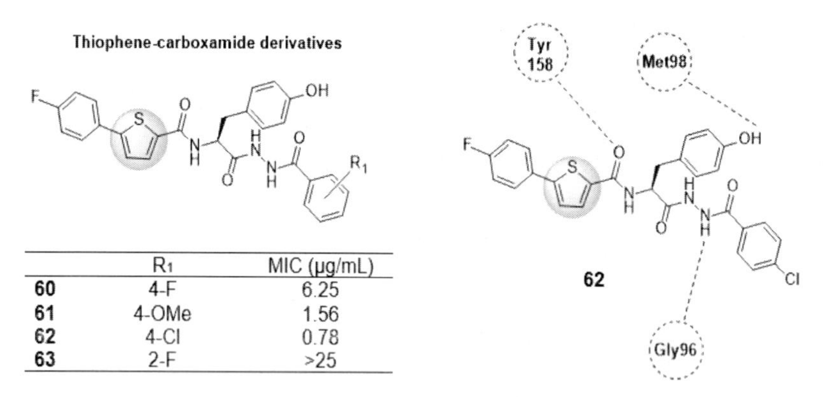

	R1	MIC (μg/mL)
60	4-F	6.25
61	4-OMe	1.56
62	4-Cl	0.78
63	2-F	>25

Figure 14. Series of 4-thiophene-carboxamide derivatives against MTB $H_{37}Rv$ and interactions of the most active compound **62** in the active site of enoyl reductase (INHA).

The same research group explored the thiophenyl-dihydroquinoline antitubercular fragment present in the structures of the active compounds **50-55** in another synthesized series of compounds, comprising a second pharmacophore 1,3,4-triazole, present in some clinically-available antitubercular drugs. Among fifteen synthesized and evaluated compounds, derivative **56** (MIC of 1.56 μg/mL) was more potent than ETB and was equipotent to ciprofloxacin. Compounds **57** and **59** (MIC of 3.12 μg/mL for both) were equipotent to ETB and compound **58** presented an MIC of 1.56 μg/mL. In general, the four most active compounds showed low cytotoxicity (Figure 13) [16].

Kulmar and coworkers [17] developed a series of 4-thiophene-carboxamide derivatives that were tested against MTB $H_{37}Rv$ **(60-63)** (Figure 14). Some substitutions at the benzene ring led to the compound **62** (R1: 4-Cl), which was 2-fold more potent than ETB. In order to obtain more information about mechanism of action and binding mode of this compound, docking and molecular dynamic (MD) studies were performed with MTB enzyme enoyl reductase (INHA). The active compound **62** exhibited good binding free energy, and their interactions with Tyr^{158}, Met^{98}, and Gly^{96} residues were similar to those of INH. MD simulation, in physiological and environmental conditions, indicated small structural re-arrangements and fewer conformational changes around the binding site than the co-crystal ligand. Furthermore, *in silico* drug-likeness ADMET properties suggest that **62** is a potential lead for anti-MTB drug development [17].

High-throughput screening (HTS), a traditional approach for discovery of hits and leads in medicinal chemistry, was applied in a large library of drug-like small molecules, leading to the discovery of the molecule **SID 92097880** (6-acetyl-2-(thiophene-2-carboxamido)-4,5,6,7-tetrahydrothieno [2,3-c]pyridine-3-carboxamide; Figure 15) that showed activity against MTB $H_{37}Rv$ with an MIC value of 7.64 μM and an SI of 13. In order to develop a more potent compound, this lead compound was used as a starting point to design 2-substituted 4,5,6,7,8,9-tetrahydrocycloocta[*b*] thiophene-3-carboxamides. Among the synthesized and screened compounds, two **(64** and **65)**, substituted with 4-tolyl and 4-phenoxyphenyl groups, respectively; inhibited MTB with MICs of 4.54 and 3.70 μM, respectively, and the new compounds were found to be more potent than ethambutol (MIC: 7.64 μM) and lead compound **SID 92097880**. In addition, derivative **65** showed SI > 10, and when tested in the presence of verapamil (an efflux pump inhibitor), the MIC value decreased 3-fold (Figure 15) [18].

Wangh and colleagues [19] demonstrated that the tetrahydrocycloocta[*b*]thiophene antitubercular fragment tethered cephalexin **(66)** showed potent activity against MTB $H_{37}Rv$ with a MIC value of 0.78 µg/mL. It is interesting to note that cephalexin substituted with ampicillin or amoxicillin generated inactive compounds. Derivative **66** showed SI ≥ 10 in three cell lines, and was considered a non-toxic, promising therapeutic agent [19].

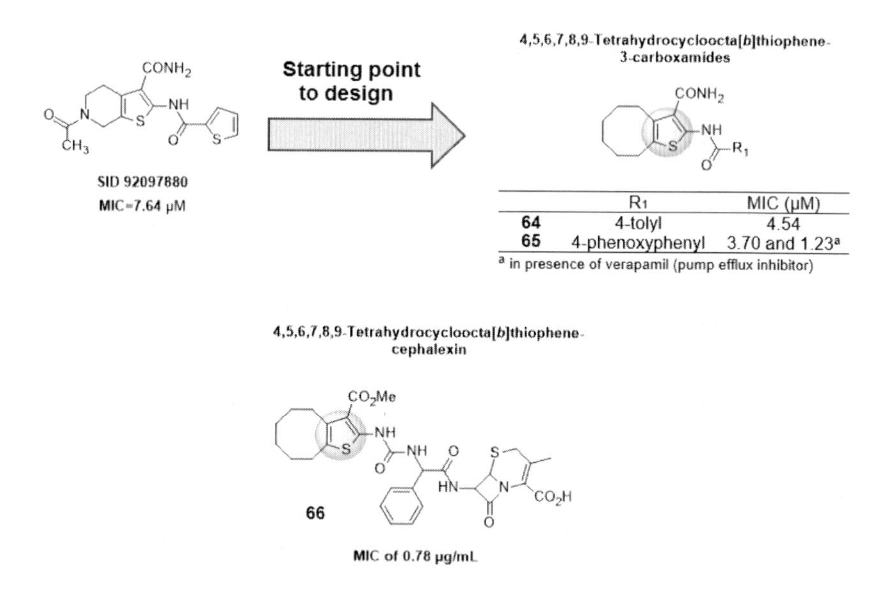

Figure 15. Lead compound discovered using HTS (High-throughput screening) as a starting point for design of active tetrahydrocycloocta[*b*]thiophene derivatives. The thiophene moiety tethered cephalexin with potent activity against MTB $H_{37}Rv$.

LEISHMANIASIS

Leishmaniasis, one of the principal NTD, is a complex of infectious diseases caused by protozoan parasites that belong to the *Leishmania* genus. These parasites are transmitted to humans and animals by the bites of the female sandfly. It affects around 12 million people in more than 98 countries distributed across the five continents. Over 20 species and subspecies of Leishmania infect humans, causing three major clinical manifestations ranging from nodular and ulcerative skin lesions (cutaneous), to progressive mucocutaneous, to visceral leishmaniasis, the most severe and potentially fatal form [20].

Currently, there is no human vaccine available against any clinical manifestation of the disease, and the chemotherapeutic treatments are very limited, including treatment with obsolete drugs with toxic side-effects, requiring long periods of treatment, and questionable effectiveness because of the emergence of drug resistance. The first-line drugs are the pentavalent antimonials ((meglumine antimonate (Glucantime®) and sodium stibogluconate (Pentostam®)). The second line includes drugs containing liposomal amphotericin B (AmBisome®), pentamidine (Pentacarinat®) and miltefosine (Impavido®) (the only available oral drug) [21, 22].

This scenario clearly emphasizes the urgent need for the development of novel drugs with antileishmanial activity. The major challenge is to discovery new drugs that meet all of the following criteria: effectiveness against the two parasitic forms, promastigotes and amastigotes (the intracellular form, living in macrophages); ability to eliminate reservoirs (decrease the parasite load of popliteal lymph nodes and spleen); possibility of oral administration; shorter treatment cycles, and avoidance of the emergence of resistant strains.

The first reports of evidence that thiophene derivatives present leishmanicidal activity were described by researchers at the Aix-Marseille University in 1993 and 1994 [23, 24]. These initial investigations with 5-nitrothiophenes demonstrated the potential of this heterocycle, and allowed various research groups around the world to begin exploring this scaffold in an attempt to identify potential hit and lead compounds.

Figure 16 shows some examples extracted from the literature, in which thiophene derivatives were shown to be active against various species and evolutionary forms of *Leishmania*. Because the assays were performed using different methodologies, and with different *Leishmania* species, comparisons of the inhibition values are not adequate; nevertheless, they allow observation of the great structural diversity that results in compounds with promising activity, including mainly 5-nitro-thiophenes and 2-amino-thiophenes.

Research groups at the Institute Pasteur Korea, after performing high-throughput (HTS) and high-content screening (HCS) of a 200.000 chemical library, identified the scaffold 2-acetamidothiophene-3-carboxamide as potential hit for the development of new anti-leishmanial agents. A series of 2-acetamidothiophene-3-carboxamide derivatives were synthesized with modifications in three parts of the scaffold: changing the cycloalkyl ring fused to thiophene, and performing modifications of the 2-acetamido and 3-carboxamide moieties with different substituents [25].

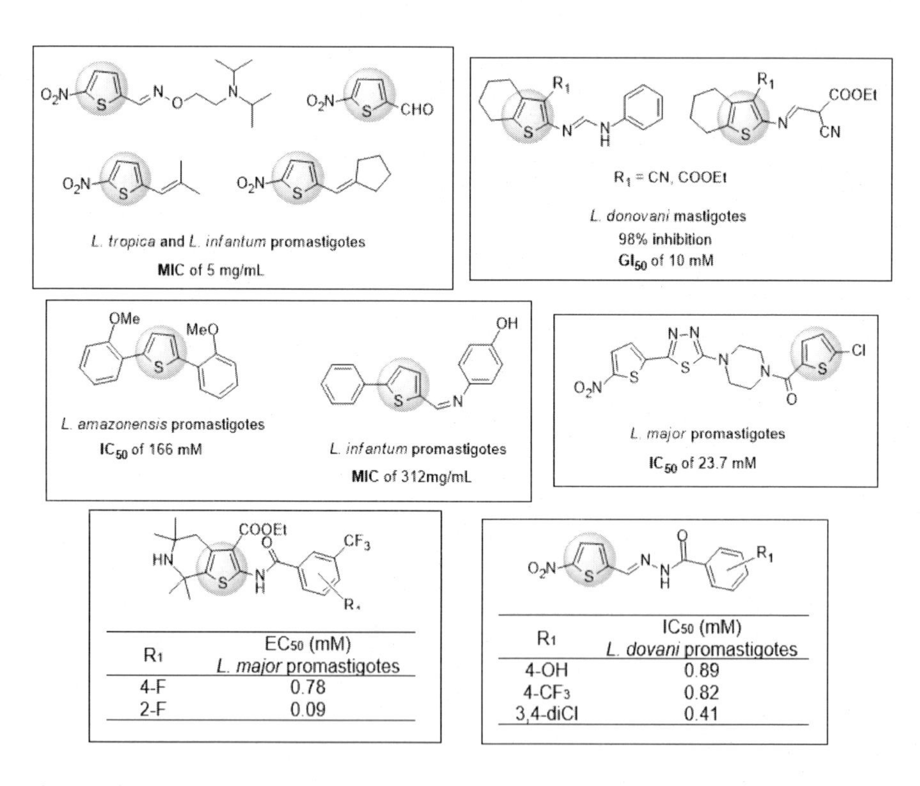

R_1	EC_{50} (mM) *L. major* promastigotes
4-F	0.78
2-F	0.09

R_1	IC_{50} (mM) *L. dovani* promastigotes
4-OH	0.89
4-CF_3	0.82
3,4-diCl	0.41

Figure 16. Examples of thiophene derivatives active against *Leishmania* species.

The data of the *in vitro* anti-leishmanial activity against amastigote and promastigote forms of *L. donovani*, and the cytotoxicity against macrophages are summarized in Table 1.

Table 1. Anti-leishmanial and cytotoxicity activities of 2-acetamidothiophene-3-carboxamides against amastigotes and promastigotes forms of *L. donovani* and THP-1 cells (values in μM)

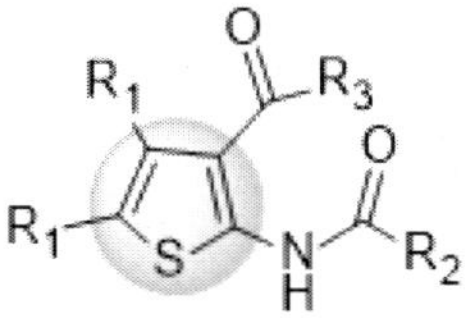

				Amast.	THP-1		Promast.
Cdpd	R1	R2	R3	EC_{50}	CC_{50}	SI	EC_{50}
67	$-(CH_2)_5-$	-	*t*-Butoxy	> 50	> 50	-	n.d
68	$-(CH_2)_5-$	CH_3	*t*-Butoxy	> 50	> 50	-	n.d
69	$-(CH_2)_5-$	CF_3	4-OCH_3-Ph-NH-	6.41	> 50	> 7.80	> 50
70	$-(CH_2)_4-$	CF_3	4-OCH_3-Ph-NH-	14.6	> 50	> 3.42	> 50
71	$-(CH_2)_3-$	CF_3	4-OCH_3-Ph-NH-	>50	> 50	-	n.d
72	CH_3	CF_3	4-OCH_3-Ph-NH-	10.1	> 50	> 4.95	> 50
73	H	CF_3	4-OCH_3-Ph-NH-	> 50	> 50	-	> 50
74	$-(CH_2)_5-$	CF_3	3-OCH_3-Ph-NH-	10.1	37.6	3.71	21.5
75	$-(CH_2)_5-$	CF_3	2-OCH_3-Ph-NH-	> 50	> 50	-	42.0
76	$-(CH_2)_5-$	CF_3	Ph-NH-	> 50	> 50	-	n.d
77	$-(CH_2)_5-$	CF_3	4-CH_3-Ph-NH-	> 50	> 50	-	n.d
78	$-(CH_2)_5-$	CF_3	4-Cl-Ph-NH-	> 50	> 50	-	n.d
79	$-(CH_2)_5-$	CF_3	4-CF_3-Ph-NH-	17.6	17.2	0.90	n.d
80	$-(CH_2)_5-$	CF_3	4-OCF_3-Ph-NH-	9.85	11.9	1.20	> 50
81	$-(CH_2)_5-$	CF_3	4-CN-Ph-NH-	8.53	9.11	1.06	> 50
82	$-(CH_2)_5-$	CF_3	4-OCH_3-Ph-CH_2-NH-	> 50	> 50	-	> 50
83	$-(CH_2)_5-$	CF_3	*t*-Butyl-NH-	> 50	> 50	-	n.d
84	$-(CH_2)_5-$	CH_3	4-OCH_3-Ph-NH-	> 50	> 50	-	> 50
85	$-(CH_2)_5-$	OCH_3	4-OCH_3-Ph-NH-	> 50	> 50	-	n.d
86	$-(CH_2)_5-$	Ph	4-OCH_3-Ph-NH-	> 50	> 50	-	n.d
Milt.	-	-	-	7.10	22.7	3.19	0.18

Cpd.: Compounds; Milt.: Miltefosine; Amast.: amastigotes; Promast.: promastigotes; n.d: not determined; EC_{50}: Effective Concentration for 50%; CC_{50}: Cytotoxicity Concentration for 50%; SI: Selectivity Index.

As can be seen in Table 1, seven compounds were active ($EC_{50} < 50$ μM) against amastigotes and only two were active against promastigote forms (compounds **74** and **75**). Compound **69** was the most active against amastigote forms (EC_{50}: 6.41 μM) being equipotent with the reference drug miltefosine (7.10 μM), with the advantage of a larger selectivity index (7.80 versus 3.19).

Concerning the structure–activity relationship (SAR), the presence of a cycloheptyl ring fused to thiophene core resulted in the most active compounds. Size reduction of the fused ring to cyclohexyl or cyclopentyl gradually decrease the activity, indicating that lipophilicity plays an important role in amastigote activity. In the 2-acetamido position, the presence of the trifluoromethyl group is essential for activity. Substitution of the three fluoro atoms by hydrogens (methyl group (**84**)) resulted in inactive compounds, suggesting that the presence of the electron withdrawing group is essential. At least, substitutions in the 3-carboxamide moiety are more tolerable. Substitutions in the para position resulted in more active compounds, especially with the presence of 4-methoxyphenylamino (**69**), 4-cyanophenylamino (**81**) and 4-trifluoromethoxyamino (**80**) groups. *Meta-* and *ortho-*substitutions gradually reduced the activity, ultimately resulting in inactive compounds (i.e., compounds **74** and **75**).

In another study published by the Eli Lilly Company and the Biomedical Sciences Institute of the São Paulo University, the researchers evaluated the antileishmanial potential of a library containing 147 2-arylbenzothiophene compounds, based on the tamoxifen chemical structure [26]. Despite the fact that tamoxifen is one of the drugs of choice for the treatment of breast cancer, recent studies demonstrate its *in vitro* and *in vivo* activity against several strains of *Leishmania* in the low micromolar range [27–29]. For the choice of the 2-arylbenzothiophene analogs, the stilbene core of tamoxifen was maintained as scaffold.

Initially, all the compounds were screened against promastigote forms of *L. chagasi.* A total of 123 compounds exhibited activity, with IC_{50} ranging from 3.6 to 121.9 μM, and 59 compounds had an IC_{50} less than or equal to that of tamoxifen ($\leq$ 17.7 μM). Table 2 presents one part of these results.

In brief, the SAR data for the 2-arylbenzothiophenes suggest that the presence of phenols (-OH linked to benzene) is not important for anti-leishmanicidal activity, and optimal potency was observed when two basic side chains were present in R1 and R2, as observed in compounds **92**, **94**, **95**, and **96**.

Table 2. Anti-leishmanial activity of benzothiophenes, added to the tamoxifen core, against promastigote forms of L. chagasi (values in μM)

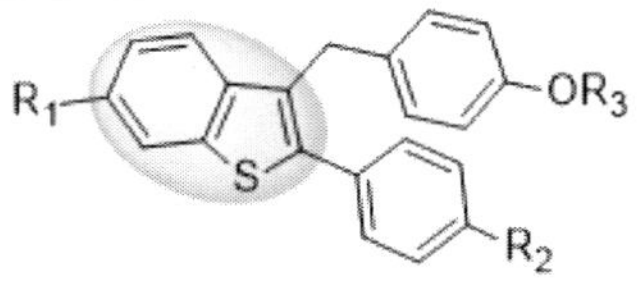

R_1, S, OR_3, R_2					
Cpd	R1	R2	R3	X	*L. chagasi* (IC_{50} ± SD)
BTP	OH	OH	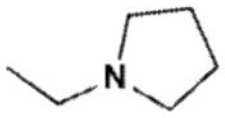 N	C=O	52.2 ± 18.2
87	H	OH	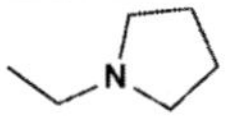 N	C=O	34.6 ± 4.1
88	H	OCH_3	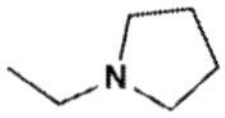 N	C=O	36.5 ± 5.7
89	H	OH	OH	C=O	88.7 ± 0.2
90	H	OCH_3	OH	C=O	79.0 ± 0.1
91	H	H	OH	C=O	65.8 ± 0.1
92	H	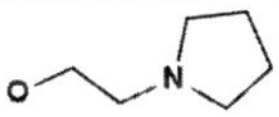 O N	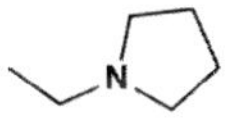 N	C=O	3.7 ± 0.3
93	H	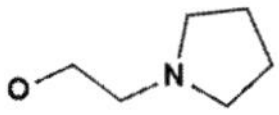 O N	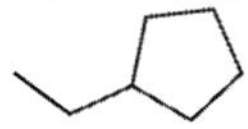	C=O	28.7 ± 1.8
94	H	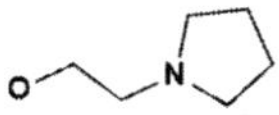 O N	 $N(i\text{-}Pr)_2$	C=O	4.3 ± 0.2
95	H	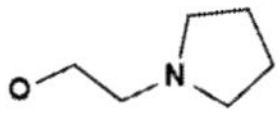 O N	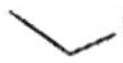 $N(n\text{-}Bu)_2$	C=O	5.4 ± 0.4
96	H	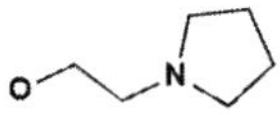 O N	 $N(i\text{-}Pr)_2$	CH_2	3.6 ± 0.2
Tam.	-	-	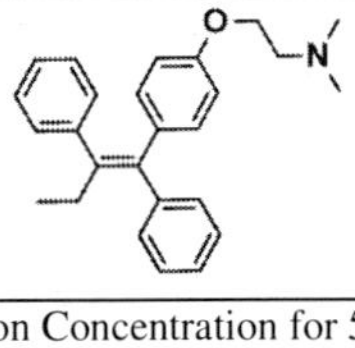 O N	-	17.7 ± 0.8

Cpd.: Compounds; Tam.: Tamoxifen; IC_{50}: Inhibition Concentration for 50%.

Another important physical property observed in the most active compounds relates to lipophilicity. As noted in the study by the Institut Pasteur Korea group, the most active benzothiophene analogs disobey one of Lipinski´s rules, and have a cLogP value higher than 5.00 (compounds **92**, **94**, **95**, and **96**, have clogP values between 6.93 – 8.26), suggesting again that lipophilicity plays an important role in the anti-leishmanicidal activity of thiophene derivatives [26].

These most promising compounds **(92**, **94** and **96)** were also active against promastigote forms of three other *Leishmania* species (with IC_{50} < 10 μM), and intracellular amastigotes of *L. chagasi*, (reducing the number of infected macrophages), being even less toxic to host cells (*Vero* cells) than the parasite (SI around 3.0) (Table 3). Taken together, these results suggest that 2-arylbenzothiophene is a promising scaffold for the development of future leishmanicidal agents.

Table 3. Anti-leishmanicidal activity of most active 2-arylbenzothiophenes against promastigotes of various *Leishmania* species (IC_{50} values in μM), amastigotes of *L. chagasi* (% of reduction of infected macrophages), and cytotoxicity to mammalian cells

	Antileishmanial activity				Cytotox.
	Promastigotes		Amastigotes		
Cpd	*L. chagasi* (IC_{50}±SD)	*L. amazonensis* (IC_{50}±SD)	*L. brasiliensis* (IC_{50}±SD)	*L. chagasi* (%infected cells±SD)	*Vero* cell (IC_{50}±SD)
92	3.7 ± 0.3	9.9 ± 0.5	5.2 ± 0.8	70.1 ± 8.1	11.7 ± 3.6
94	4.3 ± 0.2	8.0 ± 1.2	3.5 ± 0.9	64.8 ± 11.8	13.6 ± 1.6
96	3.6 ± 0.2	7.3 ± 1.8	4.2 ± 0.1	34.4 ± 6.3	13.0 ± 2.9

Cpd.: Compounds; IC_{50}: Inhibition Concentration for 50%; SD.: Standard-Deviation; Cytotox.: Cytotoxicity.

Cysteine proteases are enzymes belonging to the papain family that play important roles in the reproduction, metabolism, and intracellular survival of *Leishmania* species. Based on this knowledge, researchers from Italy and Germany searched for new synthetic inhibitors of this potential and unexplored *Leishmania* drug target. A recent study from this group reported one fused benzo[*b*]thiophene derivative **97** (Figure 17) that acts as a potent and highly selective anhydride-based inhibitor of *L. mexicana* cysteine protease CPB2.8ΔCTE, with IC_{50}: 3.7 μM and K_i: 1.3 μM. In addition, no inhibition of the highly similar human cysteine proteases (cathepsin-B and cathepsin-L) was observed, indicating high specificity [30].

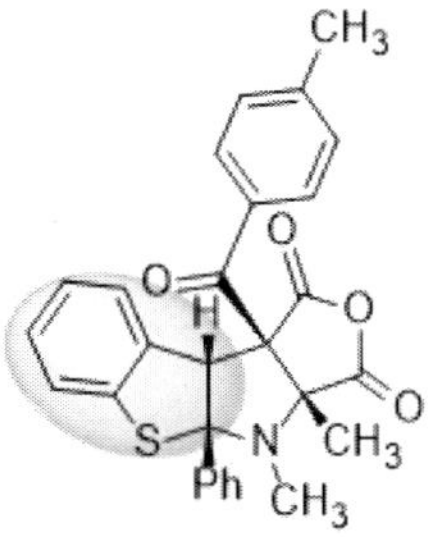

Figure 17. Structure of fused benzo[*b*]thiophene **97** selective inhibitor of *L. mexicana* cysteine protease CPB2.8ΔCTE.

Interactions of the benzo[*b*]thiophene derivative **97** specified in Figure 17 with the cysteine protease CPB2.8ΔCTE active site were investigated using Nuclear Magnetic Resonance (NMR) biomimetics and by *in silico* experiments (molecular docking, molecular dynamics and Density Functional Theory (DFT)). This study revealed evidence that the compound may act first as a competitive inhibitor, and second as an irreversible covalent inhibitor by chemo-selective attack of the residue Cys^{25} on the anhydride carbonyl, forming a covalent bond.

The last selected example of thiophene derivatives with anti-leishmania activity, compiles a series of 4 paper of our group, that already allowed to identify one new active compound with oral bioavailability.

Our group performed several steps of drug design and discovery of new and potent 2-amino-thiophene derivatives against *L. amazonensis*, including synthesis, docking, chemometric studies, leishmanicidal effects against amastigotes and promastigotes, cytotoxicity, genotoxicity, and various biological and pharmacological assays *in vitro* and *in vivo*, to elucidate the mechanism of action and toxicological effects of selected compounds.

The pharmacophore optimization steps are briefly summarized in Figure 18, also showing the chemical structures of the most active compounds against promastigote and amastigote forms of *L. amazonensis*, their IC_{50} values, and selectivity indices (SI).

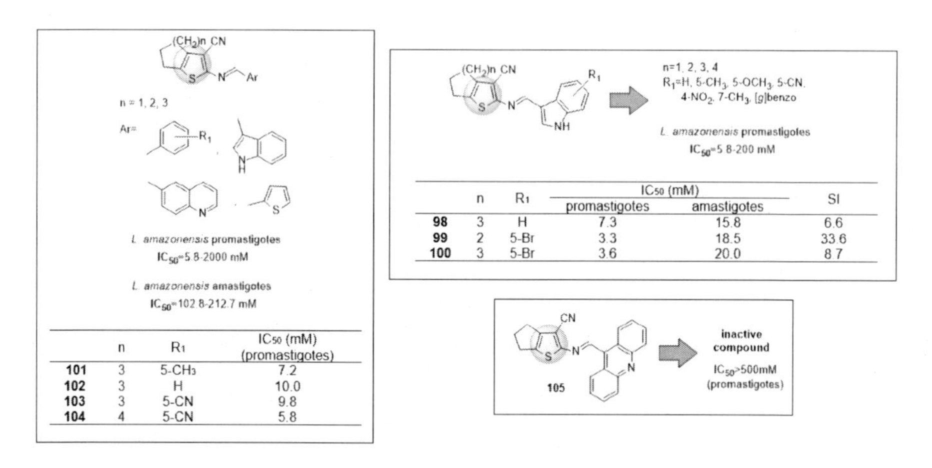

Figure 18. Pharmacophore optimization of 2-amino-thiophene derivatives indicating the most active compounds against promastigote and amastigote forms of *L. amazonensis* (IC_{50} values) and selectivity indices (SI).

In our first study, we found that 2-aminothiophene derivatives exhibit potent and selective antileishmanial activity (at low micromolar level) against both stages of *L. amazonensis* with lower levels of cytotoxicity to host cells [31].

We observed that changes in the size of the cycloalkyl ring attached to thiophene core are tolerable. The nature of the imine-bound aryl radical exerts great influence on the activity. Substitution with an indole ring was the most promising, resulting in compounds with lower IC_{50} values (3.3, 3.6, and 7.3 μM against promastigotes for compounds **99**, **100**, and **98**, respectively) and the best selectivity indices (33.6, 8.7, and 6.6, respectively for compounds **99**, **100**, and **98**, respectively) (Figure 18), showing better performance than the reference drug (meglumine antimoniate (IC_{50}: 25.7 mM, SI: 0.04).

Studies of the mechanism of action of compounds **98**, **99**, and **100** showed that these derivatives induced apoptosis in promastigotes involving phosphatidylserine externalization and DNA fragmentation, and secondary necrotic cell death. Compounds **99** and **100** also significantly reduced the parasite load in macrophages infected with the parasite, *via* an immunomodulatory mechanism associated with increased levels of TNF-α, IL-12, and NO.

With the successful identification of the new class of 2-amino-thiophenes candidates for leishmanicidal drugs, new derivatives were planned by the exploration and optimization of the pharmaco-prophylaxis, obtaining new hybrid compounds belonging to the classes of thiophene-indole and thiophene-acridine [32, 33].

The introduction of the acridine ring bond to imine (CH=N) was not positive in terms of leishmanicidal activity, resulting in compounds without promastigote activity (IC50 > 500 μM; **105**) (Figure 18). However, the thiophene-indole hybrids confirmed our previous results, resulting in many compounds with better profiles than the reference drugs (tri- and penta-valent antimonials) [32, 33].

The thiophene-indole hybrids were obtained by varying the size of the cycloalkyl ring attached to the thiophene core, and varying the substitutions in the indole ring. The compounds with the best promastigote activity were **101**, **102**, **103**, and **104**, with IC_{50} values at 7.2, 10, 9.8 and 5.8 μM, respectively (Figure 18).

As observed in previous studies [25, 26], lipophilicity appears to play an important role in the leishmanicidal activity of thiophene derivatives, because the compounds with the best activity profile were those in which the major cycloalkyl groups were bound to the thiophene ring: cycloocta[*b*]thiophenes > cyclohepta[*b*]thiophenes > cyclohexa[*b*] thiophenes > cyclopenta[*b*]thiophenes.

It was not possible to establish direct correlations between substitutions in the indole ring and biological activity; nevertheless, position 5 appears to be important for activity because the great majority of the most active compounds have substitutions in that position: 5-Br for **99** and 100, 5-CN for **103** and **104**, and 5-CH_3 for **101** [31].

Subsequently, the most active and selective 2-aminothiophene **99** was used in *in vivo* tests. Compound **99** was not toxic (LD_{50} estimated at 2500 mg/kg orally), and it did not induce genotoxicity (up to 2000 mg/kg) in a peripheral blood micronucleus assay.

The *in vivo* antileishmanial activity of compound **99** was tested in a mouse model of cutaneous leishmania (oral administration for 7 weeks). At 200 mg/kg, **99** reduced paw lesion size by 52.47 ± 5.32% (Figure 19), and decreased the parasite load of the popliteal lymph node and spleen by 42.57 ± 3.14%, and 100%, respectively, without causing weight changes or other clinically important perturbations in terms of biochemical and hematological profiles. Therefore, this agent was more active and efficient, and was less toxic than meglumine antimoniate (reference drug), the administration of which was parenteral [22].

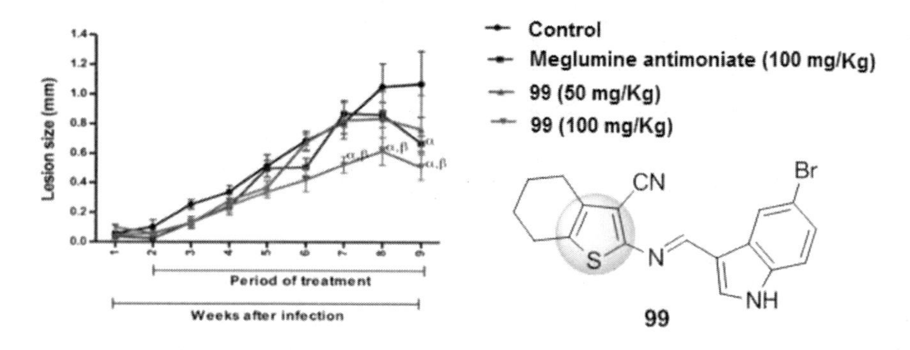

Figure 19. Lesion size growth in Swiss mice infected with *L. amazonensis* and treated orally with **99** at 50 and 200 mg/kg, and with meglumine antimoniate (reference drug) at 100 mg/Kg parenterally, for 7 weeks.

These data, coupled with the fact that compound **99** was capable of eliminating *in vitro* promastigote and amastigote forms of antimonial-resistant strains with the same IC_{50} values of the sensitive strains (suggesting no cross-resistance) demonstrates that compound **99** meets all prerequisites to become a new therapeutic alternative for the treatment of leishmaniasis. Complementary tests, including pharmacokinetic studies and activity against other *Leishmania* species still need to be conducted in order to begin the clinical trials.

HEPATITIS

The term "hepatitis" is used to designate any inflammatory process in the liver, caused by either biological agents (such as viruses), damage secondary to administration of various drugs or excessive alcohol intake, or physical agents (e.g., radiation). Among the biological agents, viruses are the primary causative agents of hepatitis. Depending on differences in composition of their genetic material, they can be classified in five subtypes: hepatitis A (HAV), B (HBV), C (HCV), D (HDV), and E (HEV) viruses. The first three of these have the highest incidence [34].

Among the viral hepatitis types with higher incidence (HAV, HBV, and HCV), there is a differentiation in terms of mode of infection. Type A is transmitted by contact with water or food contaminated with feces from infected hosts. Types B and C are transmitted parenterally. The parenteral types lead to the chronic disease and liver complications such as cirrhosis and hepatocellular carcinoma. HBV and HCV infections are still associated with systemic impairments and autoimmune manifestations or lymphoproliferative events, respectively [34, 35].

Depending on their biological characteristics, there are key aspects that differentiate infections with types B and C. While the virus causing HBV belongs to the *Hepadnaviridae* family and possesses genetic material consisting of partially double-stranded circular deoxyribonucleic acid (DNA), HCV belongs to the *Flaviviridae* family (*Hepacivirus* genus), possessing genetic material consisting of single-stranded ribonucleic acid (RNA). Infections caused by the latter can be found on all continents, remaining one of the major causes of concern in health systems worldwide [34, 36].

The hepatitis C virus is transmitted primarily through contact with contaminated blood and blood products, by organ transplantation, renal dialysis and/or intravenous drug administration, causing acute and chronic infections. HCV reaches approximately 3% of the world population, equivalent to 170 million individuals worldwide, in addition to an estimated 3 million newly infected individuals every year. This infection is five-times more prevalent than acquired immunodeficiency syndrome.

There is evidence to suggest that cirrhosis occurs in approximately 20% of cases, with hepatocellular carcinoma occurring in about 5%, besides occurrence of hepatic failure and fibrosis. It is also an important relationship with liver transplant disorders. The virus exists in six subtypes, or genotypes, distributed globally [36–44].

The causative agent of viral hepatitis C varies in size between 30 and 60 nm, consisting of a protein envelope surrounding an RNA of 9600 bases (Figure 20). The genome is subject to continuous changes in nucleotide residues owing to a high mutation rate, which helps HCV to evade immune surveillance [45, 46].

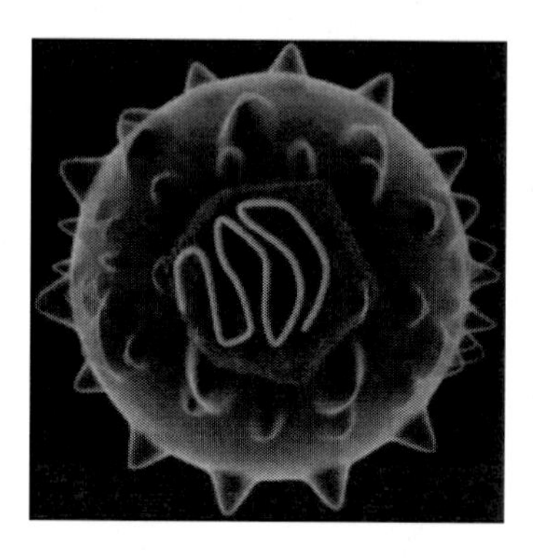

Figure 20. Structure of hepatitis C virus.

The mechanism of action of HCV remains unclear. Unfortunately, there is no available vaccine. Treatment is costly, based on the administration of interferon-α alone or in combination with ribavirin. This treatment carries considerable side-effects and has low efficacy, showing positive results in only 50%–60% of treated patients, primarily in individuals infected with genotype I (more prevalent in countries of North America, Europe, and Asia). There is also a constant increase in the development of resistance on the part of the virus [46–52].

The intensive search for anti-HCV agents has resulted in development of drugs designed for use in combination with pegylated interferon-α and ribavirin, including telaprevir and boceprevir as viral protease inhibitors, responsible for the increased success rate in the treatment of infected patients with HCV genotype 1. These discoveries represented the initiation of development of other direct-action antivirals for combination therapy, in pursuit of improved efficacy and safety. These include simeprevir and paritaprevir (inhibitors of NS3/4A viral protease), sofosbuvir and dasabuvir (inhibitors of the NS5B viral enzyme) and ledipasvir, daclatasvir and ombitasvir (NS5A inhibitors). Nevertheless, the persistence of adverse effects and development of resistance, owing to high mutation rates of the viral genetic material, clarify the urgent need for the development of efficient, affordable and safe anti-HCV agents, preferably with multitarget action [53–64].

Therefore, the search for new potent anti-HCV agents have focused on the inhibitory action of functional enzymes essential for the survival and replication of the virus within the host, including proteins involved in the coding processes of genetic information and protein expression of the infecting agent. A serine proteinase (NS3) and an RNA-dependent RNA polymerase (NS5B) are the main targets recently studied in the search for the development of inhibitors, primarily thiophene derivatives. Below we review recent developments in HCV therapeutics, that are specifically concentrates to inhibitors of NS5B [36, 38, 49, 65, 66].

The NS5B Structure

NS5B and NS3 are polyproteins encoded by the HCV genome, characterized as non-structural proteins (NS). Recalling the difference between types B and C, particularly their nucleic acids (DNA for HBV and RNA for HBC), we can understand the choices made in several recent investigations of the potential inhibitor of thiophene derivatives of NS5B, found only in the C variety, because it is an RNA-polymerase. This fact it is also important for the selective action of these inhibitors only against the viral machinery, and not against the host cells that contain genetic material based on deoxyribonucleic acid [49, 67].

The three-dimensional structure of NS5B is similar to those of other RNA polymerases that are often compared to a right hand, with its catalytic site presenting in a palm domain and the fingers and thumb domains involved in the interaction with RNA strands. Inhibitors can be classified of two ways: nucleoside (NI) and non-nucleoside (NNI) inhibitors, where the first act in terminal chains, and the latter function as non-competitive allosteric inhibitors. For NNI, because their binding sites are not within the active site, their binding points are presents in the thumb I, thumb II, palm I and palm II domains. The thumb domain II (Figure 21) is a target of interaction with thiophenes, based in the thiophene carboxylic acid structure, to which the recent inhibitory activities of NS5B have been directed. Specificities of inhibitor classes can be seen in thumb domain I, with benzimidazole and indole derivatives as known inhibitors, and to palm domains I and II, which are inhibited by benzothiadiazine and acylpyrrolidine derivatives, and by benzofurans, respectively. The existence of several inhibitor binding sites makes NS5B an advantageous target for combination therapy for HCV [52, 57, 68–71].

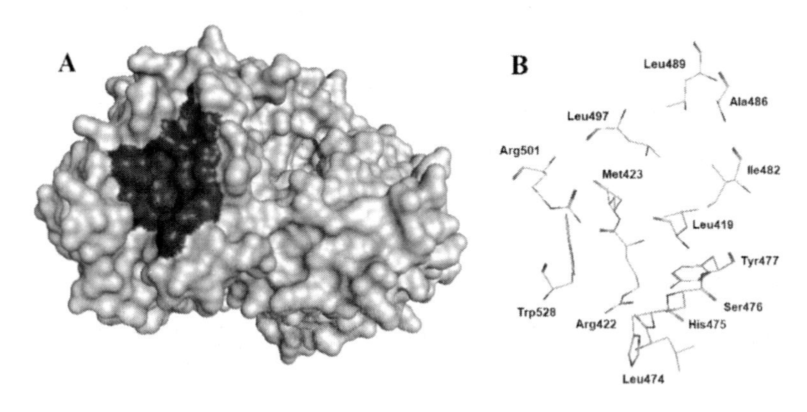

Figure 21. A – Thumb domain II surface of NS5B; B – Its contact amino acids (PDB entry: 1C2P).

The thumb domain II represents a predominantly hydrophobic depression located approximately 35 Å from the active site of NS5B. When binding to the inhibitor, it is precluded from interacting with the finger moieties of the enzyme, maintaining it in an inactive state, unable to bind to RNA. In addition, there is reduced affinity for highly energetic molecules (as GTP); this energy is necessary for the nucleic acid polymerization and elongation. Despite the fact that the mechanisms of inhibitor binding have not yet been fully clarified, computational prediction study tools can be useful in the analysis of some points of contact in the establishment of the enzyme-inhibitor complex that appear to be essential for their formation [36, 72–74].

Recent Studies of NS5B Inhibitors

As mentioned above, recent studies of NS5B inhibitors focus on the binding site of thumb domain II, based on thiophene-carboxylic acid structures (Figure 22). These are modified in three main different moieties in the search for improved inhibitor activities of RNA-dependent RNA polymerase.

This position (highlighted in R1 in the Figure 22) interacts with the hydrophobic residues in thumb domain II, specifically with Leu^{419}, Met^{423}, Ile^{482}, Ala^{486}, Leu^{489}, and Leu^{497} (shown in Figure 21). Therefore, excellent affinity for formation of enzyme-inhibitor complex depends on the lipophilic group in the 5-position of thiophene carboxylic acid derivatives. As shown in Figure 23, for compounds **107-109**, the maintenance of phenyl or thiophene groups (aromatic substituents) in this region is important for the positive effect in hydrophobic interactions with the binding site, as well as for maintenance of good inhibitory concentrations of 50% of the enzymes (IC_{50}) [36, 73].

Thiophene-carboxilic acid derivatives

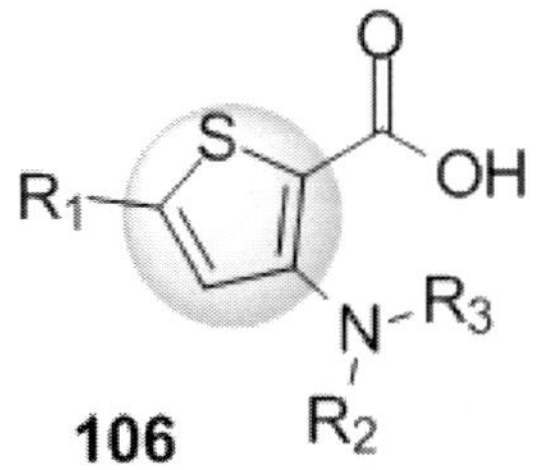

106

Figure 22. General structure of thiophene-carboxylic acid derivatives (**106**) NS5B inhibitors.

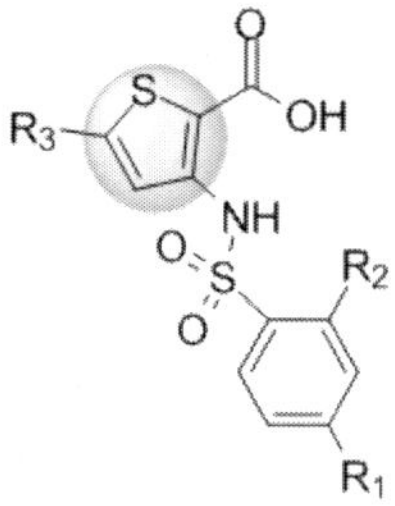

	R_1	R_2	R_3	IC_{50} (μM)
107	CH_3	H	4-FPh	0.2918
108	H	CH_3	4-CNPh	0.2663
109	CH_3	CH_3	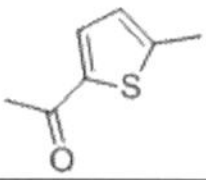	0.3073

Figure 23. Chemical structures of compounds **107**, **108**, and **109**, and its IC_{50} values for the inhibition of NS5B.

The importance of this characteristic is observed in the comparison between compounds **110** and **111** (Figure 24), where the change of a phenyl for a *tert*-butylacetylene group provides approximately 30-times greater activity against HCV, measured as the concentration of drug that induces half of the maximum effect (EC_{50}), reflecting improvement in the activity by insertion of longer, non-cyclical apolar substituents [71].

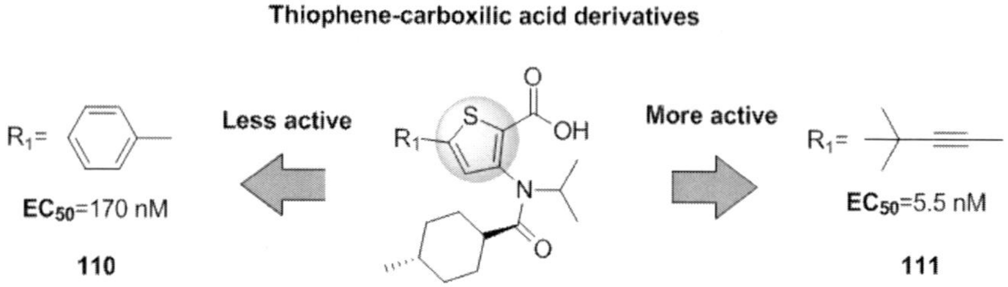

Thiophene-carboxilic acid derivatives

Less active | More active

EC_{50}=170 nM — **110** | **EC_{50}=5.5 nM** — **111**

Figure 24. Chemical structures of thiophene-carboxylic acid derivatives **110** and **111**, and its EC_{50} intrinsic values for genotype 1a of HCV.

For compounds **114** and **115** (Figure 25) we can see improvements when compared with compounds **112** and **113**, with better inhibitory interactions with the NS5B enzyme for compounds with *tert*-

butylacetylene substituents, although other radical groups can also influence the results, as will be discussed later [75].

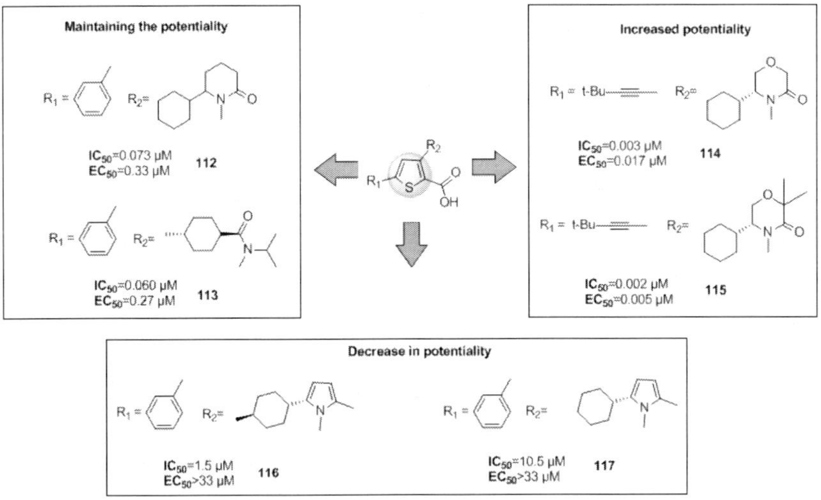

Figure 25. IC_{50} and EC_{50} values of compounds **112-115** for the genotype 1b of NS5B, and their structures.

These data make clear that hydrophobicity in the 5-position of the thiophene carboxylic acid is very important for the binding of NS5B inhibitors. Substituents with higher lipophilic contact surfaces with the contact amino acids improve these interactions, particularly the 5-*tert*-butylacetylene group, as maintained in many studies.

Modifications in the Amide Group Pf the Main Skeleton

As mentioned above, binding to thumb II pocket is known to occur in a deep site, where the groups linked to the amide group of the main skeleton (highlighted as R2 in the Figure 22) are found in the deeper regions of this field interaction, described as a very lipophilic surface.

The need for hydrophobic links in these regions can be seen when we observe the positive influence of aromatic substituent with apolar groups, with similar results for all compounds (Figure 23) tested with the 1b genotype of HCV NS5B, in concordance with the few points of structural differentiation between them. These groups *N*-linked interact with the Leu^{419}, Met^{423}, Leu^{474}, His^{475}, and Trp^{528} in the pocket site [36].

From the thumb domain II characteristics, the insertion of carbonated chains in this amide position was positively observed for compounds **113** (Figure 25) and **118–122** (Figure 26), where the 4-methylated cyclohexane ring appears to be important for the maintenance of interactions with the protein domain-ligand complex. This is an important contribution of the 4-methyl group, as demonstrated by the comparison between the potencies of compounds **113** and **118**, possibly participating in some additional interaction. In this same sense, better results are observed when an unsaturation is present in the six-membered ring, provided that it is distant from the amide group, therefore in the deeper region of the binding site, more specifically in contact with the Leu^{419}, Arg^{422}, Met^{423}, and Trp^{528} amino acids. For this series, compound **122** that maintains anti-HCV potentiality, showed better pharmacokinetic studies. Therefore, it appears to be a promising drug [71].

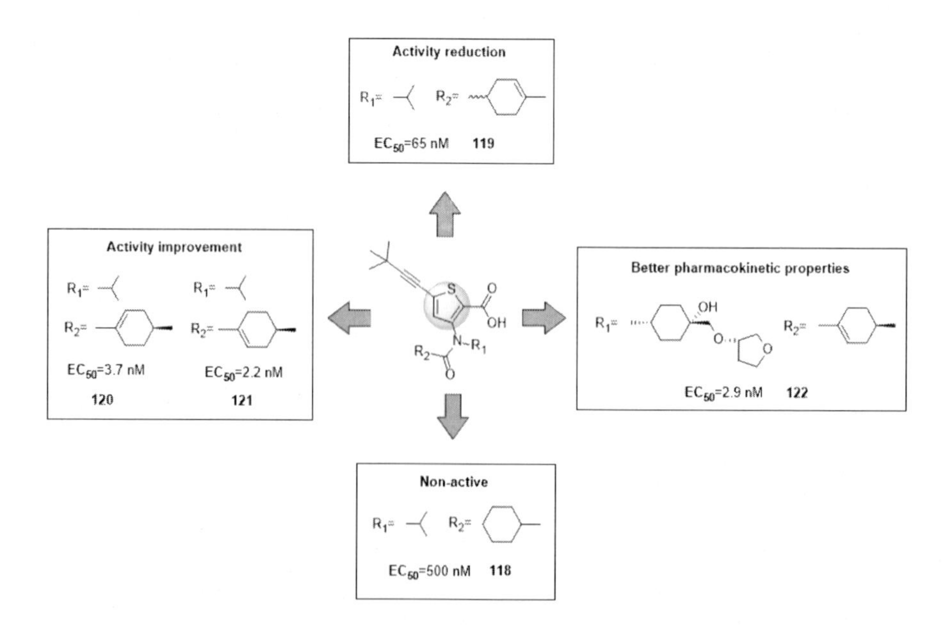

Figure 26. Structures of compounds **118–122** and their structure-activity relationships for inhibition of genotype 1a of NS5B.

This amide group, therefore, does not necessarily need to be arranged in a linear chain; rather, the cyclization in an analog lactam ring appears to maintain the potentiality of the thiophene derivatives, by contrast with compounds **112** and **113** (Figure 25), where the piperidone ring possesses a cyclohexane ring anchored to it. Furthermore, the substitution of a piperidone cycle for the morpholinone ring, which has an oxygen atom in the opposite position of the ring, improves the inhibitory potentiality of NS5B, highlighting the benefits of the additional replacement of a dimethyl group inserted on the morpholinone ring in compound **113** (Figure 25). Even with cyclization, the lipophilic contacts with thumb domain II are maintained, but with possibly positive influence of a six-membered ring present in the piperidone and morpholinone ring, confirmed when they are replaced by pyrrole heterocycle, decreasing anti-HCV potentiality [75].

For these molecules, a characteristic is important for the carbonyl group in the lactam rings. The spatial position of the carboxylic acid of the thiophene is found in a region of solvent contact, then, to tolerate polar groups, constituted by the Arg[501], Ser[476], and Tyr[477], capable of forming hydrogen bonds. Therefore, neighboring regions can be strategic points for insertion of other acceptor/donor groups of hydrogen bonds; for example, of carbonyl in the lactam ring, that can clarify good tolerance of the secondary amine and the sulfonyl oxygen in compounds **107-109** (Figure 23) [36, 73, 75].

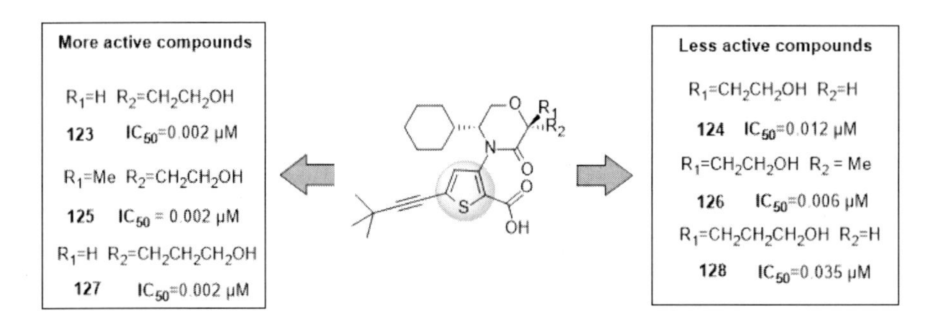

Figure 27. Conformational preference in inhibitory activity of NS5B.

Finally, for this molecule series, the positive influence of the dimethyl group in compound **115** (Figure 25) also possesses important considerations for the geometric fit in the protein-binder complex. In this position, other carbon chains are well accepted to maintain anti-HCV potential (Figure 27); however, this occurs only when they are in a *cis*-configuration, together with the cyclohexyl substituent, demonstrating the stereochemical preference for the maintenance of interactions in the complex [75].

This feature suggests the conclusion that the preferences of the deeper regions of the thumb domain II is ideal for hydrophobic interactions; furthermore, it provides a strategic point for the insertion of polar groups in their structures, that should be near the carboxylic acid moiety of the thiophene ring.

Third Possibility of Changes in Thiophene Carboxylic Acid Derivatives

The nitrogen atom of the amide group can be additionally substituted to become a tertiary group (highlighted in R3 in the Figure 22), another strategic point for molecular improvement.

The binding region between the protein and these substituents appears to be further away from the other two already mentioned, providing the possibility of inserting groups with characteristics different from the others, including polar atoms. As shown in Figure 28, compounds with more rigid groups and with a polar group (-OH) at the terminal end assist in increasing target affinity, probably creating hydrogen bonds with other amino acid residues that are also important for the maintenance of the three-dimensional structure of the enzyme. The permanency of the oxygen atom in this extremity of the formation of the ether substituent with heterocyclic extensions (compounds **133-136**, Figure 28) maintains the promising activity of this class [71].

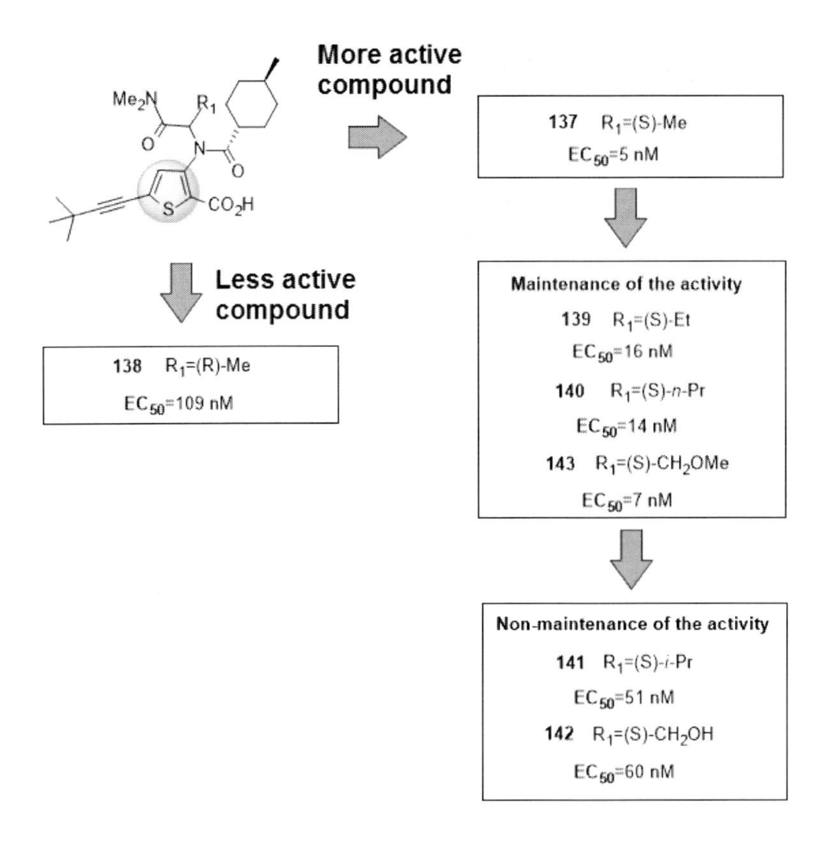

Figure 28. Structure-activity relationships for the compounds **23-30** in their inhibitory actions of NS5B genotype 1a.

For this series of compounds, the influence of other amide terminal group can be observed with the insertion of aliphatic and cyclic groups, where it is interesting to observe the need of a tertiary nitrogen (Figure 30). In this case, contrary to what was shown for compound **129** (Figure 28), the presence of a 4-hydroxyl group did not improve the inhibitory activity of compound **147** (Figure 30); nevertheless, it is interesting to note that the spacing between it and the main skeleton is different, a fact that may explain the distancing of their amino acids of contact. On the other hand, the insertion of a tertiary amine in the terminal portion appears to improve anti-HCV potency (compounds **146** and **148**). Following pattern of combination regimens for anti-HCV therapy, compound **150**, which, although is not the most active, shows better pharmacokinetic parameters, appears to be promising according to its synergistic effect with interferon-α 2a and ribavirin, or with telaprevir [76].

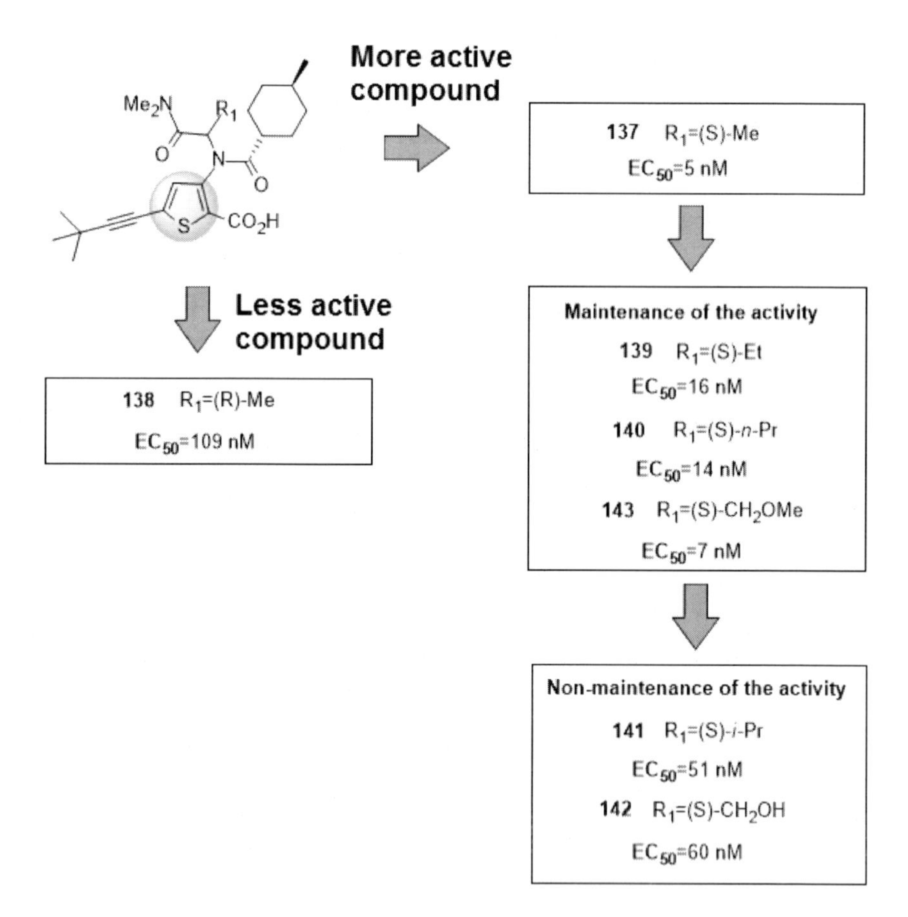

Figure 29. Stereochemical preferences for the genotype 1b of NS5B inhibitors.

The positive effect of tertiary nitrogen in this position can be also observed in compounds presented in Figure 31, as well as the stereochemical preference of this binding site for the *S*-conformation of these substituents, when compared with that of *R*-conformers [77].

As mentioned for the other regions of the thumb domain II, some specific features can be observed that improve the binding affinity between inhibitors and the NS5B enzyme. These are important for the positive influence of polar groups on the strategic amino acid Arg^{501}, as well as the better results for the derivatives with tertiary nitrogen in this position and the preference of the protein site by *S*-conformers.

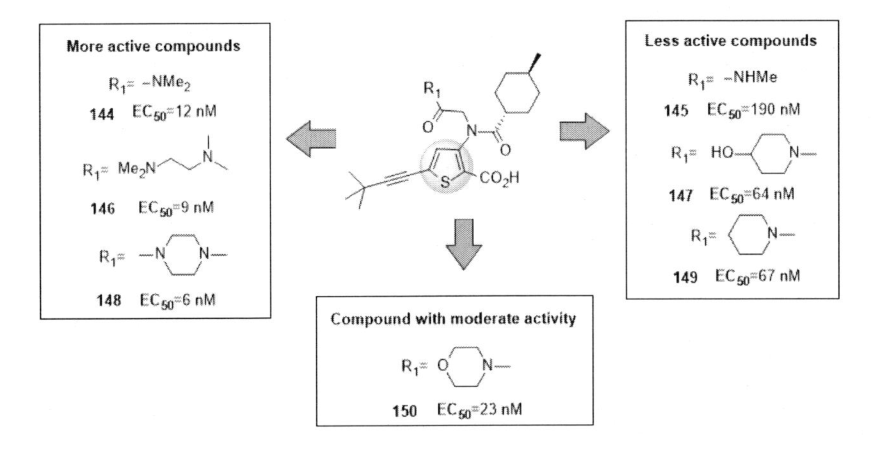

Figure 30. Structures of compounds **144-150**, with their EC_{50} values in the inhibitory activity of genotype 1b of NS5B.

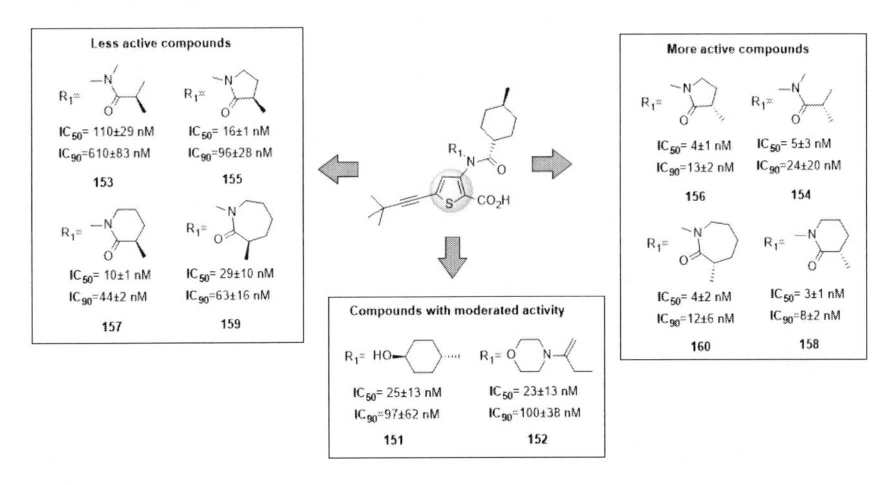

Figure 31. Structure–activity relationships of compounds **151-160** in terms of their inhibitory activity on HCV genotype 1b.

Considering that thumb domain II of NS5B is known for its excellent binding to thiophene carboxylic acid, and that this enzyme is specific to HCV, the derivations of this molecular core are important in the search of new compounds with greater affinity with their inhibitor site and, therefore, greater anti-viral activity. This binding site, as already mentioned, possesses some specific features that can be exploited for the development of new thiophene derivatives, where there is important regard for its hydrophobic surface in the interaction position with the 5-position and the main chain amide groups. This site also has strategic points of interaction with polar groups, for example the proximity of the 2-carboxylic acid substituent, where the Ser^{476} and Tyr^{477} amino acid residues are located, as well as good interactions of polar groups present in the groups adjacent to the amide group with the Arg^{501} residue.

All these characteristics have been exploited in recent studies, in the search for new thiophene inhibitors of HCV enzyme NS5B, either for monotherapy or in combination with currently available drugs. Although these studies focus on the genotype 1 type, it is known that several amino acids residues are conserved across all different genotypes; therefore, it is possible that interaction patterns seen in genotype 1 type may be maintained in other genotypes as well.

FINAL CONSIDERATIONS

Tuberculosis, one of humanity's oldest infectious diseases, remains a major public health problem in many countries around the world. Currently, it is estimated that 1/3 of the world population is infected with MTB. In order to overcome these high rates of disease incidence, new drugs need to be discovered and developed, which have well-defined criteria: (*i*) to be more potent than standard drugs, reducing the treatment period; (*ii*) present new biological targets for drugs of choice in the treatment of MDR-TB and/or XDR-TB; (*iii*) compatibility with antiretroviral drugs used in patients infected with HIV; and (*iv*) do not present antagonism with other anti-TB drugs in the standard therapeutic regimen containing at least 3 drugs.

In this chapter, several examples of synthesis and evaluation against MTB of thiophene, benzo[*b*]thiophene and nitrothiophene derivatives have been described. Some compounds showed activity against MDR-TB and

XDR-TB and were more potent than standard drugs as INH, RFM, ETB and PZA. The block of mycolic acid biosynthesis by inhibition of β-ketoacyl-CoA reductase, β-ketoacyl-ACP synthase and enoyl reductase, or reduction of the nitrothiophene and subsequently release of nitric oxide were the mechanisms of action described. In addition, drug-likeness ADMET and pharmacokinetic properties of some thiophene derivatives reinforce the potentiality of this class of compounds in anti-TB drug discovery. With regard to leishmanicidal activity, the great majority of studies emphasize the importance of lipophilicity for biological activity. Regardless of the class of thiophene derivatives used as a scaffold (2-acetamidothiophene-3-carboxamide, 2-arylbenzothiophene, benzo[*b*] thiophene, 2-amino-thiophene, and others) the most active compounds present high logP, often with significantly higher values to that established in Lipinski's rules (LogP < 5). This characteristic has shown to be important for the *in vitro* activity against the two evolutionary forms of the parasite, and against different species of *Leishmania*. However, the lack of continuity of the evaluation of the efficacy of these lead compounds in *in vivo* models provides us with great indications that the pharmacokinetic barrier prevents or at least hinders the execution of the remaining stages of development of a new drug and decreases the chance of any of these compounds becoming a new available drug. The hope is that the tools of pharmaceutical technology, such as: inclusion complexes with cyclodextrins, nanoencapsulated systems, solid dispersions, liposomes, among others, that have been used successfully in the development of new drugs against various diseases, can also be used with these thiophene derivatives helping to mask this unwanted drug feature, thus allowing in the near future that some of these lead compounds may be being evaluated in clinical study phases.

Hepatitis is a term used for designation an inflammatory process of the liver, classified inside of the group of NTD and can be caused by biological and/or physical agents, with a subdivision in five different subtypes (HAV, HBV, HCV, HDV, and HEV). Between these subtypes, the search of new anti-hepatitis agents based in thiophene core has been directed, in the vast majority, for the HCV treatment. The HCV subtype has a group of non-structural proteins characteristic of hepatitis C, in special NS5B enzyme (an RNA polymerase), which is considered a therapeutic target for thiophene derivatives. NS5B possess binding sites to non-nucleoside inhibitors and is constituted by thumb I, thumb II, palm I, and palm II domains. The thumb II domain is the binding site for thiophene-carboxylic acid derivatives inhibitors, being these the focus of the recent researches on development of new potent anti-HCV. The thumb II domain represents a depression predominantly hydrophobic, constituted by contact amino acids as Leu^{419}, Arg^{422}, Met^{423}, Leu^{474}, His^{475}, Ser^{476}, Tyr^{477}, Ile^{482}, Ala^{486}, Leu^{489}, Leu^{497}, Arg^{501}, and Trp^{528}. In this sense, synthesis strategies have aimed at the insertion of groups predominantly apolar in three regions of the thiophene-carboxylic acid core. Added to the existence of strategic positions for insertion of polar groups, as the proximity to 2-carboxylic acid moiety and groups adjacent to the amide group, that can interact with the Arg^{501} residue. As a protein of specific spatial conformation, NS5B possess a highlighted preference for *S*-conformers, demonstrating the importance of the spatial distribution of groups present in thiophene-carboxylic acid derivatives. These compounds direct search of new inhibitory alternatives of HCV NS5B, which showing promising in monotherapy regimen or in combination with available drugs.

References

[1] Bhowmik CD, Chandira RM, Jayakar B, Kumar KPS. Recent Trends of Drug Used Treatment of Tuberculosis. *J Chem Pharm Res* 2009;1:113–33.

[2] Wayne LG, Sohaskey CD. N Onreplicating P Ersistence of M Ycobacterium T Uberculosis. *Annu Rev Microbiol* 2001.

[3] Cheng AFB, Yew WW, Chan EWC, Chin ML, Hui MMM, Chan RCY. Multiplex PCR Amplimer Conformation Analysis for Rapid Detection of gyrA Mutations in Fluoroquinolone-Resistant Mycobacterium tuberculosis Clinical Isolates. *Antimicrob Agents Chemother* 2004;48:596–601. doi:10.1128/AAC.48.2.596-601.2004.

[4] Grinev AN, Kaplina N V, Pershin GN, Padeiskaya EN. *Functional Derivatives of Thiophene. XIX. Synthesis and Biological Activity of Some Esters of 2-Benzoylaminothiophene-3-carboxylic acids* 1981;14:861–3.

[5] Foroumadi A, Kargar Z, Sakhteman A, Sharifzadeh Z, Feyzmohammadi R, Kazemi M, et al. Synthesis and antimycobacterial activity of some alkyl [5-(nitroaryl)-1,3, 4-thiadiazol-2-ylthio]propionates. *Bioorganic Med Chem Lett* 2006;16:1164–7. doi:10.1016/j.bmcl.2005.11.087.

[6] Hartkoorn RC, Ryabova OB, Chiarelli LR, Riccardi GV, Makarov V, Makarov ST. Mechanism of action of 5-nitrothiophenes against mycobacterium tuberculosis. *Antimicrob Agents Chemother* 2014;58:2944–7. doi:10.1128/AAC.02693-13.

[7] Singh P, Manna SK, Jana AK, Saha T, Mishra P, Bera S, et al. Thiophene containing trisubstituted methanes [TRSMs] as identified lead against Mycobacterium tuberculosis. *Eur J Med Chem* 2015;95:357–68. doi:10.1016/j.ejmech.2015.03.036.

[8] Padam Singh, Shashi Kant Kumar, Vineet Kumar Maurya, Basant Kumar Mehta, Hafsa Ahmad, Anil Kumar Dwivedi, Vinita Chaturvedi, Tejender S. Thakur SS. S-Enantiomer of the Antitubercular Compound S006-830 Complements Activity of Frontline TB Drugs and Targets Biogenesis of Mycobacterium tuberculosis. *Cell Envelope* 2017:8453–65.

[9] Lu X, Wan B, Franzblau SG, You Q. Design, synthesis and anti-tubercular evaluation of new 2-acylated and 2-alkylated amino-5-(4-(benzyloxy)phenyl)thiophene-3-carboxylic acid derivatives. Part 1. *Eur J Med Chem* 2011;46:3551–63. doi:10.1016/j.ejmech.2011.05.018.

[10] Choi PJ, Sutherland HS, Tong AST, Blaser A, Franzblau SG, Cooper CB, et al. Synthesis and evaluation of analogues of the tuberculosis drug bedaquiline containing heterocyclic B-ring units. *Bioorganic Med Chem Lett* 2017;27:5190–6. doi:10.1016/j.bmcl.2017.10.042.

[11] Balamurugan K, Perumal S, Reddy ASK, Yogeeswari P, Sriram D. A facile domino protocol for the regioselective synthesis and discovery

of novel 2-amino-5-arylthieno-[2,3-b]thiophenes as antimycobacterial agents. *Tetrahedron Lett* 2009;50:6191–5. doi:10. 1016/j.tetlet.2009.08.085.

[12] Santhosh Reddy Patpi, Lokesh Pulipati, Perumal Yogeeswari, Dharmarajan Sriram, Nishant Jain, Balasubramanian Sridhar, Ramalinga Murthy, Anjana Devi T, Shasi Vardhan Kalivendi and SK. Design, Synthesis, and Structure−Activity Correlations of Novel Dibenzo[b,d]furan, Dibenzo[b,d]thiophene, and N-Methylcarbazole Clubbed 1,2,3-Triazoles as Potent Inhibitors of Mycobacterium tuberculosis 2012:3911–22.

[13] Pulipati L, Yogeeswari P, Sriram D, Kantevari S. Click-based synthesis and antitubercular evaluation of novel dibenzo[b,d]thiophene-1,2,3-triazoles with piperidine, piperazine, morpholine and thiomorpholine appendages. *Bioorganic Med Chem Lett* 2016;26:2649–54. doi:10.1016/j.bmcl.2016.04.015.

[14] Pulipati L, Sridevi JP, Yogeeswari P, Sriram D, Kantevari S. Synthesis and antitubercular evaluation of novel dibenzo[b,d] thiophene tethered imidazo[1,2-a]pyridine-3-carboxamides. *Bioorganic Med Chem Lett* 2016;26:3135–40. doi:10.1016/j.bmcl. 2016.04.088.

[15] Marvadi SK, Krishna VS, Sriram D, Kantevari S. Synthesis of novel morpholine, thiomorpholine and N-substituted piperazine coupled 2-(thiophen-2-yl)dihydroquinolines as potent inhibitors of Mycobacterium tuberculosis. *Eur J Med Chem* 2019;164:171–8. doi:10.1016/j.ejmech.2018.12.043.

[16] Marvadi SK, Krishna VS, Sriram D, Kantevari S. Synthesis and evaluation of novel substituted 1,2,3-triazolyldihydroquinolines as promising antitubercular agents. *Bioorganic Med Chem Lett* 2019;29:529–33. doi:10.1016/j.bmcl.2019.01.004.

[17] Kumar G, Krishna VS, Sriram D, Jachak SM. Synthesis of carbohydrazides and carboxamides as anti-tubercular agents. *Eur J Med Chem* 2018;156:871–84. doi:10.1016/j.ejmech.2018.07.047.

[18] Nallangi R, Samala G, Sridevi JP, Yogeeswari P, Sriram D. Development of antimycobacterial tetrahydrothieno[2,3-c]pyridine-3-

carboxamides and hexahydrocycloocta[b]thiophene-3-carboxamides: Molecular modification from known antimycobacterial lead. *Eur J Med Chem* 2014;76:110–7. doi:10.1016/j.ejmech.2014.02.028.

[19] Wagh MA, Baravkar SB, Jedhe GS, Borkute R, Choudhari AS, Sarkar D, et al. Design and Synthesis of 2-Amino-thiophene-Tethered Ureidopenicillin Analogs with Potent Antibacterial and Antitubercular activity. *Chemistry Select* 2018;3:3122–6. doi:10.1002/slct.201800027.

[20] Ghorbani M, Farhoudi R. Leishmaniasis in humans: Drug or vaccine therapy? *Drug Des Devel Ther* 2018;12:25–40. doi:10.2147/DDDT.S146521.

[21] Sangenito LS, da Silva Santos V, d'Avila-Levy CM, Branquinha MH, Souza dos Santos AL, de Oliveira SSC. Leishmaniasis and Chagas Disease – Neglected Tropical Diseases: Treatment Updates. *Curr Top Med Chem* 2019;19:174–7. doi:10.2174/156802661903190328155136.

[22] Rodrigues KA da F, Silva DKF, Serafim V de L, Andrade PN, Alves AF, Tafuri WL, et al. SB-83, a 2-Amino-thiophene derivative orally bioavailable candidate for the leishmaniasis treatment. *Biomed Pharmacother* 2018;108:1670–8. doi:10.1016/j.biopha.2018.10.012.

[23] Delmasl F, Gasquetl M, Madadi N, Vanellez P, Vailles A, Maldonadoz J. Synthesis and in vitro anti-protozoan activity of new 5nitrothiophene oxime ether derivatives 1993;28:23–7.

[24] Vanelle P, Ghezali S, Maldonado J, Crozet M, Delmas F, Gasquet M, et al. New 2-alkylidenemethyl-5-nitrothiophenes: preparation via SRN1 reactions and in vitro antiprotozoan activity. *Eur J Med Chem* 1994;29:41–4. doi:10.1016/0223-5234(94)90124-4.

[25] Oh S, Kwon B, Kong S, Yang G, Lee N, Han D, et al. Synthesis and biological evaluation of 2-acetamidothiophene-3-carboxamide derivatives against Leishmania donovani. *Medchemcomm* 2014;5:142–6. doi:10.1039/c3md00299c.

[26] Bonano VI, Yokoyama-Yasunaka JKU, Miguel DC, Jones SA, Dodge JA, Uliana SRB. Discovery of synthetic leishmania inhibitors

by screening of a 2-arylbenzothiophene library. *Chem Biol Drug Des* 2014;83:289–96. doi:10.1111/cbdd.12239.

[27] Miguel DC, Zauli-Nascimento RC, Yokoyama-Yasunaka JKU, Katz S, Barbiéri CL, Uliana SRB. Tamoxifen as a potential antileishmanial agent: Efficacy in the treatment of Leishmania braziliensis and Leishmania chagasi infections. *J Antimicrob Chemother* 2009;63: 365–8. doi:10.1093/jac/dkn509.

[28] Reimão JQ, Uliana SRB. Tamoxifen alters cell membrane properties in Leishmania amazonensis promastigotes. *Parasitol Open* 2018;4:4–9. doi:10.1017/pao.2018.3.

[29] Trinconi Ct, Reimão Jq, Bonano Vi, Espada Cr, Miguel Dc, Yokoyama-Yasunaka Jku, et al. Topical tamoxifen in the therapy of cutaneous leishmaniasis. *Parasitology* 2018;145:490–6. doi:10.1017/s003118201700 0130.

[30] Scala A, Micale N, Piperno A, Rescifina A, Schirmeister T, Kesselring J, et al. Targeting of the leishmania mexicana cysteine protease CPB2.8DCTE by decorated fused benzo[b] thiophene scaffold. *RSC Adv* 2016;6:30628–35. doi:10.1039/c6ra05557e.

[31] Rodrigues KADF, Dias CNDS, Neris PLDN, Rocha JDC, Scotti MT, Scotti L, et al. 2-Amino-thiophene derivatives present antileishmanial activity mediated by apoptosis and immunomodulation in vitro. *Eur J Med Chem* 2015;106:1–14. doi:10.1016/j.ejmech.2015.10.011.

[32] Félix MB, de Souza ER, de Lima M do CA, Frade DKG, Serafim V de L, Rodrigues KA da F, et al. Antileishmanial activity of new thiophene–indole hybrids: Design, synthesis, biological and cytotoxic evaluation, and chemometric studies. *Bioorganic Med Chem* 2016;24:3972–7. doi:10.1016/j.bmc.2016.04.057.

[33] de Lima Serafim V, Félix MB, Frade Silva DK, Rodrigues KA da F, Andrade PN, de Almeida SMV, et al. New thiophene–acridine compounds: Synthesis, antileishmanial activity, DNA binding, chemometric, and molecular docking studies. *Chem Biol Drug Des* 2018;91:1141–55. doi:10.1111/cbdd.13176.

[34] Walker GMLABD. *Hepatitis C Virus Infection* 2001;345:41–52.

[35] Sangare L, Diande S, Kouanda S, Dingtoumda B, Mourfou A, Ouedraogo F, et al. Mycobacterium tuberculosis drug-resistance in previously treated patients in Ouagadougou, Burkina Faso. *Ann Afr Med* 2010;9:15. doi:10.4103/1596-3519.62619.

[36] Biswal BK, Wang M, Cherney MM, Chan L, Yannopoulos CG, Bilimoria D, et al. Non-nucleoside Inhibitors Binding to Hepatitis C Virus NS5B Polymerase Reveal a Novel Mechanism of Inhibition. *J Mol Biol* 2006;361:33–45. doi:10.1016/j.jmb.2006.05.074.

[37] Wasley A, Alter MJ. Epidemiology of hepatitis C: geographic differences and temporal trends. *Semin Liver Dis* 2000;20:1–16.

[38] Chen KX, Vibulbhan B, Yang W, Nair LG, Tong X, Cheng KC, et al. Novel potent inhibitors of hepatitis C virus (HCV) NS3 protease with cyclic sulfonyl P3 cappings. *Bioorganic Med Chem Lett* 2009;19: 1105–9. doi:10.1016/j.bmcl.2008.12.111.

[39] Park JY, Lee YS, Chang BY, Kim BH, Jeon S, Park SM. Label-free impedimetric sensor for a ribonucleic acid oligomer specific to hepatitis C virus at a self-assembled monolayer-covered electrode. *Anal Chem* 2010;82:8342–8. doi:10.1021/ac1019232.

[40] Yan S, Appleby T, Gunic E, Shim JH, Tasu T, Kim H, et al. Isothiazoles as active-site inhibitors of HCV NS5B polymerase. Bioorganic *Med Chem Lett* 2007;17:28–33. doi:10.1016/j.bmcl. 2006.10.002.

[41] N. J. L, M. K. H, J. A. M, M. T. R, J. W. B, S. S. C, et al. Molecular modeling based approach to potent P2-P4 macrocyclic inhibitors of hepatitis C NS3/4A protease. *J Am Chem Soc* 2008;130:4607–9.

[42] Corbeil CR, Englebienne P, Yannopoulos CG, Chan L, Das SK, Bilimoria D, et al. Docking Ligands into Flexible and Solvated Macromolecules. 2. *Development and Application of F. Society* 2008:902–9.

[43] Negro F, Alberti A. The global health burden of hepatitis C virus infection. *Liver Int* 2011;31:1–3. doi:10.1111/j.1478-3231.2011. 02537.x.

[44] Thomas DL. Global control of hepatitis C: Where challenge meets opportunity. *Nat Med* 2013;19:850–8. doi:10.1038/nm.3184.

[45] Kieft JS, Zhou K, Grech A, Jubin R, Doudna JA. Crystal structure of an RNA tertiary domain essential to HCV IRES-mediated translation initiation. *Nat Struct Biol* 2002;9:370–4. doi:10.1038/nsb781.

[46] May MM, Brohm D, Harrenga A, Marquardt T, Riedl B, Kreuter J, et al. Discovery of substituted N-phenylbenzenesulphonamides as a novel class of non-nucleoside hepatitis C virus polymerase inhibitors. *Antiviral Res* 2012;95:182–91. doi:10.1016/j.antiviral.2012.04.008.

[47] Pearlman BL. Hepatitis C treatment update. *Am J Med* 2004;117:344–52. doi:10.1016/j.amjmed.2004.03.024.

[48] Pawlotsky JM, McHutchison JG. Hepatitis C. Development of New Drugs and Clinical Trials: Promises and Pitfalls. *Hepatology* 2004;39:554–67. doi:10.1002/hep.20065.

[49] Chen KX, Xie HY, Li ZG. QSAR analysis of 1,1-dioxoisothiazole and benzo[b]thiophene-1,1-dioxide derivatives as novel inhibitors of hepatitis C virus NS5B polymerase. *Acta Chim Slov* 2009;56:684–93.

[50] Huang Z, Murray MG, Secrist JA. Recent development of therapeutics for chronic HCV infection. *Antiviral Res* 2006;71:351–62. doi:10.1016/j.antiviral.2006.06.001.

[51] Melagraki G, Afantitis A, Sarimveis H, Koutentis PA, Markopoulos J, Igglessi-Markopoulou O. Identification of a series of novel derivatives as potent HCV inhibitors by a ligand-based virtual screening optimized procedure. *Bioorganic Med Chem* 2007;15:7237–47. doi:10.1016/j.bmc.2007.08.036.

[52] Li Y, Wang ZL, He F, Wu Y, Huang W, He Y, et al. TP-58, a novel thienopyridine derivative, protects mice from concanavalin A-induced hepatitis by suppressing inflammation. *Cell Physiol Biochem* 2012;29:31–40. doi:10.1159/000337584.

[53] Trivella JP, Gutierrez J, Martin P. Dasabuvir: a new direct antiviral agent for the treatment of hepatitis C. *Expert Opin Pharmacother* 2015;16:617–24. doi:10.1517/14656566.2015.1012493.

[54] Kwong AD, Kauffman RS, Hurter P, Mueller P. Discovery and development of telaprevir: An NS3-4A protease inhibitor for treating genotype 1 chronic hepatitis C virus. *Nat Biotechnol* 2011;29:993–1003. doi:10.1038/nbt.2020.

[55] Rosenquist Å, Samuelsson B, Johansson PO, Cummings MD, Lenz O, Raboisson P, et al. Discovery and development of simeprevir (TMC435), a HCV NS3/4A protease inhibitor. *J Med Chem* 2014;57:1673–93. doi:10.1021/jm401507s.

[56] Rizk OH, Shaaban OG, Abdel Wahab AE. Synthesis of Oxadiazolyl, Pyrazolyl and Thiazolyl Derivatives of Thiophene-2-Carboxamide as Antimicrobial and Anti-HCV Agents. *Open Med Chem J* 2017;11:38–53. doi:10.2174/1874104501711010038.

[57] Delang L, Vliegen I, Leyssen P, Neyts J. In vitro selection and characterization of HCV replicons resistant to multiple non-nucleoside polymerase inhibitors. *J Hepatol* 2012;56:41–8. doi:10.1016/j.jhep.2011.04.016.

[58] Gentile I, Borgia F, Buonomo A, Zappulo E, Castaldo G, Borgia G. ABT-450: A Novel Protease Inhibitor for the Treatment of Hepatitis C Virus Infection. *Curr Med Chem* 2014;21:3261–70. doi:10.2174/0929867321666140706125950.

[59] Sofia MJ, Bao D, Chang W, Du J, Nagarathnam D, Rachakonda S, et al. Discovery of a luoro-2′-β- C -methyluridine Nucleotide Prodrug (PSI-7977) for the treatment of hepatitis C virus. *J Med Chem* 2010;53:7202–18. doi:10.1021/jm100863x.

[60] J.O. L, J.G. T, L. X, M. M, H. G, H. L, et al. Discovery of ledipasvir (GS-5885): A potent, once-daily oral NS5A inhibitor for the treatment of hepatitis C virus infection. *J Med Chem* 2014;57:2033–46.

[61] Lemm JA, Leet JE, O'Boyle DR, Romine JL, Huang XS, Schroeder DR, et al. Discovery of Potent Hepatitis C Virus NS5A Inhibitors with Dimeric Structures. *Antimicrob Agents Chemother* 2011;55:3795–802. doi:10.1128/aac.00146-11.

[62] David Allen DeGoey, John Randolph, Dachun Liu, John Pratt, Charles Hutchins, Pamela Donner, Chris Krueger, Mark Matulenko, Sachin Patel, Christopher Motter, Lissa Nelson, Ryan Keddy, Michael Tufano, Daniel Caspi, Preethi Krishnan, Neeta Mistry, Gennadiy K and WK. Discovery of ABT-267, a pan-genotypic

inhibitor of HCV NS5A. *J Med Chem* 2014;57:2047–57. doi:10.1021/jm401398x.

[63] Manns MP, Von Hahn T. Novel therapies for hepatitis C-one pill fits all? *Nat Rev Drug Discov* 2013;12:595–610. doi:10.1038/nrd4050.

[64] Ai T, Qiu L, Xie J, Geraghty RJ, Chen L. Design and synthesis of an activity-based protein profiling probe derived from cinnamic hydroxamic acid. *Bioorganic Med Chem* 2016;24:686–92. doi:10.1016/j.bmc.2015.12.035.

[65] Grakoui A, Wychowski C, Lin C, Feinstone SM, Rice CM. Expression and identification of hepatitis C virus polyprotein cleavage products. *J Virol* 1993;67:1385–95.

[66] Kim SH, Tran MT, Ruebsam F, Xiang AX, Ayida B, McGuire H, et al. Structure-based design, synthesis, and biological evaluation of 1,1-dioxoisothiazole and benzo[b]thiophene-1,1-dioxide derivatives as novel inhibitors of hepatitis C virus NS5B polymerase. *Bioorganic Med Chem Lett* 2008;18:4181–5. doi:10.1016/j.bmcl.2008.05.083.

[67] Lindenbach BD, Rice CM. Unravelling hepatitis C virus replication from genome to function. *Nature* 2005;436:933–8. doi:10.1038/nature04077.

[68] Delang L, Coelmont L, Neyts J. *Antiviral therapy for hepatitis C virus: Beyond the standard of care. vol.* 2. 2010. doi:10.3390/v2040826.

[69] Bosse TD, Larson DP, Wagner R, Hutchinson DK, Rockway TW, Kati WM, et al. Synthesis and SAR of novel 1,1-dialkyl-2(1H)-naphthalenones as potent HCV polymerase inhibitors. *Bioorganic Med Chem Lett* 2008;18:568–70. doi:10.1016/j.bmcl.2007.11.088.

[70] Howe AYM, Cheng H, Johann S, Mullen S, Chunduru SK, Young DC, et al. Molecular mechanism of hepatitis C virus replicon variants with reduced susceptibility to a benzofuran inhibitor, HCV-796. *Antimicrob Agents Chemother* 2008;52:3327–38. doi:10.1128/AAC.00238-08.

[71] Lazerwith SE, Lew W, Zhang J, Morganelli P, Liu Q, Canales E, et al. Discovery of GS-9669, a thumb site II non-nucleoside inhibitor of

NS5B for the treatment of genotype 1 chronic hepatitis C infection. *J Med Chem* 2014;57:1893–901. doi:10.1021/jm401420j.

[72] Wang M, Ng KKS, Cherney MM, Chan L, Yannopoulos CG, Bedard J, et al. Non-nucleoside analogue inhibitors bind to an allosteric site on HCV NS5B polymerase: Crystal structures and mechanism of inhibition. *J Biol Chem* 2003;278:9489–95. doi:10.1074/jbc.M209397200.

[73] Biswal BK, Cherney MM, Wang M, Chan L, Yannopoulos CG, Bilimoria D, et al. Crystal structures of the RNA-dependent RNA polymerase genotype 2a of hepatitis C virus reveal two conformations and suggest mechanisms of inhibition by non-nucleoside inhibitors. *J Biol Chem* 2005;280:18202–10. doi:10.1074/jbc.M413410200.

[74] Di Marco S, Volpari C, Tomei L, Altamura S, Harper S, Narjes F, et al. Interdomain communication in hepatitis C virus polymerase abolished by small molecule inhibitors bound to a novel allosteric site. *J Biol Chem* 2005;280:29765–70. doi:10.1074/jbc.M505423200.

[75] Barnes-Seeman D, Boiselle C, Capacci-Daniel C, Chopra R, Hoffmaster K, Jones CT, et al. Design and synthesis of lactam-thiophene carboxylic acids as potent hepatitis C virus polymerase inhibitors. *Bioorganic Med Chem Lett* 2014;24:3979–85. doi:10.1016/j.bmcl.2014.06.031.

[76] Court JJ, Poisson C, Ardzinski A, Bilimoria D, Chan L, Chandupatla K, et al. Discovery of Novel Thiophene-Based, Thumb Pocket 2 Allosteric Inhibitors of the Hepatitis C NS5B Polymerase with Improved Potency and Physicochemical Profiles. *J Med Chem* 2016;59:6293–302. doi:10.1021/acs.jmedchem.6b00541.

[77] Li P, Dorsch W, Lauffer DJ, Bilimoria D, Chauret N, Court JJ, et al. Discovery of Novel Allosteric HCV NS5B Inhibitors. 2. Lactam-Containing Thiophene Carboxylates. *ACS Med Chem Lett* 2017;8:251–5. doi:10.1021/acsmedchemlett.6b00479.

In: Advances in Medicinal Chemistry … ISBN: 978-1-53616-368-1
Editor: E. Ferreira da Silva-Júnior

Chapter 4

DISCOVERY OF POTENT HUMAN GLUTAMINYL CYCLASE INHIBITORS AS ANTI-ALZHEIMER'S AGENTS

Phuong-Thao Tran[1,*], PhD and Van-Hai Hoang[2], PhD
[1]Department of Pharmaceutical Chemistry,
Hanoi University of Pharmacy, Hanoi, Vietnam
[2]Laboratory of Medicinal Chemistry, Research Institute
of Pharmaceutical Science, College of Pharmacy,
Seoul National University, Seoul, Republic of Korea

ABSTRACT

Alzheimer's disease (AD) is the most common dementia diseases. It leads to a decline in memory and other thinking skills that affects a person's behavior and ability to perform daily activities. Approximately 10 million patients are diagnosed with this disease every year. The financial costs of treatment and medical care for AD patients have increased significantly in recent years. While the causes of AD are not yet fully understood, it is widely known that extracellular deposition of

* Corresponding Author's E-mail: thaotp@hup.edu.vn or thaotp119@gmail.com.

amyloid-beta (Aβ) is one of two hallmark pathologies of AD. Glutaminyl cyclase (QC) enzyme has been implicated in the formation of toxic amyloid plaques by catalyzing cyclization of the *N*-terminal glutamate of β-amyloid peptides (*N*Glu-Aβ) into pyroglutamic acid (pGlu) and thus may participate in the pathogenesis of AD. Moreover, the results of a recent phase I trials indicate that QC inhibitors are well tolerated and metabolically safe for both young and elderly patients, suggesting that the inhibition of QC offers an alternative approach to conventional Aβ-lowering agents. Despite the significant potential role of QC in AD pathology, only a few QC-specific inhibitors have been developed thus far. Utilizing structural knowledge of the hQC substrates Aβ_{3E-42} and traditional screening methods for rational design, researchers from Probiodrug AG (Germany); the Laboratory of Medicinal Chemistry in the College of Pharmacy at Seoul National University (Republic of Korea); Haiquang Wu's group (China) and Phuong-Thao Tran's group from the Hanoi University of Pharmacy (Vietnam); ... have designed, synthesized, evaluated and predicted the structure-activity relationship (SAR) of QC inhibitors. In order to present the story of the discovery and development of QC inhibitors, which represents a new opportunity for the treatment of AD, this chapter is divided into two parts. The introduction in Part I presents an overview of AD and hQC, while Part II examines the discovery of potent human glutaminyl cyclase inhibitors as anti-Alzheimer's agents.

Keywords: Alzheimer's disease, AD, human glutaminyl cyclase, hQC, inhibitors

INTRODUCTION

Alzheimer's Disease

Overview

German physician and neuropathologist Alois Alzheimer first described AD in 1906 at the 37th Conference of South-West German Psychiatrists in Tübingen. At this conference, he presented the case of a patient called Auguste Deter, a 51-year-old woman who had profound memory loss, unfounded suspicions about her family, and other worsening psychological changes. An autopsy of her brain revealed a loss of neurons and abnormal deposits in and around neurons [1]. For most of the 20th century, the diagnosis of AD was reserved for individuals between the ages of 45 and 65 who developed symptoms of dementia [2]. Nowadays, however, the diagnosis of AD is independent of age, and the term "Alzheimer's disease" was formally adopted in medical nomenclature to describe individuals of all ages with a characteristic typical symptom pattern, disease course, and neuropathology [3].

In 2017, the Alzheimer's Association estimated that approximately 5.5 million Americans were living with AD [4], and that 47 million people throughout the world were affected by this disease or other forms dementia [5]. It is estimated that the number of people with AD throughout the world will exceed 70 million by 2030. The disease accounts for between 60% and 80% of dementia cases, and the majority of people with AD are 65 and older (96%). In 2015, the global cost of AD was US$818 billion, which is anticipated to increase to US$2,000 billion by 2030 [6]. Moreover, the number of deaths from AD has increased significantly, while deaths from other diseases have decreased. Of the top 10 causes of death, AD is the only one that cannot be prevented, cured, or even slowed down [4].

Unlike many other diseases, AD is a long-term process in which patients gradually lose their memory and independence. Caregiving is, therefore, necessary, which involves high physical, emotional, and financial costs.

The causes of AD are not still fully understood, although certain hypotheses have been put forward in this regard. The oldest of these is the cholinergic hypothesis, which proposes that AD is caused by reduced synthesis of neurotransmitter acetylcholine [7]. On the basis of this hypothesis, certain drugs were developed with a view to preventing acetylcholine hydrolysis (acetylcholine esterase inhibitors); drugs such as donepezil, galantamine, and rivastigmine, for example, which focus on the treatment of cognitive symptoms [7, 8]. However, analysis of the clinical efficacy of these drugs in alleviating symptoms of dementia demonstrated unsatisfactory results, and the cholinergic hypothesis has not maintained widespread support [7]. Two other hypotheses – the amyloid and tau hypotheses – enjoy full support today and are becoming potential molecular targets for AD.

Amyloid beta (A*β*) is a 36 to 43 amino acid peptide, which is a small fragment of a transmembrane protein, called the amyloid precursor protein (APP). APP is made by neurons and other brain cells. It is also found in extra-neural tissues and is especially abundant in platelets. Its function is unknown, however. The A*β* residue is derived from cleavage of APP by the enzymes *β*- and *γ*-secretase. Other enzymes further degrade Aβ monomers and oligomers. Defective clearance of A*β* from aberrant cleavage of APP and other mechanisms results in its accumulation [9, 10].

A*β*, which is a heterogeneous mixture of a peptide with different solubility, stability, biological, and toxic properties, is a significant hallmark in AD. At present, more than 20 A*β* peptides are found in the Alzheimer's brain, the majority of which are $A\beta_{42}$ and $A\beta_{40}$. When A*β* is formed, it is aggregated to produce higher-order structures. This assembly can take place in at least two ways: the soluble pathway leads to oligomers, while the other pathway leads to insoluble, plaque-forming fibrils [11]. These oligomers are suspected to be the primary toxic species in AD since they are small enough to enter synapses, highly diffuse, and able to spread to other regions of the brain [11, 12]. Soluble A*β* oligomers are aggregated into different spherical structures that range from dimers to dodecamers, and the bigger the size of the oligomeric assembly, the lower its toxicity effects are [13]. Some studies have demonstrated that A*β* oligomers impair memory function, and lead to neuron loss and synaptic dysfunction. It has been demonstrated that A*β* oligomers interact directly with cholesterol and lipid in the neuron cell membranes, forming pores and disrupting the proper permeability of membranes. They also exert their toxic effects through specific receptors. Even at a very low molecular concentration, A*β* oligomers can increase *N*-methyl-*D*-aspartic acid (NMDA) response through glutamate receptor subunit epsilon-2 (NR-2B) containing NMDA receptors. Moreover, A*β*-receptor interactions induce deterioration and loss of synapses through a redistribution of receptors, which further leads to an alteration of neuronal plasticity accompanied by oxidative stress. Recent studies have demonstrated that A*β* oligomers can induce impairment of the transduction of signal in neuronal insulin receptors, and block the activation of the insulin receptor substrate. In addition, Aβ oligomers cause an elevation in cytosolic calcium due to the creation of an ion channel. This process leads to the rapid depolarization of neurons, the release of glutamate, and excitotoxicity. Although the accumulation of Aβ inside neurons is still not well understood, some toxicity of intracellular Aβ was found. They are also affected by calcium transition on mitochondria, and levels of hyperphosphorylated tau [12-14].

Oxidative stress, which increases in Alzheimer's patients, may induce overproduced Aβ peptides. Overproduced Aβ peptides may aggregate into toxic oligomer forms, leading to the initiation of free radical processes which result in a change in protein structure and function, pathological induction condition, neuron damage, and apoptosis [12, 13].

Tau proteins are proteins that stabilize axonal microtubules. They are the product of alternative splicing from a single gene that is located on chromosome 17 and are abundant in neurons of the central nervous system, but less common elsewhere [15]. Tau proteins are highly soluble, and their primary functions are to modulate the stability of microtubules by interaction with tubulin and to promote tubulin assembly into microtubules. Tau changes its isoforms and the levels of phosphorylation to control the stability of microtubules. Typically, one protein Tau is phosphorylated with two to three phosphate groups. When over-phosphorylation occurs, Tau can be stickier and lead to the self-assembly of tangled, helical filaments and straight filaments which are involved in the pathogenesis of AD.

AD has both modifiable and unmodifiable risk factors. Age, family history, and genes are three unmodifiable factors. With regard to age, most individuals with AD are 65 or older. Moreover, individuals who have parents or siblings with AD are more likely to develop the disease than those who do not have a first-degree relative with the disease. Furthermore, having the apolipoprotein E (APOE) e4 gene and other mutated genes such as APP, PS1, and PS2 increases the risk of developing AD. Modifiable factors include smoking, midlife obesity, diabetes, hypertension, high cholesterol, and education (brain training) [5].

Glutaminyl Cyclase (QC)

Overview

Glutaminyl cyclases (EC 2.3.2.5) catalyze the intramolecular cyclization of *N*-terminal glutamine residues into pyroglutamic acid (pGlu), liberating ammoniac (Figure 1) [16, 17].

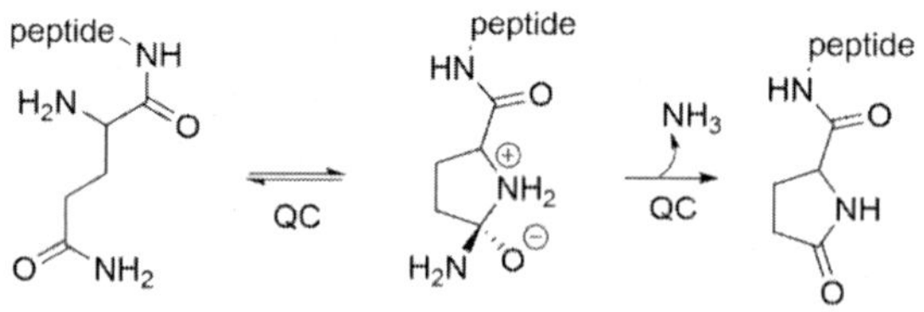

Figure 1. *N*-terminal cyclization of glutaminyl peptides by QC.

Glutaminyl cyclases have been identified in both plant and animal sources [16-18]. In mammals, they are abundant in neuroendocrine tissues, such as the hypothalamus and pituitary gland, other parts of the brain, cortex, adrenal medulla, and B lymphocytes [16, 17, 19]. Some peptide hormones and chemokines such as gastrin, TRH (thyrotropin-releasing hormone), GnRH (gonadotropin-releasing hormone), and neurotensin are substrates of QC. Although the physiological functions of QC are still ambiguous, it is believed that the formation of pyroglutamate in numerous bioactive peptides, hormones, and chemokines protects them from proteolytic degradation and can be essential for their biological activity [20-22].

Mammalian QCs are zinc-dependent glycoproteins. The structure of human QC [23], human Golgi-resident QC [24], animal glutaminyl cyclase [25], glycosylated mammalian QC [26], and their active structure site were analyzed. The structure of human QC revealed an α/β-mixed scaffold comprising a central six-stranded β-sheet surrounded by two (α7 and α9) and six (α2, α3, α4, α5, α6, and α10) α-helices on opposite sides, with two α-helices (α1 and α8) located at one edge of the β-sheet. Six loops between α3 and α4, β3 and α5, β4 and α7, β5 and α8, α8 and α9, and β6 and α10 create the active site. It is a narrow pocket near the *C*-terminal edge of central parallel strands $\beta3$, $\beta4$, and $\beta5$. The zinc ion lies at the bottom of the pocket and is coordinated by Asp^{159} O$\delta2$, Glu^{202} O$\delta1$, His^{330} N$\varepsilon2$, and a water molecule. The hydrophobic pocket is made up of residue Lys^{144}, Phe^{146}, Leu^{249}, Ile^{303}, Ile^{321}, Phe^{325}, and Trp^{329} with six water molecules inside (Figure 2) [23].

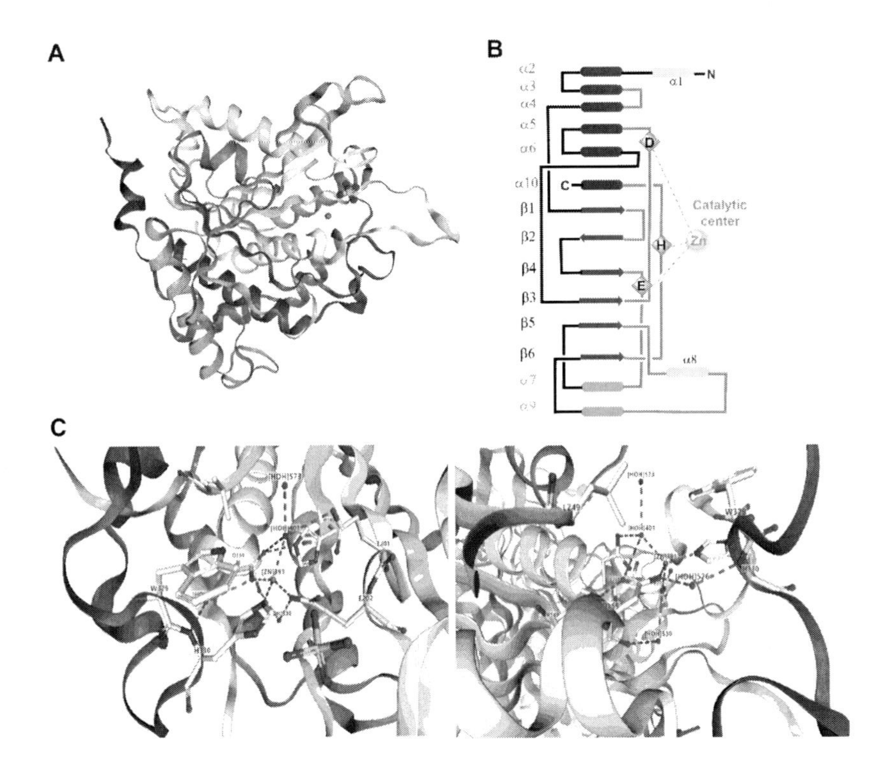

Figure 2. Structure of human QC. (A) A ribbon diagram of the overall structure of human QC. (B) A topology diagram of the human QC structure. (C) A stereo-view of the human QC catalytic region.

Kai-Fa Huang and his colleagues [23] also proposed the catalysis mechanism of human QC with Glu^{201} acting as the general base and acid to transfer a proton from the α-amino group of the substrate to the amino leaving group on the γ-amide. The functions of the zinc ion are the polarization of the γ-amide carbonyl group of the substrate and simultaneous stabilization of the oxyanion formed by the nucleophilic attack of the α-nitrogen (Figure 3).

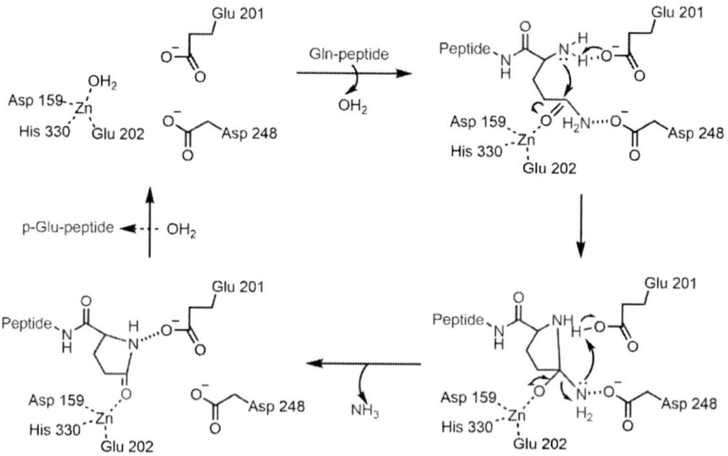

Figure 3. Proposed catalysis mechanism of human QC.

QC also catalyzes the *N*-terminal glutamate cyclization into pGlu, a reaction that is favored under acidic pH conditions [27]. While co-transfection of APP and QC leads to pGluAβ in HEK293 cells, the addition of recombinant QC in the same culture medium generated minor amounts of A$\beta_{N3(pE)}$, suggesting that the conversion is favored intracellularly [28]. This reaction is probably related to the formation of several plaque-forming peptides, such as Aβ peptides, which play a pivotal role in AD.

QC and Alzheimer's Disease

Pyroglutamate-Aβ: Formation and Pathological Functions

Formation of pyroglutamate-Aβ is a multistep process requiring the loss of the first two amino acids – aspartate and alanine – to expose the *N*-terminal glutamate at the third position of full-length Aβ, followed by intramolecular dehydration of exposed glutamate to pyroglutamate A$\beta_{(3\text{-}x)}$ peptides. However, the formation of A$\beta_{(11\text{-}x)}$ may be generated directly from the APP by β-secretase in the trans-Golgi network (Figure 4) [29].

AβxpE peptides were found in amyloid plaques [30-32], diffused amyloid plaques [31], and vascular amyloid deposits [33] of AD brain.

Pyroglutamate Aβ peptides constitute between 15% and 20% of the total Aβ and are deposited in the center of senile plaques in AD brain [30, 34]. Moreover, Aβ3pE peptides constitute the majority of the water-soluble amyloid peptides, that are thought to precede amyloid plaques in AD brain [35].

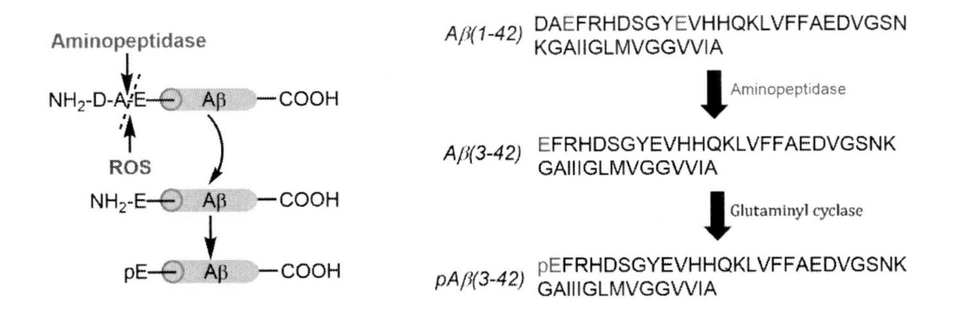

Figure 4. Scheme of pEAβ formation.

This formation of pyroglutamate-Aβ results in the loss of three charges or six charges for Aβ3pE or Aβ11pE, respectively, and leads to higher hydrophobicity and thus more stability and aggregation propensity of the AβxpE. Moreover, the formation of a lactam ring also increases the stability of peptides, due to resistance to aminopeptidase degradation. Moreover, AβxpE enhances β-sheet formation and aggregation propensity in both aqueous and hydrophobic media compared with corresponding full-length Aβ, which suggests that they are potential seeding species of aggregate formation [31, 36, 37]. Harigaya and collaborators [38] demonstrated Aβ3pE-42 oligomer formation, the most toxic form, at a concentration as low as 1.0 μg/mL, whereas this concentration is 2.5 μg/mL for $A\beta_{1-42}$ *in vitro*.

Studies on cytotoxic properties of pyroglutamate Aβ peptides have demonstrated that Aβ3pE are more toxic for neurons and astrocytes than corresponding full-length Aβ [39]. Russo and coauthors [40] demonstrated that Aβ3pE induced significantly more cell loss than other Aβ species in rat cultured hippocampal neurons and cortical astrocytes. They also suggested that Aβ3pE share similar degenerative mechanisms, i.e., apoptosis in neurons and necrosis in astrocytes with full-length Aβ. Moreover, because of early aggregation, Aβ3pE form membrane pores, which leads to change in membrane permeability [41]. They also induce redox-sensitive neuronal apoptosis [42]. Interestingly, Aβ3pE have an inhibitory effect on full-length Aβ1-42 fibrillogenesis, probably maintaining the peptides in more toxic forms, oligomers. It has recently been demonstrated that Aβ3pE form hybrid oligomers with the full-length Aβ peptides faster than self-aggregation [43]. These oligomers are more cytotoxic to cultured neurons than oligomers Aβ1-42 by tau-dependent cytotoxicity [44]. The accumulation of oligomeric aggregates of Aβ3pE42 results in a loss of lysosomal integrity, which may explain its predominant accumulation in the lysosomes [45]. Another study has demonstrated that Aβ3pE antibody increases cell survival much more significantly than other Aβ species, which also suggests the toxicity of Aβ3pE [46, 47].

Glutaminyl Cyclase Inhibition: A Possible Treatment Strategy

The expression of QC correlates with the appearance of Aβ3pE; specifically, QC is overexpressed in the brains of AD patients and animal models [48-50]. Moreover, higher levels of QC mRNA were found in the AD brain than in an age-matched healthy brain, and a correlation between QC expression and the severity of dementia was observed [48]. Studies in transgenic or QC knock-out mice have demonstrated a significant reduction in Aβ3pE levels, a decrease in plaque pathology, and a recovery the cognitive function [51]. Small molecule QC inhibitors were found and demonstrated their activity in reducing brain pGlu-Aβ levels, Aβ plaques, and gliosis, while also restoring memory deficits in AD mice [52, 53]. This suggests that inhibition of QC may prevent the formation of Aβ3pE, restore brain functions and that QC inhibitors may offer a new option for the development of novel anti-AD agents.

Discovery of Potent Human Glutaminyl Cyclase Inhibitors as Anti-Alzheimer's Agents

In the light of the QC hypothesis, research has to date focused primarily on the discovery and development of potent human QC inhibitors as anti-Alzheimer's agents. Some QC inhibitors were reported and demonstrated their biological activity in Aβ3pE and brain functions of AD mice. Moreover, one of the QC inhibitors is being studied in a current phase II trial [37, 54-60].

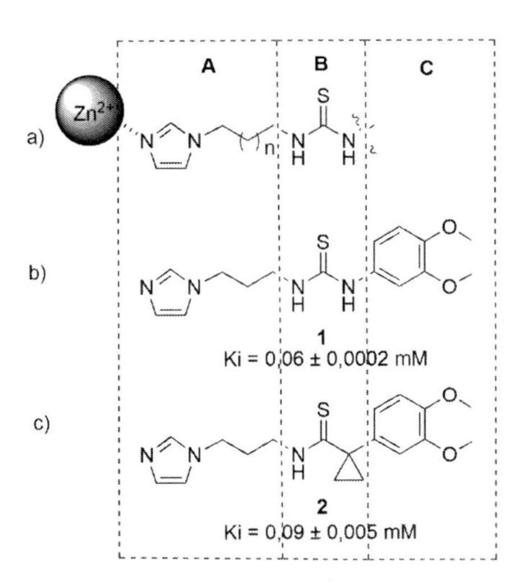

Figure 5. Structure of lead compounds **1** and **2**.

In 2006, scientists from the Probiodrug AG company (Germany) and their colleagues undertook pioneering research to find QC inhibitors by applying a ligand-based optimization screening approach of hundreds of heterocyclic compounds. They indicated that a compound of the imidazol-1-yl-alkyl thiourea type was a lead scaffold (Figure 5, a). In this study, Probiodrug reported two compounds (**1** and **2**) with the most potent QC-inhibitory effect, exhibiting K_i values in the nanomolar range (Figure 5, b and c). Consequently, these compounds were selected as lead QC inhibitors [37]. Based on the structures of specific human QC inhibitors which interacted with QC enzymes at the active site and the structures of the two lead compounds, the scientists have been able to divide the structures of QC inhibitors into three regions, as presented in Figure 5. The A-region contains imidazole conjugated as a zinc-binding moiety. The B region presents a hydrogen bond donor (Figure 5). The C-region accommodates the hydrophobic side chain penultimate to the *N*-terminus of the preferred substrates [37].

In 2009, a structure-based approach based on a homology model of the enzyme was adopted for the introduction of a methyl group in position 4 or 5 of the imidazole and the bioisosteric replacement of the thiourea moiety in the B-region was carried out. Many potent QC inhibitors were presented by Probiodrug AG (Germany) and their colleagues (Table 1) as the second generation of QC inhibitors. The structures of lead compounds **1** and **2** were replaced by the new lead scaffold **3** (Figure 6) [54].

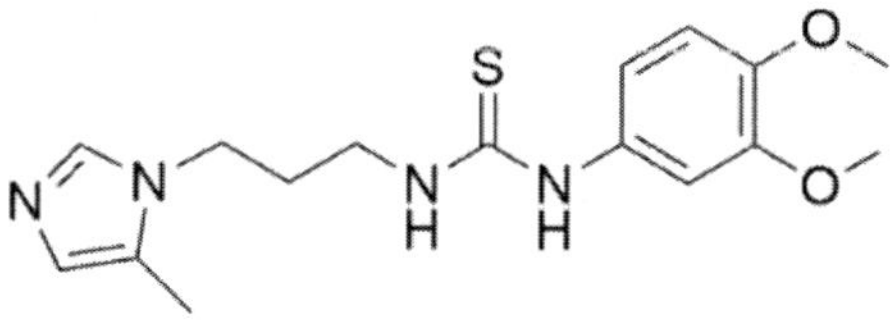

Figure 6. Structure of lead compounds **3**.

In 2013, Probiodrug AG presented a new class of inhibitors of hQC, resulting from a pharmacophore-based screening. Hit molecules were identified containing benzimidazolyl-1,3,4-oxadiazol scaffold (Figure 7, a). Benzimidazole played a role in binding to metal, replacing imidazole and 4 or 5-methylimidazole in the A-region of the previous lead compounds **1**, **2**, **3** (Figure 5 and 6), while 1,3,4-oxadiazole moiety acted as a central part of the structural framework (B-region). The Probiodrug AG scientists found that benzimidazole and benzimidazole-1,3,4-oxadiazole derivatives were the most potent inhibitors of hQC known thus far, exhibiting inhibitory constants of which ranged from 60 to 6 nM. It is significant to note that the replacement of 1,3,4-oxadiazole with 1,3,4-thiadiazole (Figure 7, b) led to the most active compound, i.e., compound **4**, with an IC_{50} value of 0.07 μM and a K_i value of 23 nM. The subsequent optimization resulted in benzimidazolyl-1,3,4-thiadiazoles and -1,2,3-triazoles with an inhibitory potency in the nanomolar range. This forms the premise for further research to identify novel QC inhibitors [56].

Table 1. The potent QC inhibitors were reported in 2009 by Probiodrug AG

	R_1	R_2	Z	K_i (ref. [58])
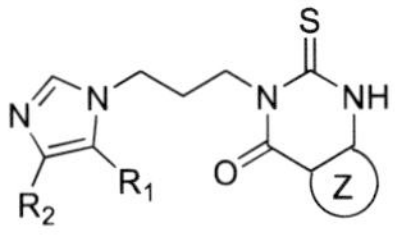	CH_3	H	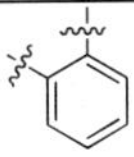	18 ± 1,9
	CH_3	H	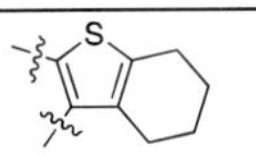	17 ± 0.3
	CH_3	H	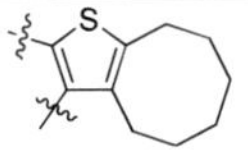	39 ± 1.0
	CH_3	H	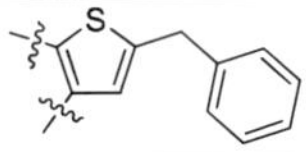	20 ± 0.6

	R_1	R_2	X	Y	R	K_i (ref. [58])
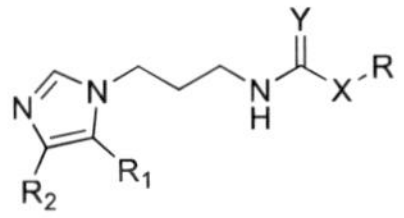	H	CH_3		S	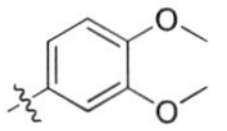	41 ± 2.0
	CH_3	H	NH	S	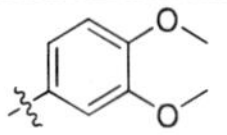	6.3 ± 0.45
	CH_3	H	S		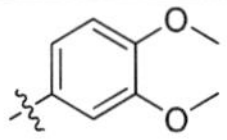	2.6 ± 0.08
	CH_3	H	NH	$-CH-NO_2$	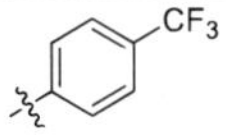 CF_3	34 ± 3.0

Furthermore, in 2013, a group of scientists from the Laboratory of Medicinal Chemistry in the College of Pharmacy at Seoul National University (Republic of Korea) published the results of their study on the structure-activity relationship (SAR) of hQC inhibitors having an *N*-(5-methyl-1*H*-imidazol-1-yl)propylthiourea template (Figure 8a) [57].

In this study, the hydrophobic dimethoxyphenyl moiety was replaced by various constrained analogs of heterocycles and aromatic rings. The results indicated that the presence of hydrogen-bond acceptors within the hydrophobic region (C-region in compounds **1**, **2**, and **3**) is essential for any inhibitory effect. Compounds with hydrogen-bond acceptors such as oxygen and nitrogen atoms appeared to be more potent than compounds lacking heteroatoms or containing delocalized aromatic rings (Figure 8, a; and Figure 9). Overall, compound **5** emerged as the most promising candidate for further study with an IC_{50} value of 58 nM (Figure 8, b) [57].

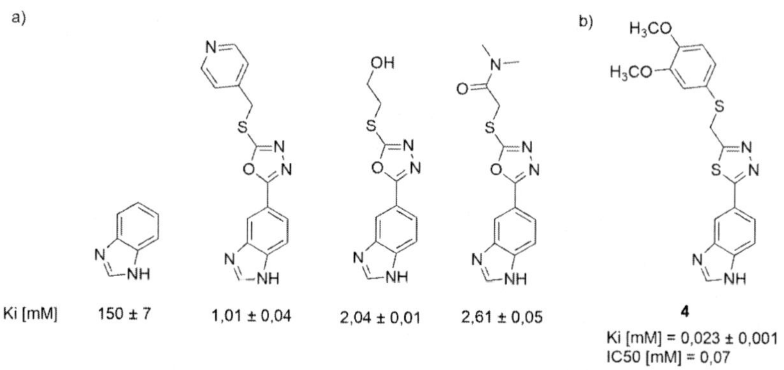

a)

Ki [mM]	150 ± 7	1,01 ± 0,04	2,04 ± 0,01	2,61 ± 0,05

b) **4**
Ki [mM] = 0,023 ± 0,001
IC50 [mM] = 0,07

Figure 7. a) Structure of benzimidazolyl-1,3,4-oxadiazole derivatives; b) Structure of compound **4**.

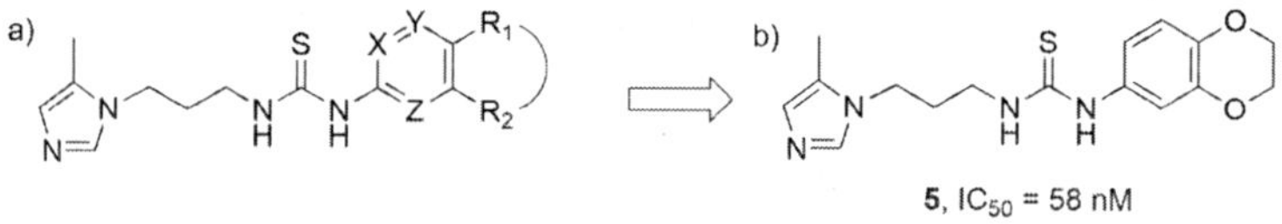

a) b) **5**, IC_{50} = 58 nM

Figure 8. Structure of *N*-aryl *N*-(5-methyl-1*H*-imidazol-1-yl)propylthiourea derivatives.

In 2015, Probiodrug AG [60] reported compound **PQ912** (Figure 10) as a competitive QC inhibitor, essential for the formation of pE-Aβ peptides. **PQ912** was considered safe and well-tolerated with dose-proportional pharmacokinetics up to doses of 200 mg. At higher doses – up to 1800 mg – exposure was super-proportional, and exposure in elderly subjects was approximately 1.5- to 2.1-fold higher. Exposure in cerebrospinal fluid (CSF) was approximately 20% of the unbound drug in plasma, and both serum and CSF QC activity was inhibited in a dose-related manner [60].

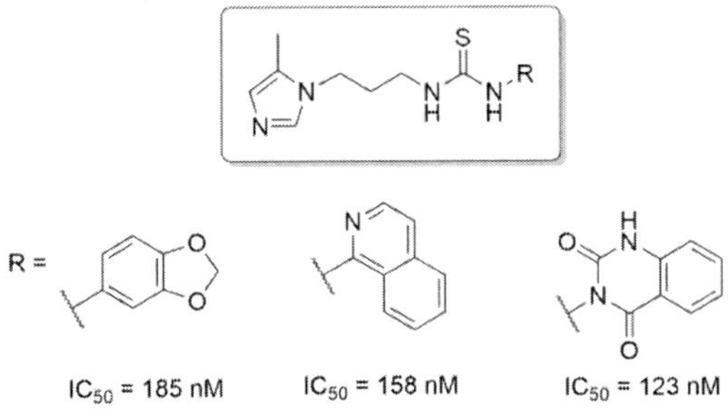

R =

IC_{50} = 185 nM IC_{50} = 158 nM IC_{50} = 123 nM

Figure 9. Structure of compounds containing dioxane, 1-isoquinoline or quinazoline moiety in C-region.

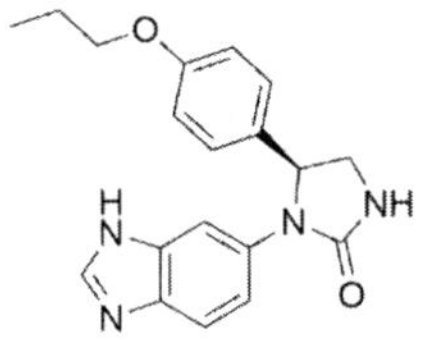

Figure 10. Structure of **PQ912**.

In 2016, a new lead structure containing the sulfolipids scaffold for QC inhibitors was presented by Hielscher-Michael S. and his colleagues [61]. Three QC inhibitors from microalgae belonging to the family of sulfolipids (Figure 11) were identified by using a new reverse metabolomics technique, including an activity-correlation analysis (AcorA), which was based on the correlation of bioactivities to mass spectral data with the aid of mathematical informatics deconvolution. The compounds demonstrated a QC inhibition of 81% and 76%, both at 0.25 mg/mL concentrations [61].

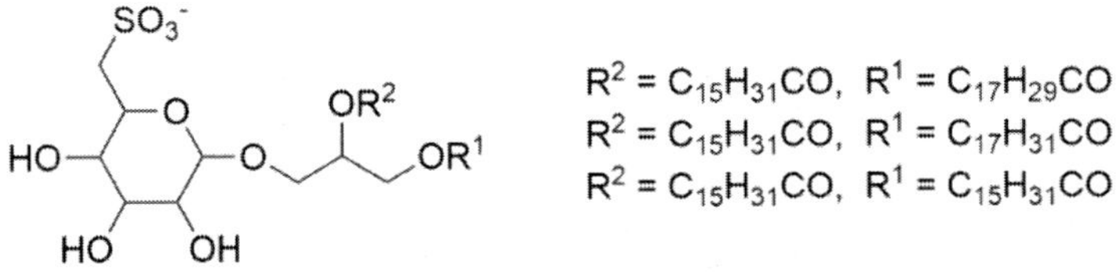

$R^2 = C_{15}H_{31}CO$, $R^1 = C_{17}H_{29}CO$
$R^2 = C_{15}H_{31}CO$, $R^1 = C_{17}H_{31}CO$
$R^2 = C_{15}H_{31}CO$, $R^1 = C_{15}H_{31}CO$

Figure 11. Structure of QC inhibitors from microalgae.

Moreover, in 2016, Li M., Dong Y. and their colleagues [55] investigated the inhibitory effect of flavonoids on hQC. Three series of apigenin derivatives – phenol-4′ (R^1), C5–OH (R^2) and C7–OH (R^3) – were modified, synthesized and evaluated the QC inhibition by spectrophotometric assessment (Figure 12). In this study, an analysis of the SAR and molecular docking was performed to analyze the binding mode of the synthesized flavonoids to the active site of hQC. Based on the research results, it could be concluded that the effect of R^1 on the inhibitory activity of apigenin derivatives is slight, but R^2 (C5–OH) is favored and R^3 (C7–OH), in particular, is essential for the inhibitory potency. These results therefore strongly implied that apigenin derivatives containing C5–OH and C7–OH might present a new class of QC inhibitors and flavonoids that should be thoroughly investigated for the treatment of AD and other QC related diseases [55].

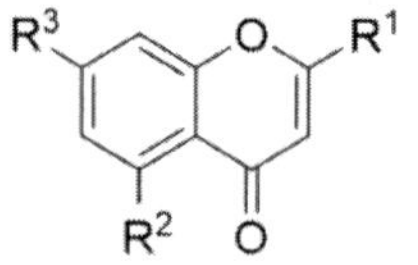

R^1 = phenol, benzaldehyde, imidazole, furane, thiophene, pyridine and their derivatives
R^2 = -H, -OH
R^3 = -H, -OH

Figure 12. Structure of apigenin derivatives as QC inhibitors.

In 2017, the scientists from the Laboratory of Medicinal Chemistry in the College of Pharmacy at Seoul National University (Republic of Korea) and their colleagues [62], designed a library of QC inhibitors based on the proposed binding mode of the preferred substrate, $A\beta_{3E-42}$. Among the preferred substrates of hQC, these scientists focused on the antepenultimate Arg of $A\beta_{3E-42}$. While the substrate specificity of this specific position has not been clearly elucidated, it was speculated that the additional side-chain might be able to act as a positional anchor to provide enhanced binding interactions. Based on the structure of former lead compound **3** and the above expectation, a novel scaffold with an additional pharmacophoric D-region was designed, as presented in Figure 13. These new series of compounds contain various amine-type functional groups in the D-region, which mimic the guanidine moiety of Arg. In this work, newly designed QC inhibitors were focused on the D-region, and their SAR investigated based on *in vitro* inhibition for human QC. Furthermore, the efficacy of the selected inhibitors was evaluated by measuring their ability to reduce the formation of $A\beta_{3(pE)\text{-}42}$ and to prevent memory impairment in AD model mice [62].

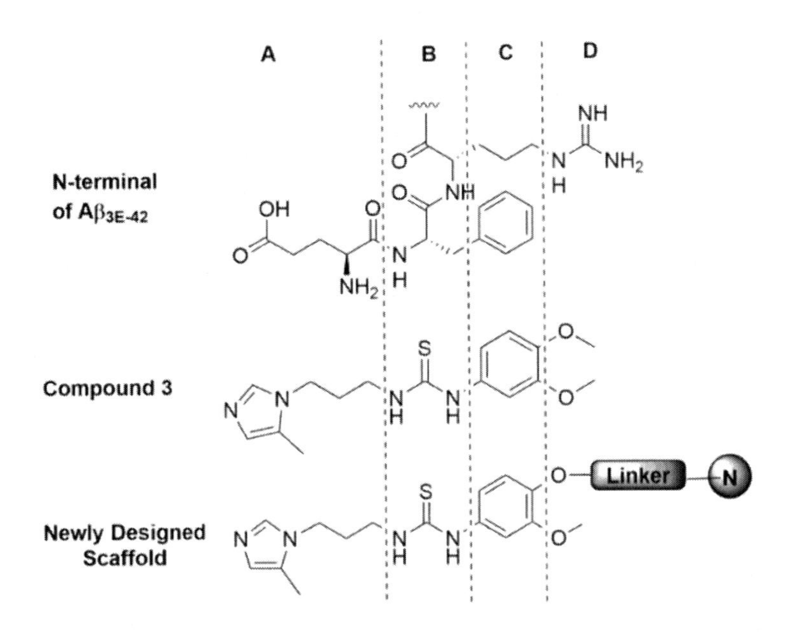

Figure 13. Newly designed scaffold for QC inhibitors.

Finally, the specific binding interactions between the selected inhibitors and the QC active site were analyzed by performing molecular docking studies. An *in vitro* SAR study identified several excellent QC inhibitors demonstrating 5- to 40-fold increases in potency compared to any known QC inhibitor (Table 2). When tested in mouse models of AD, compound **6** significantly reduced the brain concentrations of pyroform $A\beta$ and total $A\beta$ and restored cognitive functions. This potent $A\beta$-lowering effect was achieved by incorporating an additional binding region into a previously established pharmacophoric model, resulting in strong interactions with the carboxylate group of Glu^{327} in the QC binding site (Figure 14). The results of this study reveal useful insights into designing novel QC inhibitors as a potential treatment option for AD [62].

Table 2. hQC inhibition *in vitro* and in acute model studies *in vivo*[a]

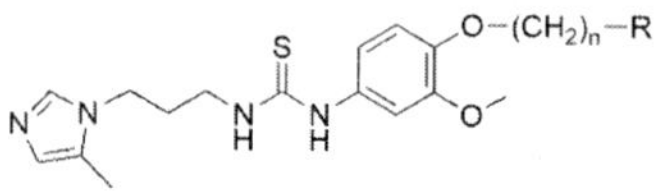

Cpd	R	n	*In vitro* IC_{50}	% inhibition of human $A\beta_{N3pE-42}$ (i.c.v. injected)	% inhibition of human $A\beta_{N3pE-42}$ (i.p. injected)	PAMPA (-logPe)
3	H	1	29.2	41.6	NE	6.3
7	*-NH_2	3	5.3	64.6	NE	6.2
8	*-N<	4	3.7	35.7		
9	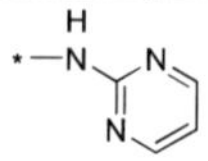 *—NH-(pyrimidin-2-yl)	4	5.7	27.3		
10	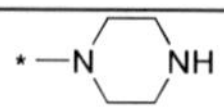 *—N(piperazine)NH	2	0.7	77.2	3.6	6.1
11		4	4.8	29.2		
12	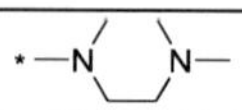 *—N(piperazine)N—	2	4.6	18.7		
13	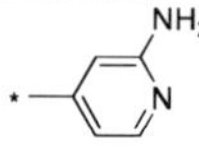 *—(3-aminopyridin-2-yl), NH_2	2	5.5	38.0		
14		4	4.5	66.1	54.7	5.0

[a]5 µL of human $A\beta_{3\text{-}40}$ in PBS (1 µg/µL) was injected into the deep cortical/hippocampus to 5 weeks old ICR mice (25 g, n = 4, male) using a stereotaxic frame to induce acute $A\beta$ toxicity. Test compounds were administrated via icv injection. Sandwich ELISA was performed for the quantification of the brain $A\beta_{N3pE-40}$.

In 2018, Van T. H. Ngo and her colleagues [63, 64] presented the SAR of analogs with modifications in the D-region as continuous series of compounds containing various amine-type functional groups in the D-region and evaluated their biological activity (Template A, Figure 15). Most compounds in this series exhibited potent activity *in vitro*. Moreover, the SAR analysis and the molecular docking studies identified compound **15** as a potential candidate since it forms an additional hydrophobic interaction in the hQC active site (Figure 16) [63, 64].

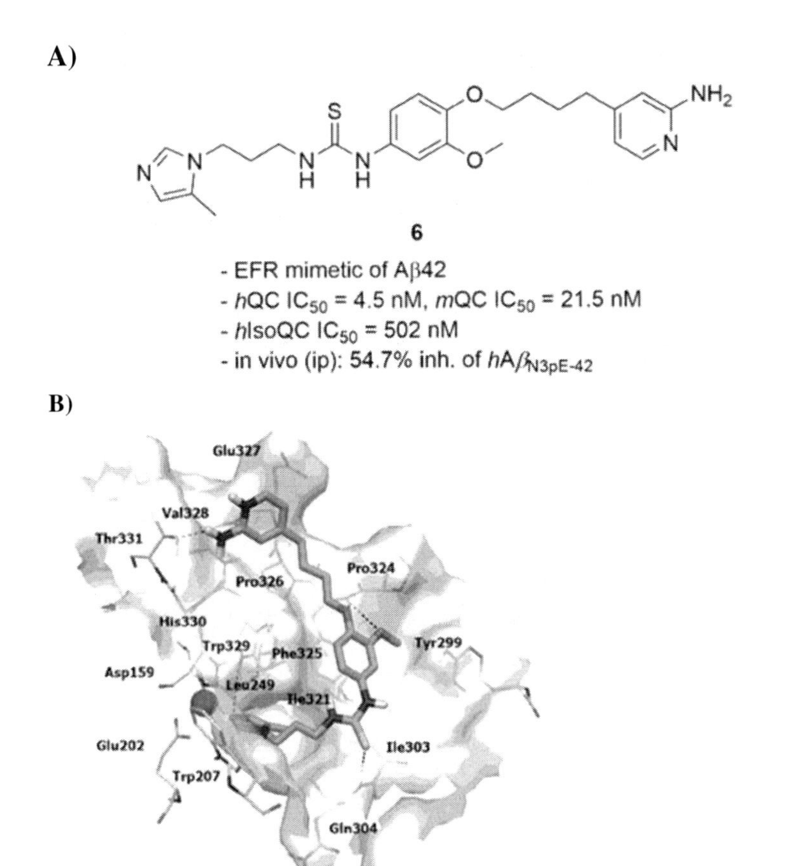

Figure 14. A) Compound **6** and its activities; B) Docked of **6** in hQC (*continued*).

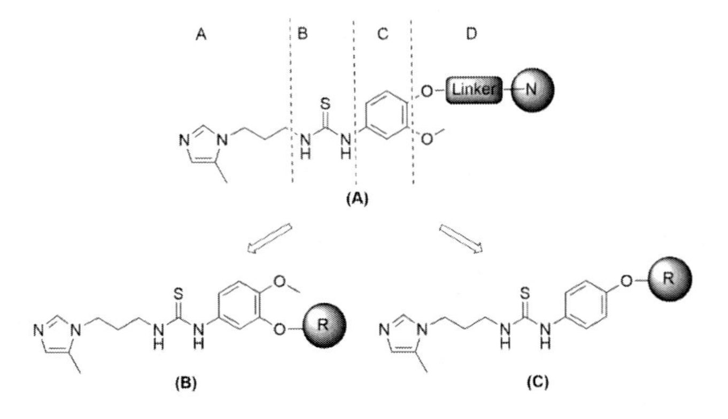

Figure 15. A) Template A; (B) Template B; (C) Template C.

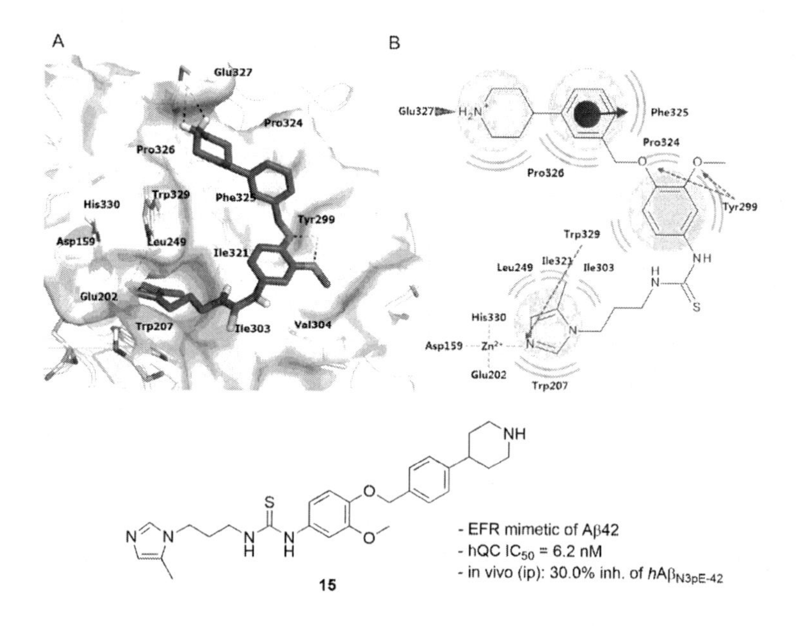

Figure 16. Refined structure of compound 15 docked with hQC.

Table 3. hQC inhibition *in vitro* and in acute model studies *in vivo*[a] of potent 3-aminoalkyloxy-4-methoxyphenyl derivatives

Cpd	R	n	*In vitro* IC50	Cyto-toxicity at 10 µM (% of control)	hERG assay (10 µM, % inhibition)	% inhibition of human AβN3pE-42 (i.c.v. injected)[a]	PAMPA (-logPe)
16	*-NH2	2	7.9	~100	15.3	35.3	5.7
17		3	9.0	~100	8.8	12.8	5.6
18		4	8.8	~100	2.8	36.6	5.9
19	* NH	2	7.3	~100	14.9	1.8	5.3
20		3	8.8	~100	25.6	5.5	5.0
21		4	7.9	~100	35.1	14.4	5.7

[a]5 µL of human Aβ3–40 in PBS (1 µg/µL) was injected into the deep cortical/hippocampus to 5 weeks old ICR mice (25 g, n = 4, male) using a stereotaxic frame to induce acute Aβ toxicity. Test compounds were administrated via icv injection. Sandwich ELISA was performed for the quantification of the brain AβN3pE-40.

Table 4. hQC inhibition *in vitro* and in acute model studies *in vivo*[a] of potent 4-aminoalkyloxyphenyl derivatives

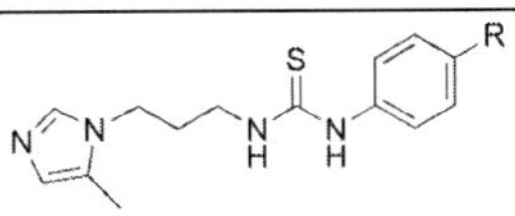

Cpd	R	*In vitro* IC_{50}	Cyto-toxicity at 10 μM (% of control)	hERG assay (10 μM, % inhibition)	% inhibition of human $A\beta_{N3pE-42}$ (i.c.v. injected)[a]	PAMPA (-logPe)
22	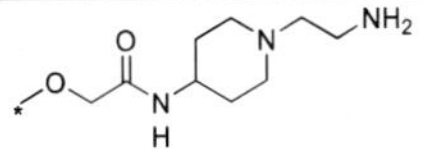	7.9	~100	19.9	NE	10.0
23	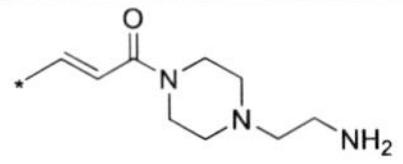	6.4	~100	7.2	22.9	9.0

[a]5 μL of human Aβ3–40 in PBS (1 μg/μL) was injected into the deep cortical/hippocampus to 5 weeks old ICR mice (25 g, n = 4, male) using a stereotaxic frame to induce acute Aβ toxicity. Test compounds were administrated via icv injection. Sandwich ELISA was performed for the quantification of the brain AβN3pE-40.

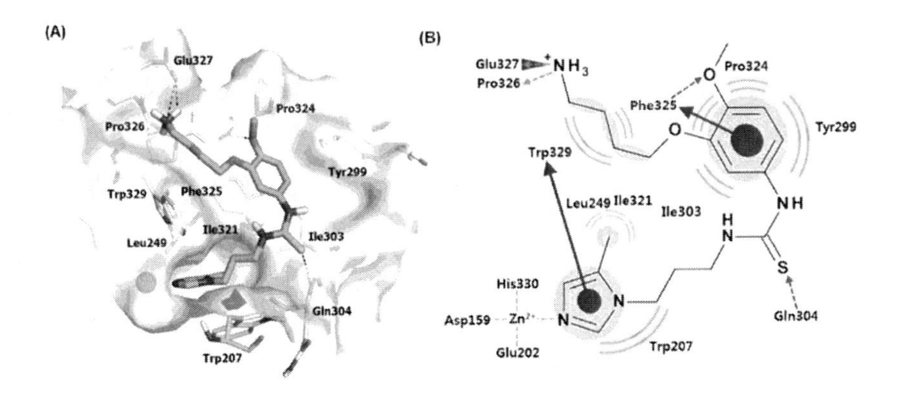

Figure 17. Docked and refined structure of **18** in hQC.

In 2018, Van T. H. Ngo and her colleagues also undertook a SAR investigation of the Phe-Arg memetic region (D-region) of hQC inhibitors. In this study, novel QC inhibitors that contain 3-aminoalkyloxy-4-methoxyphenyl (Template B, Figure 15) and 4-aminoalkyloxyphenyl groups (Template C, Figure 15) replaced the amine-type functional groups in the D-region of QC inhibitors' pharmacophore in previous reported (Template A, Figure 15). Several potent inhibitors were identified, demonstrating IC_{50} values in a low nanomolar range, and were further studied for *in vitro* toxicity and *in vivo* activity (Tables 3 and 4). Of these, inhibitors **16** and **18** displayed the most potent $A\beta_{N3pE-40}$-lowering effects in an *in vivo* acute model with reasonable BBB penetration, without exhibiting cytotoxicity or hERG inhibition. The molecular modeling analysis of **18** indicated that the salt bridge interaction and the hydrogen bonding in the active site provided a high potency (Figure 17). Given the potent activity and favorable BBB penetration with low cytotoxicity, compound **18** may serve as a potential candidate for anti-AD agents.

Final Considerations

Over the past twelve years or more, many small-molecule QC inhibitors have been synthesized, and their biological activities evaluated. Many synthesized compounds demonstrated potential activities in relation to Aβ3pE in the brain functions of AD mice. One of these compounds, **PQ912**, is being studied in a phase II trial at present. It can, therefore, be seen that the study of QC inhibitors is a "hot" topic; one that has attracted the attention of scientists throughout the world with a view to the discovery and development of novel potential agents for the prevention and treatment of AD.

References

[1] Maurer K, Volk S, Gerbaldo H. Auguste D and Alzheimer's disease. *The Lancet*. 1997;349(9064):1546-49. doi: 10.1016/S0140-6736(96)10203-8.

[2] Boller F, Forbes MM. History of dementia and dementia in history: An overview. *J Neurol Sci.* 1998;158(2):125-33. doi: 10.1016/S0022-510X(98)00128-2.

[3] Amaducci LA, Rocca WA, Schoenberg BS. Origin of the distinction between Alzheimer's disease and senile dementia. *Neurology.* 1986;36(11):1497. doi: 10.1212/WNL.36.11.1497.

[4] Association As. 2017 *Alzheimer's disease facts and figures*. 2017. Report No.: 1552-5260 Contract No.: 4.

[5] Prince M, Comas-Herrera A, Knapp M, Guerchet M, Karagiannidou M. World Alzheimer report 2016: improving healthcare for people living with dementia: coverage, quality and costs now and in the future. 2016.

[6] McDade E, Bateman RJ. Stop Alzheimer's before it starts. *Nature* News. 2017.

[7] Martorana A, Esposito Z, Koch G. Beyond the Cholinergic Hypothesis: Do Current Drugs Work in Alzheimer's Disease? *CNS Neurosci Ther.* 2010;16(4):235-45. doi: 10.1111/j.1755-5949.2010.00175.x.

[8] Wilkins. LW. *Alzheimer Disease Drugs. Nursing 2010 Drug Handbook*: Philadelphia, Pa.: Wolters Kluwer/Lippincott Williams & Wilkins, ©2010.; 2010. p. 546-51.

[9] O'Brien RJ, Wong PC. Amyloid precursor protein processing and Alzheimer's disease. *Annu Rev Neurosci.* 2011;34:185-204. doi: 10.1146/annurev-neuro-061010-113613.

[10] Zheng H, Koo EH. Biology and pathophysiology of the amyloid precursor protein. *Mol Neurodegener.* 2011;6(1):27-27. doi: 10.1186/1750-1326-6-27.

[11] Schnabel J. Amyloid: Little proteins, big clues. *Nature.* 2011;475: S12. doi: 10.1038/475S12a.

[12] Salahuddin P, Fatima MT, Abdelhameed AS, Nusrat S, Khan RH. Structure of amyloid oligomers and their mechanisms of toxicities: Targeting amyloid oligomers using novel therapeutic approaches. *Eur J Med Chem.* 2016;114:41-58. doi: 10.1016/j.ejmech.2016.02.065.

[13] Sengupta U, Nilson AN, Kayed R. The Role of Amyloid-β Oligomers in Toxicity, Propagation, and Immunotherapy. *EBioMedicine*. 2016;6:42-49. doi: 10.1016/j.ebiom.2016.03.035.

[14] Zhao LN, Long HW, Mu Y, Chew LY. The Toxicity of Amyloid ß Oligomers. *Int J Mol Sci*. 2012;13(6). doi: 10.3390/ijms13067303.

[15] Goedert M, Wischik C, Crowther R, Walker J, Klug A. Cloning and sequencing of the cDNA encoding a core protein of the paired helical filament of Alzheimer disease: identification as the microtubule-associated protein tau. *Proc Natl Acad of Sci USA*. 1988;85(11): 4051-55. doi: 10.1073/pnas.85.11.4051.

[16] Busby WH, Quackenbush GE, Humm J, Youngblood WW, Kizer J. An enzyme (s) that converts glutaminyl-peptides into pyroglutamyl-peptides. Presence in pituitary, brain, adrenal medulla, and lymphocytes. *J Biol Chem*. 1987;262(18):8532-36. doi.

[17] Fischer WH, Spiess J. Identification of a mammalian glutaminyl cyclase converting glutaminyl into pyroglutamyl peptides. *Proc Natl Acad of Sci USA*. 1987;84(11):3628-32. doi: 10.1073/pnas.84.11.3628.

[18] Oberg KA, Ruysschaert JM, Azarkan M, Smolders N, Zerhouni S, Wintjens R, Amrani A, Looze Y. Papaya glutamine cyclase, a plant enzyme highly resistant to proteolysis, adopts an all-β conformation. *Eur J Biochem*. 1998;258(1):214-22. doi: 10.1046/j.1432-1327.1998.2580214.x.

[19] 1-residue ovine hypothalamic peptide that stimulates secretion of corticotropin and beta-endorphin. *Science*. 1981;213(4514):1394. doi: 10.1126/science.6267699.

[20] [Böckers TM, Kreutz MR, Pohl T. Glutaminyl-Cyclase Expression in the Bovine/Porcine Hypothalamus and Pituitary. *J Neuroendocrinol*. 1995;7(6):445-53. doi: 10.1111/j.1365-2826.1995.tb00780.x.

[21] Pohl T, Zimmer M, Mugele K, Spiess J. Primary structure and functional expression of a glutaminyl cyclase. *Proc Natl Acad Sci USA*. 1991;88(22):10059-63. doi: 10.1073/pnas.88.22.10059.

[22] Waniek A, Hartlage-Rübsamen M, Höfling C, Kehlen A, Schilling S, Demuth H-U, Roßner S. Identification of thyrotropin-releasing

hormone as hippocampal glutaminyl cyclase substrate in neurons and reactive astrocytes. Biochimica et Biophysica Acta (BBA) - *Molecular Basis of Disease*. 2015;1852(1):146-55. doi: 10.1016/j.bbadis.2014.11.011.

[23] Huang K-F, Liu Y-L, Cheng W-J, Ko T-P, Wang AHJ. Crystal structures of human glutaminyl cyclase, an enzyme responsible for protein N-terminal pyroglutamate formation. *Proc Natl Acad Sci USA*. 2005;102(37):13117-22. doi: 10.1073/pnas.0504184102.

[24] Huang K-F, Liaw S-S, Huang W-L, Chia C-Y, Lo Y-C, Chen Y-L, Wang AHJ. Structures of human Golgi-resident glutaminyl cyclase and its complexes with inhibitors reveal a large loop movement upon inhibitor binding. *The Journal of biological chemistry*. 2011;286(14): 12439-49. doi: 10.1074/jbc.M110.208595.

[25] Huang K-F, Wang Y-R, Chang E-C, Chou T-L, Wang Andrew HJ. A conserved hydrogen-bond network in the catalytic centre of animal glutaminyl cyclases is critical for catalysis. *Biochem J*. 2008;411(1): 181. doi: 10.1042/BJ20071073.

[26] Ruiz-Carrillo D, Koch B, Parthier C, Wermann M, Dambe T, Buchholz M, Ludwig H-H, Heiser U, Rahfeld J-U, Stubbs MT, Schilling S, Demuth H-U. Structures of Glycosylated Mammalian Glutaminyl Cyclases Reveal Conformational Variability near the Active Center. *Biochemistry*. 2011;50(28):6280-88. doi: 10.1021/bi200249h.

[27] Schilling S, Hoffmann T, Manhart S, Hoffmann M, Demuth H-U. Glutaminyl cyclases unfold glutamyl cyclase activity under mild acid conditions. *FEBS Lett*. 2004;563(1-3):191-96. doi: 10.1016/S0014-5793(04)00300-X.

[28] Cynis H, Scheel E, Saido TC, Schilling S, Demuth H-U. Amyloidogenic processing of amyloid precursor protein: evidence of a pivotal role of glutaminyl cyclase in generation of pyroglutamate-modified amyloid-β. *Biochemistry*. 2008;47(28):7405-13. doi: 10.1021/bi800250p.

[29] Gunn AP, Masters CL, Cherny RA. Pyroglutamate-Aβ: Role in the natural history of Alzheimer's disease. *The Intl J Bio Cell Biol.* 2010;42(12):1915-18. doi: 10.1016/j.biocel.2010.08.015.

[30] Kawashima S. Dominant and differential deposition of distinct β-amyloid peptide species, AβN3(pE), in senile plaques. *Neuron.* 1995;14(2):457-66. doi: 10.1016/0896-6273(95)90301-1.

[31] Iwatsubo T, Saido TC, Mann DM, Lee VM, Trojanowski JQ. Full-length amyloid-beta (1-42(43)) and amino-terminally modified and truncated amyloid-beta 42(43) deposit in diffuse plaques. *Amer J Path.* 1996;149(6):1823-30. doi.

[32] Sullivan CP, Berg EA, Elliott-Bryant R, Fishman JB, McKee AC, Morin PJ, Shia MA, Fine RE. Pyroglutamate-Aβ 3 and 11 colocalize in amyloid plaques in Alzheimer's disease cerebral cortex with pyroglutamate-Aβ 11 forming the central core. *Neurosci Lett.* 2011;505(2):109-12. doi: 10.1016/j.neulet.2011.09.071.

[33] Kuo Y-M, Emmerling MR, Woods AS, Cotter RJ, Roher AE. Isolation, Chemical Characterization, and Quantitation of Aβ 3-Pyroglutamyl Peptide from Neuritic Plaques and Vascular Amyloid Deposits. *Biochem Biophys Res Commun.* 1997;237(1):188-91. doi: 10.1006/bbrc.1997.7083.

[34] Mori H, Takio K, Ogawara M, Selkoe DJ. Mass spectrometry of purified amyloid beta protein in Alzheimer's disease. *J Biol Chem.* 1992;267(24):17082-86. doi.

[35] Russo C, Saido TC, DeBusk LM, Tabaton M, Gambetti P, Teller JK. Heterogeneity of water-soluble amyloid β-peptide in Alzheimer's disease and Down's syndrome brains. *FEBS Lett.* 1997;409(3):411-16. doi.

[36] He W, Barrow CJ. The Aβ 3-Pyroglutamyl and 11-Pyroglutamyl Peptides Found in Senile Plaque Have Greater β-Sheet Forming and Aggregation Propensities *in vitro* than Full-Length Aβ. *Biochemistry.* 1999;38(33):10871-77. doi: 10.1021/bi990563r.

[37] Buchholz M, Heiser U, Schilling S, Niestroj AJ, Zunkel K, Demuth H-U. The First Potent Inhibitors for Human Glutaminyl Cyclase:

Synthesis and Structure–Activity Relationship. *J Med Chem.* 2006;49(2):664-77. doi: 10.1021/jm050756e.

[38] Harigaya Y, Saido TC, Eckman CB, Prada C-M, Shoji M, Younkin SG. Amyloid β Protein Starting Pyroglutamate at Position 3 Is a Major Component of the Amyloid Deposits in the Alzheimer's Disease Brain. *Biochem Biophys Res Commun.* 2000;276(2):422-27. doi: 10.1006/bbrc.2000.3490.

[39] Russo C, Salis S, Dolcini V, Venezia V, Song X, Teller JK, Schettini G. Identification of Amino-Terminally and Phosphotyrosine-Modified Carboxy-Terminal Fragments of the Amyloid Precursor Protein in Alzheimer's Disease and Down's Syndrome Brain. *Neurobiol Dis.* 2001;8(1):173-80. doi: 10.1006/nbdi.2000.0357.

[40] Russo C, Violani E, Salis S, Venezia V, Dolcini V, Damonte G, Benatti U, D'Arrigo C, Patrone E, Carlo P, Schettini G. Pyroglutamate-modified amyloid β-peptides – AβN3(pE) – strongly affect cultured neuron and astrocyte survival. *J Neurochem.* 2002;82(6):1480-89. doi: 10.1046/j.1471-4159.2002.01107.x.

[41] Piccini A, Russo C, Gliozzi A, Relini A, Vitali A, Borghi R, Giliberto L, Armirotti A, D'Arrigo C, Bachi A, Cattaneo A, Canale C, Torrassa S, Saido TC, Markesbery W, Gambetti P, Tabaton M. β-Amyloid Is Different in Normal Aging and in Alzheimer Disease. *J Biol Chem.* 2005;280(40):34186-92. doi: 10.1074/jbc.M501694200.

[42] Youssef I, Florent-Béchard S, Malaplate-Armand C, Koziel V, Bihain B, Olivier J-L, Leininger-Muller B, Kriem B, Oster T, Pillot T. N-truncated amyloid-β oligomers induce learning impairment and neuronal apoptosis. *Neurobiol Aging.* 2008;29(9):1319-33. doi: 10.1016/j.neurobiolaging.2007.03.005.

[43] Pyroglutamate-modified Aβ(3-42) affects aggregation kinetics of Aβ(1-42) by accelerating primary and secondary pathways. *Chem Sci.* 2017;8(7):4996-5004. doi: 10.1039/C6SC04797A.

[44] Nussbaum JM, Schilling S, Cynis H, Silva A, Swanson E, Wangsanut T, Tayler K, Wiltgen B, Hatami A, Rönicke R, Reymann K, Hutter-Paier B, Alexandru A, Jagla W, Graubner S, Glabe CG, Demuth H-U, Bloom GS. Prion-like behaviour and tau-dependent

cytotoxicity of pyroglutamylated amyloid-β. *Nature*. 2012;485:651. doi: 10.1038/nature11060.

[45] De Kimpe L, van Haastert ES, Kaminari A, Zwart R, Rutjes H, Hoozemans JJM, Scheper W. Intracellular accumulation of aggregated pyroglutamate amyloid beta: convergence of aging and Aβ pathology at the lysosome. *Age* (Dordrecht, Netherlands). 2013;35(3):673-87. doi: 10.1007/s11357-012-9403-0.

[46] Piechotta A, Parthier C, Kleinschmidt M, Gnoth K, Pillot T, Lues I, Demuth H-U, Schilling S, Rahfeld J-U, Stubbs MT. Structural and functional analyses of pyroglutamate-amyloid-β-specific antibodies as a basis for Alzheimer immunotherapy. *J Biol Chem.* 2017;292(30):12713-24. doi: 10.1074/jbc.M117.777839.

[47] Antonios G, Borgers H, Richard BC, Brauß A, Meißner J, Weggen S, Pena V, Pillot T, Davies SL, Bakrania P, Matthews D, Brownlees J, Bouter Y, Bayer TA. Alzheimer therapy with an antibody against N-terminal Abeta 4-X and pyroglutamate Abeta 3-X. *Sci Rep.* 2015;5:17338-38. doi: 10.1038/srep17338.

[48] Valenti MT, Bolognin S, Zanatta C, Donatelli L, Innamorati G, Pampanin M, Zanusso G, Zatta P, Carbonare LD. Increased glutaminyl cyclase expression in peripheral blood of Alzheimer's disease patients. *J Alzheimers Dis.* 2013;34(1):263-71. doi: 10.3233/JAD-120517.

[49] Morawski M, Schilling S, Kreuzberger M, Waniek A, Jäger C, Koch B, Cynis H, Kehlen A, Arendt T, Hartlage-Rübsamen M. Glutaminyl cyclase in human cortex: correlation with (pGlu)-amyloid-β load and cognitive decline in Alzheimer's disease. *J Alzheimers Dis.* 2014;39(2):385-400. doi: 10.3233/JAD-131535.

[50] Schilling S, Zeitschel U, Hoffmann T, Heiser U, Francke M, Kehlen A, Holzer M, Hutter-Paier B, Prokesch M, Windisch M. Glutaminyl cyclase inhibition attenuates pyroglutamate Aβ and Alzheimer's disease–like pathology. *Nat Med.* 2008;14(10):1106. doi.

[51] Jawhar S, Wirths O, Schilling S, Graubner S, Demuth H-U, Bayer TA. Overexpression of glutaminyl cyclase, the enzyme responsible for pyroglutamate Aβ formation, induces behavioral deficits, and

glutaminyl cyclase knock-out rescues the behavioral phenotype in 5XFAD mice. *J Biol Chem*. 2011;286(6):4454-60. doi.

[52] Schilling S, Zeitschel U, Hoffmann T, Heiser U, Francke M, Kehlen A, Holzer M, Hutter-Paier B, Prokesch M, Windisch M, Jagla W, Schlenzig D, Lindner C, Rudolph T, Reuter G, Cynis H, Montag D, Demuth H-U, Rossner S. Glutaminyl cyclase inhibition attenuates pyroglutamate Aβ and Alzheimer's disease–like pathology. *Nat Med*. 2008;14:1106. doi: 10.1038/nm.1872

[53] Schilling S, Appl T, Hoffmann T, Cynis H, Schulz K, Jagla W, Friedrich D, Wermann M, Buchholz M, Heiser U, Von Hörsten S, Demuth H-U. Inhibition of glutaminyl cyclase prevents pGlu-Aβ formation after intracortical/hippocampal microinjection *in vivo/in situ*. *J Neurochem*. 2008;106(3):1225-36. doi: 10.1111/j.1471-4159.2008.05471.x.

[54] Buchholz M, Hamann A, Aust S, Brandt W, Böhme L, Hoffmann T, Schilling S, Demuth H-U, Heiser U. Inhibitors for Human Glutaminyl Cyclase by Structure Based Design and Bioisosteric Replacement. *J Med Chem*. 2009;52(22):7069-80. doi: 10.1021/jm900969p.

[55] Li M, Dong Y, Yu X, Zou Y, Zheng Y, Bu X, Quan J, He Z, Wu H. Inhibitory effect of flavonoids on human glutaminyl cyclase. *Bioorg Med Chem*. 2016;24(10):2280-86. doi: 10.1016/j.bmc.2016.03.064.

[56] Ramsbeck D, Buchholz M, Koch B, Böhme L, Hoffmann T, Demuth H-U, Heiser U. Structure–Activity Relationships of Benzimidazole-Based Glutaminyl Cyclase Inhibitors Featuring a Heteroaryl Scaffold. *J Med Chem*. 2013;56(17):6613-25. doi: 10.1021/jm4001709.

[57] Tran PT, Hoang VH, Thorat SA, Kim SE, Ann J, Chang YJ, Nam DW, Song H, Mook-Jung I, Lee J, Lee J. Structure–activity relationship of human glutaminyl cyclase inhibitors having an N-(5-methyl-1H-imidazol-1-yl)propyl thiourea template. *Bioorg Med Chem*. 2013;21(13):3821-30. doi: 10.1016/j.bmc.2013.04.005.

[58] Li M, Dong Y, Yu X, Li Y, Zou Y, Zheng Y, He Z, Liu Z, Quan J, Bu X, Wu H. Synthesis and Evaluation of Diphenyl Conjugated

Imidazole Derivatives as Potential Glutaminyl Cyclase Inhibitors for Treatment of Alzheimer's Disease. *J Med Chem.* 2017;60(15):6664-77. doi: 10.1021/acs.jmedchem.7b00648.

[59] Lues I, Weber F, Meyer A, Bühring U, Hoffmann T, Kühn-Wache K, Manhart S, Heiser U, Pokorny R, Chiesa J, Glund K. A phase 1 study to evaluate the safety and pharmacokinetics of PQ912, a glutaminyl cyclase inhibitor, in healthy subjects. *Alzheimers Dement.* 2015;1(3):182-95. doi: 10.1016/j.trci.2015.08.002.

[60] AG P. *Safety and Tolerability of PQ912 in Subjects with Early Alzheimer's Disease* (SAPHIR) 2015 [updated June 1, 2017. Available from: https://clinicaltrials.gov/ct2/show/NCT02389413?term=PQ912.

[61] Hielscher-Michael S, Griehl C, Buchholz M, Demuth H-U, Arnold N, Wessjohann AL. Natural Products from Microalgae with Potential against Alzheimer's Disease: Sulfolipids Are Potent Glutaminyl Cyclase Inhibitors. *Mar Drugs.* 2016;14(11). doi: 10.3390/md14110203.

[62] Hoang VH, Tran PT, Cui M, Ngo VTH, Ann J, Park J, Lee J, Choi K, Cho H, Kim H, Ha H-J, Hong H-S, Choi S, Kim Y-H, Lee J. Discovery of Potent Human Glutaminyl Cyclase Inhibitors as Anti-Alzheimer's Agents Based on Rational Design. *J Med Chem.* 2017;60(6):2573-90. doi: 10.1021/acs.jmedchem.7b00098.

[63] Ngo VTH, Hoang VH, Tran PT, Van Manh N, Ann J, Kim E, Cui M, Choi S, Lee J, Kim H, Ha H-J, Choi K, Kim Y-H, Lee J. Structure-activity relationship investigation of Phe-Arg mimetic region of human glutaminyl cyclase inhibitors. *Bioorg Med Chem.* 2018;26(12):3133-44. doi: 10.1016/j.bmc.2018.04.040.

[64] Ngo VTH, Hoang VH, Tran PT, Ann J, Cui M, Park G, Choi S, Lee J, Kim H, Ha H-J, Choi K, Kim Y-H, Lee J. Potent human glutaminyl cyclase inhibitors as potential anti-Alzheimer's agents: Structure-activity relationship study of Arg-mimetic region. *Bioorg Med Chem.* 2018;26(5):1035-49. doi: 10.1016/j.bmc.2018.01.015.

In: Advances in Medicinal Chemistry … ISBN: 978-1-53616-368-1
Editor: E. Ferreira da Silva-Júnior

Chapter 5

CANCER AND COMPUTATIONAL MEDICINAL CHEMISTRY: ADVANCES AND PERSPECTIVES IN DRUG DISCOVERY AND DESIGN

Rafaela Molina de Angelo[1], MD, Heberth de Paula[2], PhD, Sheila Cruz Araujo[3], PhD, Michell Oliveira Almeida[3,4], PhD, Simone Queiroz Pantaleão[3], PhD and Kathia Maria Honorio[1,3,*], PhD

[1]School of Arts, Sciences and Humanities, University of São Paulo (USP), São Paulo, Brazil
[2]Federal University of Espírito Santo (UFES), Alegre, Espírito Santo, Brazil
[3]Center of Human and Natural Sciences, Federal University of ABC (UFABC), Santo André, Brazil
[4]São Carlos Institute of Chemistry, University of São Paulo (USP), São Carlos, Brazil

[*] Corresponding Author's E-mail: kmhonorio@usp.br.

ABSTRACT

Cancer is considered a global problem, with 18.1 million new cancer cases and 9.6 million deaths due to cancer estimated in 2018. Currently, cancer involves a pool of genetic disorders that alter and promote indiscriminate cell reproduction, and its causes range from infectious agents to environmental factors, behavioral, and lifestyle. Since cancer is a complex disease, discovering, and designing a new drug to treat this illness is challenging. In this scenario, there are several strategies to plan a new drug candidate, and the central focus is the identification of new chemical entities (NCE) with a potential therapeutic application, including established and validated macromolecular targets, as well as new biological targets. Studies that determine the properties of a chemical compound with potential therapeutic interest are quite complex and require a large amount of resources and time. In computational drug design, there are two main approaches: ligand-based drug design (LBDD) and structure-based drug design (SBDD). Therefore, in this chapter, we will briefly outline the main advances and challenges related to the design and discovery of anti-cancer drugs employing LBDD and SBDD techniques, presenting successful studies that integrate experimental and computational tools.

Keywords: cancer, computational medicinal chemistry, drug design, SBDD, LBDD

INTRODUCTION

Cancer is recognized as a global problem and is not limited only to industrialized countries [1, 2]. Some reports have estimated that the number of deaths due to cancer surpassed the number of deaths caused by any cardiac event (heart attack, coronary artery disease, or other) in 2011. This illness is the leading cause of death in the world [3]. Cancer is a pool of genetic disorders that alter and promote indiscriminate cell reproduction [4], and its causes range from infectious agents to environmental factors, behavioral factors, and lifestyle [2].

In the early 20th century, Boveri proposed that chromosomal changes would lead cells to reproduce uncontrollably [5]. However, these ideas were only tested in the late '70s and early '80s when it was demonstrated that the existence of mutant genes caused cancer [6, 7]. A few years later, it was established that there would be two main types of cancer-causing genes: oncogenes and tumor suppressor genes. Also, there would be changes that would trigger the process of genomic disease, like the replacement of nucleotides, changes in the number of copies of chromosomes, DNA rearrangements, and others [8]. In recent years, several innovative techniques have emerged and made possible the rise of a biological understanding of this disease, like next-generation sequencing [9] and other genomic technologies, genome-wide association studies [10], proteomics analysis [11], RNA interference studies [12], and methods related to biological chemistry [13].

These studies have identified and continued to produce new insights on the relationships between specific mutations and clinical response, as well as new approaches that are useful for diagnosis and prognosis of the disease [14]. With the advent of these techniques, it has been shown that less than 1% of all human genes appear to have the potential to confer advantages of growth to a cell if changed. A typical tumor contains between 2 to 8 of these mutated genes and mutations in other regions do not confer advantages of growth [15]. It is interesting to highlight that cancer cells are dependent on abnormal proteins that their modified genome encodes [16]. These changes result in constitutive signaling that is not governed by the usual mechanisms, leading to growth, survival, and unregulated cell division [17]. Although there is a considerable variety of mutations in the genome of different cancer cells, it seems that all modified genetic products affect a common set of biological pathways. These proteins become attractive targets for the development of new specific therapies for cancer [16].

The first generation of drugs, initiated in the mid-20th century, used for the treatment of cancer contained almost all cytotoxic agents, which act by damaging DNA, blocking topoisomerases, or attaching to microtubules [18]. These therapies had a remarkable success, leading to an increase in the survival of patients, especially in the treatment of leukemias, lymphomas, testicular cancer, and children's malignancies. On the other hand, with a few exceptions, the success was only modest in epithelial tumors. In the late '90s, it became clear that more significant gains in the survival of patients would be achieved only with small chemical changes in classical cytotoxic agents [19]. These "new" approaches are still being made now. The use of drugs designed to interfere specifically with biological targets that have a critical role in the growth and progression of tumor is known as "targeted therapy" [20] mark the new paradigm of the early twenty-first century [21]. This requires detailed knowledge of the molecular basis of cancer [20]. This made possible an essential change in the classification of this disease: its histopathological characteristics are no longer purely analyzed, and it is initialized by a search for molecular features responsible for the clinical response [22]. Today, conventional chemotherapy has been combined and even replaced by monoclonal antibodies, protein kinases inhibitors, inhibitors of cellular differentiation, or immunomodulatory agents [21]. The US Food and Drug Administration (FDA) has released 109 medications with these features for oncological use in the last ten years (http://www.centerwatch.com/).

Despite the success of the targeted therapy, several cases of resistance have arisen. For this type of patient, there is no adequate therapy. Currently, a second choice therapy in these cases is the "old" cytotoxic agents that are nonspecific and minimally effective, as well as highly toxic [23]. A common mechanism of resistance is genomic alteration, such as amplification or second site mutation [24]. New studies have been made in order to identify potential agents to combat resistant cells. One of the different ways to achieve this goal is by looking for highly selective drugs for specific mutant proteins that do not respond to traditional treatment (e.g., [23, 25, 26]).

The cost of developing new drugs is high, and despite the significant increase in investments, the number of new drugs that are released for consumption has been dropping [27]. It is estimated that the time it takes for a substance to reach the market, since its identification and/or synthesis to the clinical trials, is about 9 to 12 years and consumes $800 million [28]. Since the '80s, computers have ceased to be just a simple apparatus and have played an essential role in the discovery and design of new drugs, with several *in silico* tools being developed [29]. These tools essentially allow researchers to prioritize some compounds over others. Consequently, it is possible to increase speed and to reduce the costs of discovering target-specific drugs [22, 30].

There are several strategies to study new drug candidates, and the central focus of these studies is the identification of new chemical structures (NCS) with potential therapeutic use, both for established and validated macromolecular targets and for new biological targets. Studies that determine the properties of a chemical compound with potential therapeutic interest are quite complex and require a large amount of resources and time. In the computational planning of drug candidates, various approaches can be employed, such as LBDD (Ligand-Based Drug Design) and SBDD (Structure-Based Drug Design) techniques [31].

Knowledge of macromolecular target structures or ligand-receptor complexes allows the application of SBDD strategies. In contrast, when the structure of the target chosen is not known, LBDD methods can be utilized by exploring the properties and characteristics of a series of bioactive ligands. In many cases, the integrated use of SBDD and LBDD strategies can generate useful information in NCS planning [32]. Today, the qualified management of information is a crucial factor, making possible the organization and the analysis of a vast volume of data available on bioactive ligands and biological targets.

Bioactive molecules (or hits) can be identified from real (e.g., biological, biochemical) or virtual (e.g., computational) screenings of natural products, synthetic compounds or combinatorial collections, or by rational planning. It should be emphasized, however, that in all cases the biological properties must be experimentally determined, and the development of standardized and validated assays is necessary. Generally, in the initial stages of drug design, molecules with low potency and affinity are identified and must be optimized for many pharmacodynamic (e.g., potency, affinity, and selectivity) and pharmacokinetic properties (e.g., absorption, metabolism, bioavailability, and others). Optimized compounds are selected as lead compounds for further development as drug candidates [33].

Within the drug design process, the study of pharmacokinetic properties has enormous value and is a critical and hard point. The development of *in silico* predictive models of ADMET (administration, distribution, metabolism, excretion, and toxicity) properties gained greater importance in the '90s [34], and the perspectives point to an even more promising future. Application of *in silico* techniques in large scale is another clear advantage of these methodologies when compared to *in vitro* and *in vivo* methods used in the determination of ADMET properties. Information management is another extremely advantageous component of *in silico* techniques, especially in the treatment of extensive collections of compounds and the organization of targeted databases.

The major challenge that translates into the significant benefit of *in silico* methods lies in their ability to predict pharmacokinetic properties of compounds not yet synthesized. The generation of more robust models for prediction studies of ADMET properties has increased, favoring the development of models based on different strategies [35].

Another critical point in this process involves the possibility of exploring the large chemical space by delineating the work of identifying, selecting and optimizing molecules capable of interacting (presenting high affinity and selectivity) with the selected molecular target (e.g., enzymes and receptors), which represents the biological space. Several strategies can be employed for the investigation of the chemical-biological space, such as techniques related to the organization of databases, use of different molecular filters, high-throughput screening (HTS), and virtual screening (VS) [36]. This chapter will present the main advances and challenges related to the design and discovery of anti-cancer drugs employing some LBDD and SBDD techniques, as well as successful studies that integrate experimental and computational tools.

DRUG-RECEPTOR INTERACTIONS

The affinity of a drug by its receptor, whether enzyme, nucleic acid or another biomacromolecule, is determined by the free energy difference

(ΔG) involved in the formation of the drug-receptor complex, taking into account the energy of the drug and the receptor alone [37]. The biological activity of the drug can be related to its affinity by the receptor at the equilibrium of the formation reaction of the drug-receptor complex, which is measured by the dissociation constant. The interactions involved in the formation of the drug-receptor complex (intermolecular interactions) are usually weak and, for this reason, only happen when the molecular surfaces are closely related and complementary to each other. As a result, reversible interactions are usually observed, which is one of the criteria needed for most of the drugs used in the clinical trials. The forces involved in the drug-receptor complex commonly are ionic, ion-dipole, dipole-dipole, hydrogen bonding, charge transfer interaction, van der Waals, and hydrophobic interaction [38]. An example of a method that assesses the interactions between a receptor and a drug involves the calculation of binding energy, as illustrated in Figure 1.

The next sections will conduct a discussion on different methodologies that analyze the molecular interactions and their applications in studies related to cancer, for example, molecular docking.

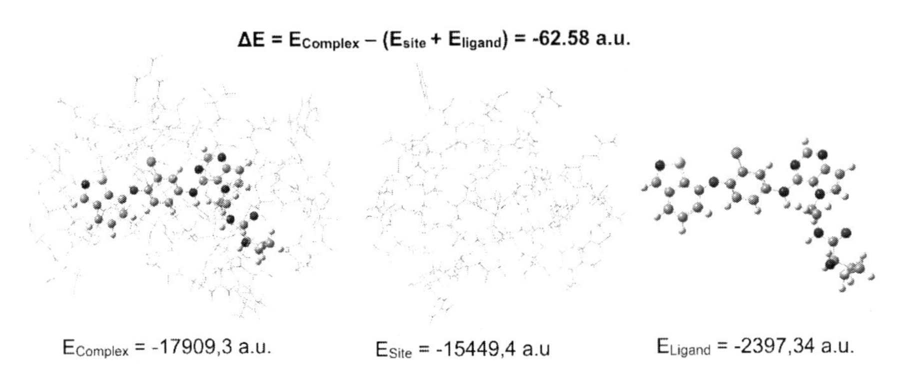

$\Delta E = E_{Complex} - (E_{site} + E_{ligand}) = -62.58$ a.u.
$E_{Complex}$ = -17909,3 a.u.
E_{Site} = -15449,4 a.u
E_{Ligand} = -2397,34 a.u.

Figure 1. Calculation of binding energy taking into account HER-2 (a biological target related to cancer) and one of its inhibitors.

MOLECULAR DOCKING

Molecular docking is used in drug design projects to refer to methods that attempt to identify compounds capable of interacting with a given macromolecular target and to identify the geometry adopted by both in the process of fitting [39]. The central problem in docking is to generate a number of molecular structures for bioactive substances, to evaluate the binding potential of each structure, and to classify the candidates according to some criterion, which is usually the application of a scoring function. Given the complexity involved in finding viable ligands, the docking procedure can be performed at multiple levels. At its lowest level, the search for ligands and the identification of their binding mode can be done manually. Having the 3D-structure of the binding site, the candidate molecule is subjected to translations and rotations in such a way to adjust it to the available space in the site and check the spatial arrangement and complementarity of the functional groups of the candidate concerning the binding site. The viability of the binding mode obtained from docking can be tested by calculating binding energy (rigid fitting), minimizing potential system energy (partially flexible fitting), or from molecular dynamics simulation (flexible fitting), which may include the calculation of thermodynamic properties as evaluation parameters. This procedure has limited application and is restricted to cases where experienced researchers already have a prior knowledge of the interaction mode from similar structures to the candidate molecule and its respective target. It is worth noting that the knowledge on the mode of analog binding does not guarantee good results. There are cases in the literature where similar structures assume very different binding modes [40]. Many cancer studies use docking techniques to understand the interaction between receptors and drugs, helping the design of new inhibitors against several biological targets related to cancer. For example, Kang and coworkers [42] showed a combination of ligand-based naïve Bayesian (NB) models and structure-based molecular docking to develop a virtual screening (VS) pipeline to identify potential VEGFR2 inhibitors from FDA-approved drugs [41].

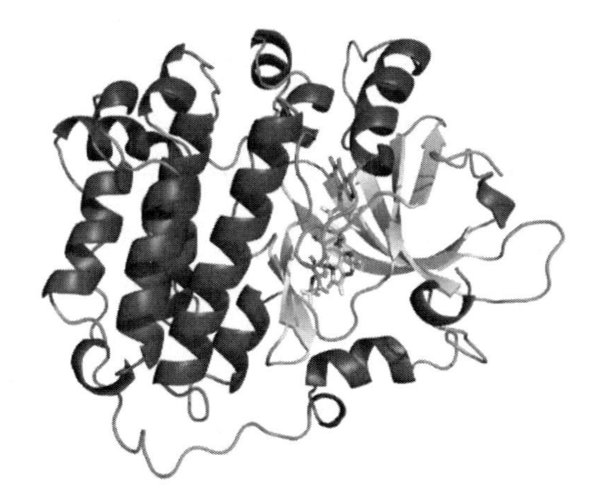

Figure 2. Docking result illustrating an inhibitor at the binding site of HER-2.

By engaging vast amounts of information (chemical and biological data), docking is essentially a computational procedure that can generate hundreds and/or millions of poses (conformations) of ligands in a short time. The main aspects about docking procedure involve the generation of molecular poses, the fitting of each structure (pose) at the binding site, and the evaluation of poses by a scoring function. Some computational programs have been developed to generate large numbers and a variety of molecular structures which can be tested immediately [42]. An example of a result obtained from molecular docking is displayed in Figure 2, which illustrates HER-2 interacting with an inhibitor.

Docking programs use various types of algorithms to obtain the conformations of the ligands at the binding site. For example, there are those that use three-dimensional mesh-based methods for molecular fitting. The first algorithms used rigid docking while searching by ligands. However, several algorithms that consider molecular flexibility in the pose search have emerged in recent years [43].

The evaluation of the pose fitting is done utilizing a scoring function capable of classifying the molecular candidates according to the interaction degree with the binding site. This function generally measures potential energy or binding free energy as a sum of specific contributions that occur during the drug-receptor interaction such as hydrophobic interactions, van der Waals, hydrogen bonds, electrostatic interactions, and others. These, in turn, are described in terms of geometric parameters such as distances and interatomic angles. Functions of this type are called a force field, and several have been developed. The need for speed in the evaluation of the "binding energy" from millions of fittings makes scoring functions rather simple. More elaborate functions are often used in more advanced stages of the search process when a small number of candidates have already been selected. At this point, force fields for proteins and nucleic acids can be used [44]. Alongside docking analyses, it is also possible to check the hydrophobic maps of the drug-receptor complex, as presented in Figure 3, where the hydrophobic surfaces of HER-2 are displayed in the presence of a ligand. Understanding the molecular fitting and finding out the bioactive pose of the inhibitor, it is possible to analyze the binding site and to identify regions containing residues with hydrophobic and hydrophilic characteristics from hydrophobic maps.

There are methods described in the literature used for the identification of binding sites and cavities that may complement the visual identification of suitable poses obtained from docking simulations. One of them was developed by Connolly [45] to measure the local curvature from the macromolecular surface. This method is based on the generation of a sphere centered on the protein surface, followed by the fraction measurement from sphere volume that intercepts the macromolecule. If the fraction is greater than one-half, the surface is concave, and if it is smaller than one-half, it is convex. A significant advance in the detection of macromolecular sites was the construction of methods used to predict favorable sites for interaction with small molecules by using three-dimensional meshes. [46].

The safest method to identify binding sites of a bioactive ligand is the crystallization of the macromolecule in complex with the substance, followed by 3D-structural resolution by means of X-ray diffraction, for example. The increasing availability of 3D-macromolecular structures in crystallographic databases increases the chance to find out the macromolecular structure of interest, eventually complexed with some ligand. Even when the macromolecular system is not found in databases, homologous systems are available and can be used in the homology modeling, for example. Considering the high correspondence degree between the primary structures of enzymes, a high similarity is expected among their tertiary structures. Much information about known binding sites of structures should also be valid for the unknown targets [47].

Docking simulations have been performed for biological targets related to cancer. Rai and collaborators [48] synthesized potential inhibitors of prostate cancer-related proteins pro-caspase 8 and 9. From molecular docking, the authors have chosen a potential anticancer compound lead, reinforcing their results obtained *in vitro* experiments. In the studies of Angelo and coworkers [49], molecular docking was used to validate modifications in dual molecules against HER-2 and EGFR, as illustrated in Figure 4.

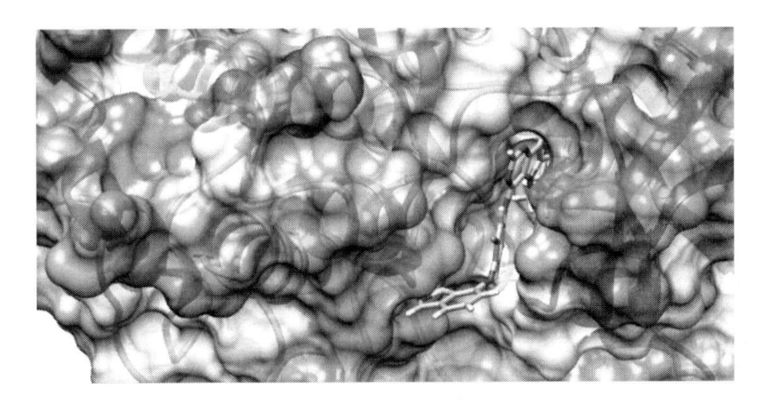

Figure 3. Hydrophobic surface of HER-2 with a ligand docked at its binding site.

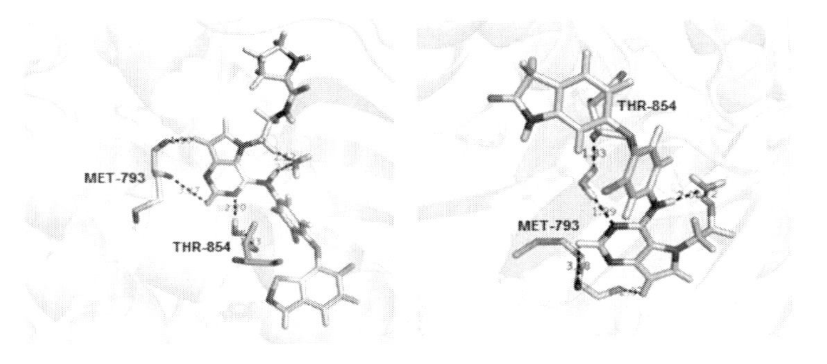

Figure 4. Intermolecular interactions of two ligands and HER-2.

Therefore, molecular docking can be considered a powerful tool to understand interactions between ligands and biological targets, as well assisting the model's construction that correlate molecular structure and biological data, as will be discussed in the next section.

Quantitative Structure-Activity Relationship (QSAR)

In drug design, it is crucial to understand the relationships between chemical structure and the biological activity of a compound series. A qualitative description of this relationship is known as structure-activity relationships (SAR). However, if quantitative biological data is available, a quantitative analysis can be performed (QSAR - Quantitative Structure-Activity Relationship) [50-52].

The biological activity of a compound is an essential feature in drug design research. Other fundamental aspects involve toxicology, physicochemical features, and ADMET, as well as selectivity against other targets [53]. It is viable to quantitatively describe the relationships between chemical structures and each of these biological data. Therefore, different terminologies for these analyses are found in the literature, for example, QSAR (Quantitative Structure-Activity Relationships), QSPR (Quantitative Structure-Property Relationships), and QSTR (Quantitative Toxicology-Property Relationships and/or Quantitative Structure-Selectivity Relationships). To obtain the structure-activity relationships, it is necessary to employ multivariate statistical methods that are the mathematic basis for the QSAR models. The use of the multivariate analysis, data description, classification, and regression modeling involved in QSAR studies are combined with the final objective of interpreting and predicting the target-property of interest, for example, values of EC_{50} and IC_{50}. Figure 5 shows the main steps involved in the construction of predictive QSAR models [51, 52, 54].

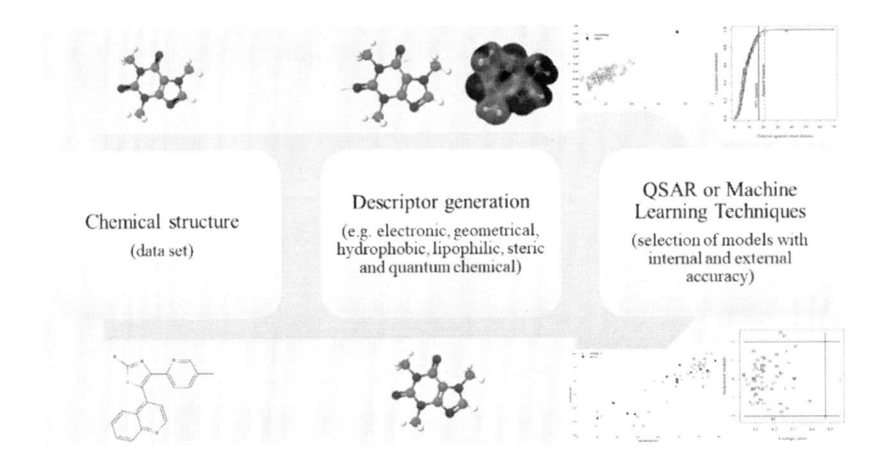

Figure 5. Workflow used to construct QSAR models.

Some QSAR techniques have been widely used over the past decades, for example, Comparative Molecular Field Analysis (CoMFA) and Comparative molecular similarity index analysis (CoMSIA). CoMFA is a grid-based technique, introduced in 1988, and broadly used to construct 3D-QSAR models. It is based on the supposition that the ligand-receptor interactions are non-covalent and changes in biological activities may correlate with changes in the steric and electrostatic fields of the bioactive ligands. Field values are correlated with biological data by using a multivariate statistical technique, such as Partial Least Square (PLS). CoMSIA is an extension of the CoMFA methodology whereby molecular similarity indices are considered as a set of field descriptors that will be related to the biological data. Another powerful QSAR method is 3D-pharmacophore modeling, because it may identify new potential drugs using hypothesis on the 3D-arrangement of structural properties such as hydrogen bond donors and acceptors, hydrophobic groups, and aromatic rings of compounds that bind to the target. Pharmacophore concept takes into account that structurally diverse compounds bind to the binding sites of biological targets in a similar way according to pharmacophoric elements interacting with the same functional groups of the receptor [55-58]. Next, some applications of QSAR approaches in the drug design aiming cancer cases will be presented.

A QSAR study on EGFR and HER-2 inhibitors was performed to discover and/or develop a dual inhibitor as a strategy to design more potent cancer drug candidates. In this study, analyses of the chemical properties of a group of substances having affinity by both HER-2 and EGFR were carried out to understand the main factors involved in the interaction between these inhibitors and their targets. CoMFA and CoMSIA techniques were applied to 63 compounds, and the results indicated important molecular characteristics related to the inhibition of the biological targets. Finally, new compounds were proposed as candidates to inhibit both biological targets [59].

Other example of QSAR study selected activin-like kinase 5 (ALK-5) as an attractive target to treat cancer and QSAR models were constructed to explore the relationships between the molecular structure of 1,5-naphthyridine, pyrazole, and quinazoline derivatives and the ALK-5 inhibition. From a dataset containing 59 compounds, several electronic descriptors, polar surface area, logP, and topological descriptors were also calculated. The ordered predictor selection (OPS) algorithm, weighted principal component analysis (PCA) and Fisher's weights (FW), combined with sequential forward selection, were employed to select the most relevant descriptors to be employed in all PLS regressions. From this procedure, a predictive model was obtained and can be used to predict the biological activity of new compounds as drug candidates to treat cancer [60].

Another relevant aspect of QSAR modeling is the accuracy estimation and applicability of QSAR and QSPR models. This aspect was addressed in a study that used biological and physicochemical properties for an Ame mutagenicity set. In this study, a parameter was developed and called as "distance to model," which was defined as a metric of similarity between the training and test compounds that have been subjected to QSAR modeling. This approach was applied in 30 QSAR models that employed the Ames mutagenicity data. This analysis identifies 30-60% of compounds having a prediction accuracy based on the Ames data. In sense, *in silico* predictions can be used to reduce by fifty percent the cost of experimental measurements by providing a similar prediction accuracy [61].

The last study presented in this section on QSAR models employed a forward stepwise multiple linear regression method to predict the activity of withanolides analogs against breast cancer. The most effective QSAR model for anticancer activity, using data against the SK-Br3 cell, showed a high correlation with the biological activity. Besides molecular docking simulations, a study on the binding conformations and different interaction behaviors was performed in order to reveal a plausible mechanism of interaction between active with anolides and β-tubulin [62]. In short, some aspects of QSAR modeling were presented in this section, as well as their application in case studies. It is possible to use QSAR analyses as additional information for virtual screening, which will be discussed hereafter.

VIRTUAL SCREENING

Virtual screening (VS) can be defined as a drug discovery strategy that allows identification of potential bioactive ligands in a considerable compound database [63, 64]. This can be performed from two approaches: structure-based virtual screening (SBVS) or ligand-based virtual screening (LBVS). In both cases, it is essential to select virtual libraries with compounds that respect drug-like and ADMET characteristics to minimize errors and discontinuity in subsequent steps related to biological assays until its approval as drugs. One of the most popular databases is ZINC [64], which contains over 35 million purchasable chemical compounds.

When the 3D-structure of the biological target and its action mechanism are known, it is possible to carry out the screening based on the receptor, and the focus is to find compounds that interact at the binding sites of the biological target. This is possible by selecting docking programs that adequately respond to the system of interest, regarding the different search algorithms, such as (*i*) random (genetic algorithm or Monte Carlo) and (*ii*) systematic (incremental, database of conformers or non-stochastic). Another essential parameter to be selected is related to the scoring function, which is used to rank the suitable poses of the ligands based on calculations of the binding energy from force fields. The main docking programs used in virtual screening studies are FlexX [65], DOCK [66], GOLD [67], GLIDE [68], AutoDock [69], and AutoDock Vina [70], which present different algorithms to obtain poses (conformations) of the ligands and different scoring functions.

In an LBVS strategy, compounds with known biological activity can be employed to screen a compound database by using a measure of similarity, for example. Thus, we can search by pharmacophoric patterns in the compound databases, such as hydrogen bonding donors or acceptors, charged groups, hydrophobic regions, substituents containing aromatic rings, and others. Also, it is possible to search for molecular similarity by calculating similarity parameters such as Tanimoto, Jaccard, Sorensen, and others [64]. Currently, the increase in the number of compounds and crystallographic structures of biological targets deposited in the databases led to a new category of studies: virtual reverse docking. The main objective of this approach is to perform a virtual search for biological receptors of known bioactive substances, where the molecular interactions in different biological targets are assessed, and a possible target is selected. Another critical aspect in VS is related to the techniques to validate the obtained results. The best-known validations of a VS study involve the use of enrichment curves with decoys (non-ligands containing physicochemical properties similar to true ligands) and the construction of ROC (Receiver Operating Characteristics) curve and area under the curve (AUC) [71].

From a fast search in the literature, we can find many studies that employ VS to identify potential ligands of biological targets related to cancer. For example, Yousuf and coauthors [63] have employed a structure-based virtual screening to identify multi-targeted inhibitors against breast cancer, where the biological targets were the HSP90 co-chaperone and the human epidermal growth factor EGFR and HER2/neu receptor. For this, the authors previously carried out a study on binding sites using the CASTp (Surface Protein Surface Atlas) server.

AutoDock Vina was used for analyzing the possible interactions between the 50 selected compounds and EGFR, HER2, and HSP90. The PubChem and ZINC databases were fused, summing 3 million compounds. Binding energy cutoffs were applied (-8.9, -8.5, and -8.3 kcal/mol for EGFR, HER2, and HSP90, respectively), resulting in 71 compounds. Next, after a careful ADMET analysis with visual inspection from interactions in the regions of interest of the biological targets, five potential compounds were selected for future assays. So, from this study, we can observe that besides molecular docking and VS, it is essential to understand the dynamics of the drug-receptor complex. To do this, an essential computational technique used is molecular dynamics, which will be presented following.

Molecular Dynamics

Molecular dynamics (MD) simulations use classical mechanics as a principle and have the objective to provide information about the microscopic dynamic behavior of the system, where this behavior is calculated as a function of time (Newton equations), and all atoms of the system under study are considered [72]. Then, the computational simulations by molecular dynamics perform calculations of the positions of the atoms of the complex at a certain time interval. A widely used time interval is the nanosecond order (1ns = 10^{-9} s) [73]. In order to perform these calculations, it is necessary to insert a set of interaction potentials known as force fields, which are empirical and potential energy functions. Thus, with the use of the force field it is possible to calculate the total potential energy of the system, $V(r)$, from the 3D-structures of the system. The potential energy $V(r)$ is described as the sum of various energy terms, such as bound atoms (bond lengths and angles, as well as dihedral angles) and atoms (interactions of van der Waals and Coulomb). An example of a force field is displayed in equation 1 [74]:

$$V_{(r)} = \sum \frac{1}{2}\, kl\,(l - l_o)^2 + \sum \frac{1}{2}\, k\theta\,(\theta - \theta_o)^2 +$$

$$\sum \frac{1}{2} K_\varphi\,(1\cos(n_\varphi - \varphi_o))^2 + \sum 4\varepsilon \left(\frac{A}{r_{ij}^{12}}\right) - \left(\frac{B}{r_{ij}^{6}}\right) + \sum 4\varepsilon \left(\frac{q_i\, q_j}{4\pi\varepsilon_o\, r_{ij}}\right) \quad \text{eq. (1)}$$

Equation (1) illustrates the force field used to calculate the total energy of a system, and this equation is divided into five energy terms: bond lengths and angles, potential energy of torsion, Lennard-Jones potential, and Coulomb potential. Computational programs such as GROMACS, AMBER, CHARMM, and NAMD are widely used in the MD simulations. Several applications of DM in cancer studies can be found in the literature. One example of this was performed by Almeida and coworkers [75], where the authors studied the inhibition of ALK-5 (activin receptor-like kinase 5) by six inhibitors. One of the techniques used in this study was MD simulations using AMBER 12. The analysis was carried out to evaluate the dynamic behavior of the inhibitors in the binding site of ALK-5 and the stability of the biological target in the presence of these inhibitors [75]. Uchibori and colleagues [76] performed molecular dynamics simulations to analyze the action of EGFR (biological target related to lung cancer) inhibitors. From the results obtained, it was possible to verify the efficacy of brigatinib concerning the triple-mutant EGFR [76]. In 2018, Kaboli and coauthors [77] carried out a study employing molecular dynamics to understand the interactions between the BRAF target (related to skin cancer) and berberine derivatives [77]. Trejo-Soto and colleagues [78] also used molecular dynamics simulations to analyze the interactions between ATK (autologous tumor killing) and some inhibitors. The results obtained from MD simulations, besides providing valuable information about the stability of the kinase binding site, also showed that selective inhibitors could be promising [78]. So, from the previously cited articles, it is possible to verify the importance of the use of molecular dynamics in studies related to cancer and the results obtained from this technique can also be used in the calculation of the binding free energy involved in the protein-ligand interaction, as will be discussed in the next section.

Calculation of Binding Free Energies

From the analysis of a molecular dynamics simulation [79], it is possible to combine the values of energies obtained from the continuous solvent methods by Poisson-Boltzmann surface area or generalized Born method (MM-PBSA and MM-GBSA, respectively) and the method based on solvation interaction energy (SIE) [80-82]. These approaches are popular in calculations to estimate the binding energy of small molecules (inhibitors or ligands) with macromolecules (proteins, enzymes, DNA, or RNA). These methodologies use the implicit solvation and are based on the more stable trajectories obtained from the molecular dynamics simulations, because in this type of simulation the calculations performed, besides estimating the energy, seek a more stable conformation of the target under study. Concerning MM-GBSA and MM-PBSA techniques, the free energy calculation is based on the following equation [83]:

$$\Delta G = E_{MM} - E_{El} - E_{vdW} - G_{Pol} - G_{Np} - TS \qquad \text{eq. (2)}$$

where E_{MM} is the standard energy terms obtained from molecular dynamics (bond, angle, and dihedral), E_{El} is the value for electrostatic interactions, E_{vdW} corresponds to the van der Waals interactions, G_{Pol} and G_{Np} are the polar and nonpolar contributions to the free energy of solvation, T is the absolute temperature, and S is the entropy of the system. Figure 6 shows a representation of the free energy calculations from MM-GBSA and MM-PBSA techniques [84].

The main difference between the two methodologies is, for MM-PBSA, electrostatic energy is calculated by the finite Poisson-Boltzmann difference and, for MM-GBSA, this energy is calculated from the generalized Born method. About SIE, this method treats the protein-ligand system with atomistic details and effects of implicit solvation. The binding free energy between the ligand and the protein is calculated by: [82, 85]

$$\Delta G_{bind} = \alpha\,[E_{inter}^{Coul}\,(Din) + \Delta G_{desolv}^{R}\,(Din) + E_{inter}^{vdW} + \gamma\,(\rho, Din)\,\Delta MSA\,(\rho)] + C \qquad \text{eq. (3)}$$

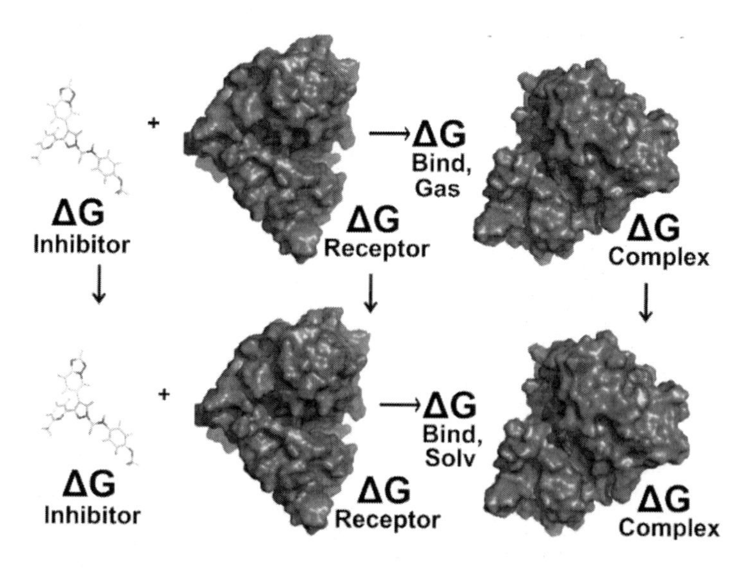

Figure 6. Schema illustrating the calculation of free energy from MM-GBSA and MM-PBSA approaches.

The equation (3) indicates that the approximate calculation of the binding energy between protein and inhibitor (in aqueous solvent) is made from the contribution of interaction energy (E_{inter}) and desolvation free energy (ΔG_{desolv}). The electrostatic portion is included in the SIE methodology, which is formed by Coulomb intermolecular energy (E_{inter}^{Coul}) and desolvation free energy (ΔG_{desolv}^{R}). Also, a non-apolar portion is added, which involves the intermolecular energy of van der Waals (E_{inter}^{vdW}) and the free non-polar desolvation energy (ΔG_{desolv}^{np}). Finally, "ρ" is the derivation factor of the atomic radius of Born, Din is the inner dielectric constant of the solute, "γ" is the molecular surface tension coefficient, and ΔMSA is the molecular surface area of the solute in the bond [82, 85].

Methods that estimate binding free energy have been employed in cancer studies. For example, the study performed by Wan and collaborators [86] uses MM-GBSA and MM-GBSA to estimate the free energy of allosteric inhibitors at the binding site of the EGFR target, which is related to lung cancer cases [87].

Zhao and coworkers [88] published another example of a study that uses free energy calculations (MM-GBSA) to estimate the free energy of

BRAF inhibitors, which are related to bowel cancer. The results showed that the calculation of free energy was essential to understand the selectivity of these inhibitors concerning the BRAF target. In relation to the SIE method, Chen and coauthors [88] performed a study that uses this methodology to estimate the binding free energy of the p53-MDM2 protein with several inhibitors, and the obtained results can help the design of more potent inhibitors of p53-MDM2. Almeida and collaborators [75] performed a study that, in addition to molecular dynamics simulations, used free energy calculations for ALK-5 and a series of bioactive substances. In this work, the three methods mentioned (MM-GBSA, MM-PBSA, and SIE) were used to determine the free energy of six inhibitors in relation to ALK-5. The complex containing the most active molecule presented the lower value of free energy from the calculations using the three methodologies. The results showed that this inhibitor increases the stability of the target, and this finding can be used in the design of new molecules [75]. Therefore, this section showed the applications of calculations related to binding free energy in studies related to the design of anticancer substances. Another advantageous approach in drug design and discovery is the combination of quantum mechanics and molecular mechanics, i.e., hybrid methods, which make the drug-receptor analyses using quantum mechanics to study the regions of significant interactions and molecular mechanics to analyze the environment. Some details and applications of this methodology will be discussed in the next section.

Hybrid Methods: Quantum Mechanics/Molecular Mechanics (QM/MM)

Enzymes present a high capacity to perform biochemical catalysis. Understanding enzymatic reactions is a critical process in technological applications, drug design, and studies of new catalysts. Therefore, one of the tools able to assess these aspects is a computational simulation (computational chemistry, bioinformatics) [89]. In the previous sections, several studies of medicinal and computational chemistry were presented. Now, to study condensed phase systems such as an enzymatic mechanism, hybrid methods using quantum mechanics and molecular mechanics (QM/MM) [89-91] will be discussed. This type of hybrid simulation is increasingly used in enzymatic systems, since the technique that uses molecular mechanics (molecular dynamics simulations) does not perform analysis of the electrons of the system, and this type of problem is solved by using quantum mechanics, which involves high computational costs, being unfeasible for large molecular systems. Thus, the hybrid QM/MM approach is extremely useful in biomolecular studies, since it is possible to perform a quantum treatment only for the most important region of the system, i.e., where reactions occur chemically (QM), and the rest of the environment is classically treated (MM) [90]. This hybrid simulation involves the calculation of the Hamiltonian of the system, as can be seen in equation 4 [92]:

$$H = H_{QM} + H_{QM/MM} + H_{MM} \qquad \text{eq. (4)}$$

where H_{QM} is the quantum mechanics Hamiltonian, $H_{QM/MM}$ is the Hamiltonian that couples QM and MM regions, and H_{MM} is related to the MM region. Figure 7 shows an example on the definition of the QM and MM regions in one biological system (ALK-5 complexed with an inhibitor).

There also are more two approaches besides the conventional QM/MM techniques: subtractive QM/MM and additive QM/MM. In relation to the subtractive approach, the most used method is ONIOM, which was developed by Dapprich and colleagues, and Vreven and coauthors [93, 94] and can be performed with up to 3 layers. Equation (5) illustrates an example of ONIOM calculation with two layers.

$$E^{ONIOM\,(QM/MM)} = E^{QM}_{Model} + E^{MM}_{Real} - E^{MM}_{Model} = E^{High}_{Model} + E^{Low}_{Real} - E^{Low}_{Modelo} \qquad \text{eq. (5)}$$

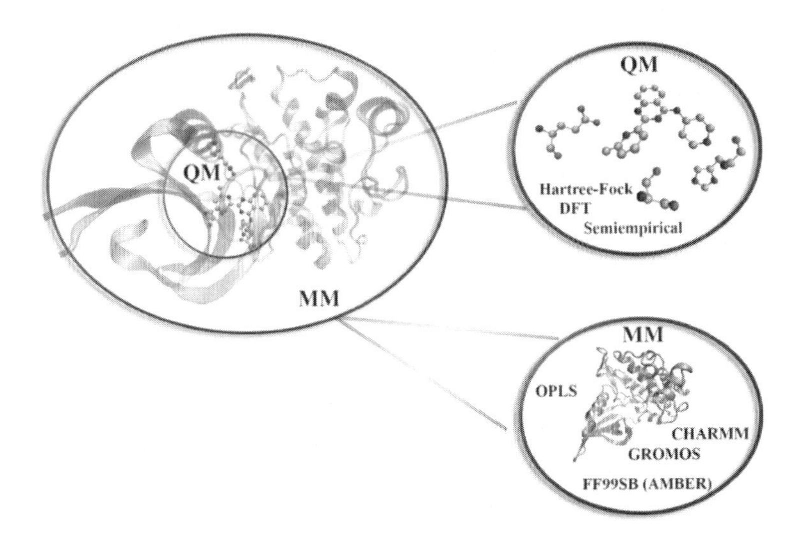

Figure 7. QM and MM regions used in quantum mechanics/molecular mechanics calculations.

The equation (5) reports the real system (contains all the atoms of the system and those calculated only by the level MM) and the model system, which contains the atoms of the QM region, the link atoms that are used to limit the pending bonds caused by the disconnection of regions with covalent bonds between QM and MM regions [93, 94].

In the additive QM/MM, the QM region is embedded in the largest MM system, and the total potential energy is the sum of QM, MM and QM/MM, as can be seen in equation (6) [91]:

$$V_{QM/MM} = V_{QM}(QM) + V_{MM}(MM) + V_{QM-MM}(QM + MM) \quad \text{eq. (6)}$$

The QM/MM methods have also been widely used in cancer studies. For example, Callegari and coauthors [95] employed the hybrid method to study the reaction mechanism of the molecule osimertinib, which is an EGFR inhibitor. The results obtained from QM/MM showed a possible mechanism of covalent inhibition of this cancer-related target [95]. There are other studies related to cancer and QM/MM calculations [96, 97].

Another study using the QM/MM methodology was performed by Śliwa and coworkers [98], in which the authors used ONIOM (implemented in Gaussian) to study the interactions of a negative allosteric modulator with the metabotropic glutamate receptor 1 (mGluR1), which is also related to cancer. The results obtained in this study were relevant to understanding better the interactions that occur between this molecule and the allosteric binding pocket of mGluR1 [98]. Another example of a QM/MM application was published in 2019 by Lu and coauthors [99], where the authors used this approach to study ErbB2 inhibitors [99]. Therefore, from the examples presented previously, it is possible to note the importance of using QM/MM method in cancer studies, since this methodology gives electronic information on ligands, biological targets and inhibitor-receptor complex related to cancer.

FINAL CONSIDERATIONS

A rational approach in drug design is needed to maximize the chances of finding out more potent and safer drug candidates to treat different illnesses, in particular, cancer. *In silico* techniques have been widely employed to exploit the potential opportunities related to new drug targets and extensive libraries of small compounds are now readily available through combinatorial chemistry. Computer-aided drug design techniques can be useful in reducing costs and speeding up drug design and discovery. This has become possible because of the development of more accurate and reliable algorithms, the use of more thoughtfully planned strategies, and the incredible increase in the computer power (hardware) to allow studies with more accuracy and speed. Therefore, this study focused on the main aspects and some applications of *in silico* techniques applied in cancer studies, where QSAR modeling, molecular docking, and molecular dynamics (MD) simulations and calculations of binding free energy were used to study small molecules into biological targets related to cancer.

ACKNOWLEDGMENTS

The authors would like to thank FAPESP (2016/18840-3, 2014/27189-9, 2015/20314-5, and 2018/06680-7), CAPES, and CNPq for funding.

REFERENCES

[1] Parkin DM, Stjernswärd J, Muir CS. Estimates of the worldwide frequency of twelve major cancers. *Bull World Health Organization* 1984;62:163-82.

[2] Vineis P, Wild CP. Global cancer patterns: causes and prevention. *The Lancet* 2014;383:549-57. doi:10.1016/S0140-6736(13)62224-2.

[3] Jacques F, Isabelle S, Rajesh D, Sultan E, Colin M, Marise R, et al. Cancer incidence and mortality worldwide: Sources, methods and major patterns in GLOBOCAN 2012. *Int J Cancer* 2015;136:E359-E86. doi:10.1002/ijc.29210.

[4] Hollstein M, Alexandrov LB, Wild CP, Ardin M, Zavadil J. Base changes in tumour DNA have the power to reveal the causes and evolution of cancer. *Oncogene* 2016;36:158. doi:10.1038/onc.2016.192.

[5] Boveri T. Concerning the Origin of Malignant Tumours by Theodor Boveri. Translated and annotated by Henry Harris. *J Cell Sci* 2008;121:1-84. doi:10.1242/jcs.025742.

[6] Stehelin D, Varmus HE, Bishop JM, Vogt PK. DNA related to the transforming gene(s) of avian sarcoma viruses is present in normal avian DNA. *Nature* 1976;260:170. doi:10.1038/260170a0.

[7] Tabin CJ, Bradley SM, Bargmann CI, Weinberg RA, Papageorge AG, Scolnick EM, et al. Mechanism of activation of a human oncogene. *Nature* 1982;300:143. doi:10.1038/300143a0.

[8] MacConaill LE, Garraway LA. Clinical Implications of the Cancer Genome. *Journal of Clinical Oncology* 2010;28:5219-28. doi:10.1200/jco.2009.27.4944.

[9] Meyerson M, Gabriel S, Getz G. Advances in understanding cancer genomes through second-generation sequencing. *Nat Rev Genetics* 2010;11:685. doi:10.1038/nrg2841.

[10] Freedman ML, Monteiro ANA, Gayther SA, Coetzee GA, Risch A, Plass C, et al. Principles for the post-GWAS functional characterization of cancer risk loci. *Nat Genetics* 2011;43:513. doi:10.1038/ng.840.

[11] Hanash S, Taguchi A. The grand challenge to decipher the cancer proteome. *Nat Rev Cancer* 2010;10:652. doi:10.1038/nrc2918.

[12] Brough R, Frankum JR, Sims D, Mackay A, Mendes-Pereira AM, Bajrami I, et al. Functional Viability Profiles of Breast Cancer. *Cancer Discov* 2011;1:260-73. doi:10.1158/2159-8290.cd-11-0107.

[13] Workman P, Collins I. Probing the Probes: Fitness Factors For Small Molecule Tools. *Chem Bio* 2010;17:561-77. doi:10.1016/j.chembiol.2010.05.013.

[14] Chin L, Hahn WC, Getz G, Meyerson M. Making sense of cancer genomic data. *Genes & Development* 2011;25:534-55. doi:10.1101/gad.2017311.

[15] Vogelstein B, Papadopoulos N, Velculescu VE, Zhou S, Diaz LA, Kinzler KW. Cancer Genome Landscapes. *Science* 2013;339:1546-58. doi:10.1126/science.1235122.

[16] Stratton MR, Campbell PJ, Futreal PA. The cancer genome. *Nature* 2009;458:719. doi:10.1038/nature07943.

[17] Hanahan D, Weinberg Robert A. Hallmarks of Cancer: The Next Generation. *Cell* 2011;144:646-74. doi:10.1016/j.cell.2011.02.013.

[18] Chabner BA, Roberts Jr TG. Chemotherapy and the war on cancer. *Nat Rev Cancer* 2005;5:65. doi:10.1038/nrc1529.

[19] Pezaro CJ, Mukherji D, De Bono JS. Abiraterone acetate: redefining hormone treatment for advanced prostate cancer. *Drug Discov Today* 2012;17:221-6. doi:10.1016/j.drudis.2011.12.012.

[20] Sawyers C. Targeted cancer therapy. *Nature* 2004;432:294. doi:10.1038/nature03095.

[21] Neidle S. *Cancer Drug Design Discov.* Elsevier; 2011.

[22] Azuaje F. Computational models for predicting drug responses in cancer research. *Brief Bioinfo* 2017;18:820-9. doi:10.1093/bib/bbw065.

[23] Walter AO, Sjin RTT, Haringsma HJ, Ohashi K, Sun J, Lee K, et al. Discovery of a mutant-selective covalent inhibitor of EGFR that overcomes T790M-mediated resistance in NSCLC. *Cancer Discov* 2013;3:1404-15. doi:10.1158/2159-8290.CD-13-0314.

[24] Lovly CM. Combating acquired resistance to tyrosine kinase inhibitors in lung cancer. American Society of Clinical Oncology educational book/*ASCO Am Soc Clin Oncol Meet* 2015:e165-e73. doi:10.14694/EdBook_AM.2015.35.e165.

[25] Cross DAE, Ashton SE, Ghiorghiu S, Eberlein C, Nebhan CA, Spitzler PJ, et al. AZD9291, an irreversible EGFR TKI, overcomes T790M-mediated resistance to EGFR inhibitors in lung cancer. *Cancer Discov* 2014;4:1046-61. doi:10.1158/2159-8290.CD-14-0337.

[26] Zhou W, Ercan D, Chen L, Yun C-h, Li D, Capelletti M, et al. Novel mutant-selective EGFR kinase inhibitors against EGFR T790M. *Nat* 2009;462:1070-4. doi:10.1038/nature08622.

[27] Scannell JW, Blanckley A, Boldon H, Warrington B. Diagnosing the decline in pharmaceutical R&D efficiency. *Nat Rev Drug Discov* 2012;11:191. doi:10.1038/nrd3681.

[28] Dickson M, Gagnon JP. The cost of new drug discovery and development. *Discov Med* 2004;4:172-9.

[29] Song CM, Lim SJ, Tong JC. Recent advances in computer-aided drug design. *Briefings in Bioinformatics* 2009;10:579-91. doi:10.1093/bib/bbp023.

[30] Rao VS, Srinivas K. Modern drug discovery process: an in silico approach. *J Bioinfo Seq Anal* 2011;3:89-94.

[31] Bhogal N, Balls M. Translation of new technologies: from basic research to drug discovery and development. *Curr Drug Discov Technol* 2008;5:250-62. doi:10.2174/157016308785739839.

[32] Guido RV, Oliva G, Andricopulo AD. Virtual screening and its integration with modern drug design technologies. *Curr Med Chem* 2008;15:37-46.

[33] Andricopulo AD, Salum LB, Abraham DJ. Structure-based drug design strategies in medicinal chemistry. *Curr Top Med Che*m 2009;9:771-90.

[34] Whittaker PA. What is the relevance of bioinformatics to pharmacology? *Trends Pharmacol Sci* 2003;24:434-9. doi:10.1016/S0165-6147(03)00197-4.

[35] Tetko IV, Bruneau P, Mewes HW, Rohrer DC, Poda GI. Can we estimate the accuracy of ADME-Tox predictions? *Drug Discov Today* 2006;11:700-7. doi:10.1016/j.drudis.2006.06.013.

[36] Lombardino JG, Lowe JA, 3rd. The role of the medicinal chemist in drug discovery--then and now. *Nat Rev Drug Discov* 2004;3:853-62. doi:10.1038/nrd1523.

[37] Almeida MO, Costa CHS, Gomes GC, Lameira J, Alves CN, Honorio KM. Computational analyses of interactions between ALK-5 and bioactive ligands: insights for the design of potential anticancer agents. *J Biomol Struct Dyn* 2017:1-13. doi:10.1080/07391102.2017.1404938.

[38] Montanari C. *Química medicinal: métodos e fundamentos em planejamento de fármacos* [*Medicinal Chemistry: Methods and Fundamentals of Drug Planning*]. São Paulo: Edusp 2011;720

[39] Kuntz ID, Meng EC, Shoichet BK. Structure-based molecular design. *Acc Chem Res* 1994;27:117-23.

[40] Brown DG, Visse R, Sandhu G, Davies A, Rizkallah PJ, Melitz C, et al. Crystal structures of the thymidine kinase from herpes simplex virus type-1 in complex with deoxythymidine and ganciclovir. *Nat Struct Biol* 1995;2:876-81.

[41] Kang D, Pang X, Lian W, Xu L, Wang J, Jia H, et al. Discovery of VEGFR2 inhibitors by integrating naïve Bayesian classification, molecular docking and drug screening approaches. *RSC Advances* 2018;8:5286-97.

[42] Pearlman R. Rapid generation of high quality approximate 3D molecular structures. *Chem Des Auto News* 1987;2:5-6.

[43] Magalhães CS. *Algoritmos genéticos para o problema de docking proteína-ligante* [*Genetic Algorithms for Protein-Binder Docking Problem*]. Petrópolis: Laboratório Nacional de Computação Científica 2006.

[44] Rodrigues RP, Mantoani SP, de Almeida JR, Pinsetta FR, Semighini EP, da Silva VB, et al. Estratégias de triagem virtual no planejamento de fármacos [Virtual Screening Strategies in Drug Planning]. *Revista Virtual de Química* 2012;4:739-76.

[45] L Connolly M. Measurement of protein surface shape by solid angles. *Journal of Molecular Graphics* 1986;4:3-6. doi:10.1016/0263-7855(86)80086-8.

[46] Goodford PJ. A computational procedure for determining energetically favorable binding sites on biologically important macromolecules. *J Med Chem* 1985;28:849-57.

[47] Al-Lazikani B, Jung J, Xiang Z, Honig B. Protein structure prediction. *Curr Opin Chem Biol* 2001;5:51-6.

[48] Rai R, Dutta RK, Singh S, Yadav DK, Kumari S, Singh H, et al. Synthesis, biological evaluation and molecular docking study of 1-amino-2-aroylnaphthalenes against prostate cancer. *Bioorg Med Chem Lett* 2018;28:1574-80. doi:10.1016/j.bmcl.2018.03.057.

[49] De Angelo RM, Almeida MdO, De Paula H, Honorio KM. Studies on the Dual Activity of EGFR and HER-2 Inhibitors Using Structure-Based Drug Design Techniques. *Int J Mol Sci* 2018;19:3728.

[50] Bunin BA. Increasing the efficiency of small-molecule drug discovery. *Drug Discov Today* 2003;8:823-6.

[51] Cherkasov A, Muratov EN, Fourches D, Varnek A, Baskin, II, Cronin M, et al. QSAR modeling: where have you been? Where are you going to? *J Med Chem* 2014;57:4977-5010. doi:10.1021/jm4004285.

[52] He T, Heidemeyer M, Ban FQ, Cherkasov A, Ester M. SimBoost: a read-across approach for predicting drug-target binding affinities

using gradient boosting machines. *J Cheminfo* 2017;9 doi:10.1186/s13321-017-0209-z.

[53] Ying SL, Du XJ, Fu WT, Yun D, Chen LP, Cai YP, et al. Synthesis, biological evaluation, QSAR and molecular dynamics simulation studies of potential fibroblast growth factor receptor 1 inhibitors for the treatment of gastric cancer. *European J Med Chem* 2017; 127:885-99. doi:10.1016/j.ejmech.2016.10.066.

[54] Li H, Hassona MDH, Lack NA, Axerio-Cilies P, Leblanc E, Tavassoli P, et al. Characterization of a New Class of Androgen Receptor Antagonists with Potential Therapeutic Application in Advanced Prostate Cancer. *Mol Cancer Therap* 2013;12:2425-35. doi:10.1158/1535-7163.mct-13-0267.

[55] Cramer RD, Patterson DE, Bunce JD. Comparative molecular field analysis (CoMFA). 1. Effect of shape on binding of steroids to carrier proteins. *J Am Chem Soc* 1988;110:5959-67. doi:10.1021/ja00226a005.

[56] Gao XL, Liu DH, Wang Z, Dai K. Quantitative Structure Tribo-Ability Relationship for Organic Compounds as Lubricant Base Oils Using CoMFA and CoMSIA. *J Tribol Transac Asme* 2016;138:7. doi:10.1115/1.4033191.

[57] Caballero J. 3D-QSAR (CoMFA and CoMSIA) and pharmacophore (GALAHAD) studies on the differential inhibition of aldose reductase by flavonoid compounds. *J Mol Graph Model* 2010;29:363-71. doi:10.1016/j.jmgm.2010.08.005.

[58] Araujo SC, Maltarollo VG, Honorio KM. Computational studies of TGF-betaRI (ALK-5) inhibitors: analysis of the binding interactions between ligand-receptor using 2D and 3D techniques. *European Journal of Pharmaceutical Sciences: Official Journal of the European Federation for Pharmaceutical Sciences* 2013;49:542-9. doi:10.1016/j.ejps.2013.05.015.

[59] de Angelo R, Almeida M, de Paula H, Honorio K. Studies on the Dual Activity of EGFR and HER-2 Inhibitors Using Structure-Based Drug Design Techniques. *Int J Mol Sci* 2018;19:3728.

[60] Araujo SC, Maltarollo VG, Silva DC, Gertrudes JC, Honorio KM. ALK-5 Inhibition: A Molecular Interpretation of the Main Physicochemical Properties Related to Bioactive Ligands. *J Braz Chem Soc* 2015;26:1936-46. doi:10.5935/0103-5053.20150172.

[61] Sushko I, Novotarskyi S, Körner R, Pandey AK, Cherkasov A, Li J, et al. Applicability domains for classification problems: benchmarking of distance to models for Ames mutagenicity set. *J Chem Info Model* 2010;50:2094-111.

[62] Yadav DK, Kumar S, Saloni HS, Kim M-h, Sharma P, Misra S, et al. Molecular docking, QSAR and ADMET studies of withanolide analogs against breast cancer. *Drug Design Develop Therapy* 2017;11:1859.

[63] Yousuf Z, Iman K, Iftikhar N, Mirza MU. Structure-based virtual screening and molecular docking for the identification of potential multi-targeted inhibitors against breast cancer. *Breast cancer* (Dove Medical Press) 2017;9:447-59. doi:10.2147/bctt.s132074.

[64] Reddy S, Reddy KT, Kumari V. Ligand based virtual screening to identify potential anti cancer ligands similar to Withaferin A targeting indoleamine 2, 3-dioxygenase. *Biosciences Biotechnology Research Asia* 2014;11:887-93.

[65] Kramer B, Rarey M, Lengauer T. Evaluation of the FLEXX incremental construction algorithm for protein–ligand docking. Proteins: *Struc Func Bioinf* 1999;37:228-41.

[66] Ewing TJ, Makino S, Skillman AG, Kuntz ID. DOCK 4.0: search strategies for automated molecular docking of flexible molecule databases. *J Comp Aided Mol Design* 2001;15:411-28.

[67] Verdonk ML, Cole JC, Hartshorn MJ, Murray CW, Taylor RD. Improved protein–ligand docking using GOLD. *Proteins: Struc Func Bioinfo* 2003;52:609-23.

[68] Friesner RA, Banks JL, Murphy RB, Halgren TA, Klicic JJ, Mainz DT, et al. Glide: a new approach for rapid, accurate docking and scoring. 1. Method and assessment of docking accuracy. *J Med Chem.* 2004;47:1739-49.

[69] Goodsell DS, Morris GM, Olson AJ. Automated docking of flexible ligands: applications of AutoDock. *Journal of Molecular Recognition* 1996;9:1-5.

[70] Trott O, Olson AJ. AutoDock Vina: improving the speed and accuracy of docking with a new scoring function, efficient optimization, and multithreading. *J Comp Chem.* 2010;31:455-61.

[71] Ferreira RS, Oliva G, Andricopulo AD. Integração das técnicas de triagem virtual e triagem biológica automatizada em alta escala: oportunidades e desafios em P&D de fármacos [Integration of virtual screening and high-level automated biological screening techniques: opportunities and challenges in drug R&D]. *Quim Nova* 2011;34:1770-8.

[72] Namba AM, Silva VBd, Silva CHTPd. Dinâmica molecular: teoria e aplicações em planejamento de fármacos [Molecular dynamics: theory and applications in drug planning]. *Eclética Química* 2008;33:13-24.

[73] Bockmann RA, Grubmuller H. Nanoseconds molecular dynamics simulation of primary mechanical energy transfer steps in F1-ATP synthase. *Nat Struct Mol Biol* 2002;9:198-202.

[74] Ackermann T. C. L. Brooks III, M. Karplus, B. M. Pettitt. Proteins: A Theoretical Perspective of Dynamics, Structure and Thermodynamics, Volume LXXI, in: *Advances in Chemical Physics*, John Wiley & Sons, New York 1988. 259 Seiten, Preis: US $ 65.25. Berichte der Bunsengesellschaft für physikalische Chemie 1990;94:96-. doi:10.1002/bbpc.19900940129.

[75] Almeida MO, Costa CHS, Gomes GC, Lameira J, Alves CN, Honorio KM. Computational analyses of interactions between ALK-5 and bioactive ligands: insights for the design of potential anticancer agents. *J Biomol Struc Dynamics* 2018;36:4010-22. doi:10.1080/07391102.2017.1404938.

[76] Uchibori K, Inase N, Araki M, Kamada M, Sato S, Okuno Y, et al. Brigatinib combined with anti-EGFR antibody overcomes osimertinib resistance in EGFR-mutated non-small-cell lung cancer. *Nat Commun* 2017;8:14768. doi:10.1038/ncomms14768.

[77] Jabbarzadeh Kaboli P, Ismail P, Ling K-H. Molecular modeling, dynamics simulations, and binding efficiency of berberine derivatives: A new group of RAF inhibitors for cancer treatment. *PLOS ONE* 2018;13:e0193941. doi:10.1371/journal.pone.0193941.

[78] Hernández-Campos A, Romo-Mancillas A, Medina-Franco JL, Castillo R. In search of AKT kinase inhibitors as anticancer agents: structure-based design, docking, and molecular dynamics studies of 2,4,6-trisubstituted pyridines AU - Trejo-Soto, Pedro Josué. *J Biomol Struc Dynamics* 2018;36:423-42. doi:10.1080/07391102.2017.1285724.

[79] Meller Ja. *Mol Dynamics*. eLS. John Wiley & Sons, Ltd; 2001.

[80] Wang J, Hou T, Xu X. Recent Advances in Free Energy Calculations with a Combination of Molecular Mechanics and Continuum Models. *Curr Comp Aided Drug Des*. 2006;2:287-306. doi:10.2174/157340906778226454.

[81] Su P-C, Tsai C-C, Mehboob S, Hevener KE, Johnson ME. Comparison of Radii Sets, Entropy, QM Methods, and Sampling on MM-PBSA, MM-GBSA, and QM/MM-GBSA Ligand Binding Energies of F. tularensis Enoyl-ACP Reductase (FabI). *J Comp Chem*. 2015;36:1859-73. doi:10.1002/jcc.24011.

[82] Sulea T, Cui Q, Purisima EO. Solvated Interaction Energy (SIE) for Scoring Protein–Ligand Binding Affinities. 2. Benchmark in the CSAR-2010 Scoring Exercise. *J Chem Info Model*. 2011;51:2066-81. doi:10.1021/ci2000242.

[83] Genheden S, Ryde U. The MM/PBSA and MM/GBSA methods to estimate ligand-binding affinities. *Exp Op Drug Discov*. 2015;10:449-61. doi:10.1517/17460441.2015.1032936.

[84] Zoete V, Irving MB, Michielin O. MM–GBSA binding free energy decomposition and T cell receptor engineering. *J Mol Recog*. 2010;23:142-52. doi:10.1002/jmr.1005.

[85] Naïm M, Bhat S, Rankin KN, Dennis S, Chowdhury SF, Siddiqi I, et al. Solvated Interaction Energy (SIE) for Scoring Protein−Ligand Binding Affinities. 1. Exploring the Parameter Space. *J Chem Info Model*. 2007;47:122-33. doi:10.1021/ci600406v.

[86] Wan S, Bhati AP, Zasada SJ, Wall I, Green D, Bamborough P, et al. Rapid and Reliable Binding Affinity Prediction of Bromodomain Inhibitors: A Computational Study. *J Chem Theory Comput* 2017;13:784-95. doi:10.1021/acs.jctc.6b00794.

[87] Acuña J, Piermattey J, Caro D, Bannwitz S, Barrios L, López J, et al. Synthesis, Anti-Proliferative Activity Evaluation and 3D-QSAR Study of Naphthoquinone Derivatives as Potential Anti-Colorectal Cancer Agents. *Molecules.* 2018;23 doi:10.3390/molecules23010186.

[88] Zhao K, Zhou X, Ding M. Molecular insight into mutation-induced conformational change in metastasic bowel cancer BRAF kinase domain and its implications for selective inhibitor design. *J Mol Graph Model.* 2018;79:59-64. doi:10.1016/j.jmgm.2017.11.005.

[89] van der Kamp MW, Mulholland AJ. Combined Quantum Mechanics/Molecular Mechanics (QM/MM) Methods in Computational Enzymology. *Biochem.* 2013;52:2708-28. doi:10. 1021/bi400215w.

[90] Warshel A, Levitt M. Theoretical studies of enzymic reactions: Dielectric, electrostatic and steric stabilization of the carbonium ion in the reaction of lysozyme. *J Mol Bio.* 1976;103:227-49. doi:10.1016/0022-2836(76)90311-9.

[91] Groenhof G. Introduction to QM/MM Simulations. In: Monticelli L, Salonen E, editors. *Biomolecular Simulations: Methods and Protocols.* Totowa, NJ: Humana Press; 2013, p. 43-66.

[92] Warshel A. Computer Simulations of Enzyme Catalysis: Methods, Progress, and Insights. *Ann Rev Biophys Biomol Struc.* 2003;32:425-43. doi:10.1146/annurev.biophys.32.110601.141807.

[93] Dapprich S, Komáromi I, Byun KS, Morokuma K, Frisch MJ. A new ONIOM implementation in Gaussian98. Part I. The calculation of energies, gradients, vibrational frequencies and electric field derivatives. *Journal of Molecular Structure: THEOCHEM* 1999; 461–462:1-21. doi:10.1016/S0166-1280(98)00475-8.

[94] Vreven T, Byun KS, Komáromi I, Dapprich S, Montgomery JA, Morokuma K, et al. Combining Quantum Mechanics Methods with

Molecular Mechanics Methods in ONIOM. *J Chem Theo Comp.* 2006;2:815-26. doi:10.1021/ct050289g.

[95] Callegari D, Ranaghan KE, Woods CJ, Minari R, Tiseo M, Mor M, et al. L718Q mutant EGFR escapes covalent inhibition by stabilizing a non-reactive conformation of the lung cancer drug osimertinib. *Chem Sci.* 2018;9:2740-9. doi:10.1039/C7SC04761D.

[96] Koulgi S, Achalere A, Sonavane U, Joshi R. Investigating DNA Binding and Conformational Variation in Temperature Sensitive p53 Cancer Mutants Using QM-MM Simulations. *PLOS ONE.* 2015;10:e0143065. doi:10.1371/journal.pone.0143065.

[97] Wityk P, Wieczór M, Makurat S, Chomicz-Mańka L, Czub J, Rak J. Dominant Pathways of Adenosyl Radical-Induced DNA Damage Revealed by QM/MM Metadynamics. *J Chem Theo Comp.* 2017;13:6415-23. doi:10.1021/acs.jctc.7b00978.

[98] Śliwa P, Kurczab R, Bojarski AJ. ONIOM and FMO-EDA study of metabotropic glutamate receptor 1: Quantum insights into the allosteric binding site. *Int J Quantum Chem* 2018;118:e25617. doi:10.1002/qua.25617.

[99] Lu J, Zhou K, Yin X, Xu H, Ma B. Molecular insight into the T798M gatekeeper mutation-caused acquired resistance to tyrosine kinase inhibitors in ErbB2-positive breast cancer. *Comp Bio Chem.* 2019;78:290-6. doi:10.1016/j.compbiolchem.2018.12.007.

In: Advances in Medicinal Chemistry … ISBN: 978-1-53616-368-1
Editor: E. Ferreira da Silva-Júnior

Chapter 6

INDOLEAMINE 2,3-DIOXYGENASE 1 INHIBITORS: DISCOVERY, DEVELOPMENT, AND PROMISE IN CANCER IMMUNOTHERAPY

Van-Hai Hoang[1,*], PhD and Phuong-Thao Tran[2], PhD
[1]Laboratory of Medicinal Chemistry, Research Institute of Pharmaceutical Science, College of Pharmacy, Seoul National University, Seoul, Republic of Korea
[2]Department of Pharmaceutical Chemistry, Hanoi University of Pharmacy, Hanoi, Vietnam

ABSTRACT

Success in cancer immunotherapy, including FDA-approved biological drugs targeting-immune checkpoint process, has led the therapy to become a promising exciting new era. It is recognized that this therapy, which fights against cancer by activating T cell-mediated adaptive immunity, provides more benefits than traditional approaches

* Corresponding Author's Email: hoanghai@snu.ac.kr.

such as chemotherapy and radiotherapy. The therapy can combine with traditional therapies to improve anti-cancer, resulting in a higher probability of survival, compared to single-agent approaches. Tryptophan is an essential amino acid, which is mostly metabolized through the kynurenine pathway. The pathway has three "gatekeepers": indoleamine 2,3-dioxygenase 1 (IDO1), indoleamine 2,3-dioxygenase 2 (IDO2), and tryptophan 2,3- dioxygenase (TDO) to control levels of tryptophan. Because of its primary function in the kynurenine pathway, IDO1 can cause tryptophan starvation, which inhibits the proliferation of immune cells; and induce immunosuppression through metabolites of kynurenine. In cancer cells, the overexpression of IDO1 makes a microenvironment around cancer cells to help them escape the immune system. Therefore, IDO1 has been considered as an anticancer molecule and become a favorite drug discovery program. Uniquely, unlike other cancer immune checkpoint targets, most inhibitors of which are biological macromolecules, IDO1 inhibitors are small molecules, marking them become the most attractive issue in the immuno-oncology area. In only ten years, thousands of IDO1 inhibitors have discovered, and at least five compounds are going on trial studies. In order to provide the story of the discovery and development of IDO1 inhibitors, this chapter is divided into two parts. The first part provides the much-needed summary of the relationship between the immune system and cancer as well as the structure and role of IDO1 in cancer treatment. Moreover, the last part of the chapter traces the discovery and development IDO1 inhibitors from beginning to now, which provide lead identification, SAR study, lead optimization, considers issues, and outlook of IDO1 inhibitors.

Keywords: immuno-oncology, kynurenine pathway, indoleamine 2,3-dioxygenase 1, IDO1 inhibitors

INTRODUCTION

Cancer and Immune System

Cancer is a large group of diseases characterized by the uncontrolled growth and spread of abnormal cells. Cancer is the second leading cause of death in the world, is an estimated 18.1 million new cases and 9.6 million deaths in 2018 [1]. Nowadays, some causes of cancer were determined, such as smoking, alcohol, pollution, diet, body weight issue, infectious agents, UV light, radiation, and others. Everyday everybody is exposed to these causes, but they do not catch clinical cancer. The human body has mechanisms to identify and eliminate cancer cells, and one of the most useful things is controlled by the immune system.

As a host defense system, the immune system determines whether a factor is "self" and "non-self," then attach and eliminate the "non-self" one. It plays the most critical role in the system. Cancer cells are abnormal cells and are recognized as "non-self." However, the process is not simple, but it is a complex and dynamic process between the immune system and cancer cells. The role of the immune system was proposed for the first time by Burnet and Thomas in the 1950s and developed in several following decades. It is named immunoediting and composed of three phases: elimination, equilibrium, and escape [2].

- Phase 1: Elimination. The elimination phase also is known as immunosurveillance, starting when cancer cells have arisen. The process includes of innate and adaptive immune responses of the immune system to eradicate cancer cells that are formed in the body. Firstly, macrophages and stromal cells surrounding the tumor cells release some inflammatory agents which induce inflammatory signals. The signals are necessary for the recruitment of innate cells such as natural killer cells (NK), macrophages, dendritic cells (DCs) and natural killer T cells (NKT) to the tumor site. At this time, NK and NKT are stimulated to produce interferon-gamma (IFN-γ). The released IFN-γ causes tumor deaths and promotes the synthesis of chemokines CXCL9, CXCL10, CXCL11 which block the formation of new blood vessels. Also, NK and macrophage produce interleukin 12 (IL-12) promoting to kill more tumor cells through apoptosis and reactive oxygen. Whereas, DCs ingest the tumor cells and their debris. After ingestion, DCs migrate to lymph nodes, presenting the tumor antigens (TA) to naive $CD4^{+}$T cells in order to differentiate and develop TA-specific T cells. The TA-specific T cells move to the tumor site and eliminate antigen-bearing tumor cells. During the process, chemokines are still produced to enhance tumor cell elimination [2-5].

- Phase 2: Equilibrium. It is the next step in the cancer immunoediting process. Under the immune selection pressure, some cancer cells are unstable and rapidly mutate, grow, and acuminate, while others are detected and eliminated. During this period, the tumor cells try to find methods to survive. This is a long fight between the immune system and tumor cells, and frequently, it takes many years. Lymphocytes and IFN-γ are two primary agents for tumor cell variants selection [2-5].
- Phase 3: Escape. The escape starts when tumor cells beat the elimination of the immune system. They are able to fight against the immune system, grow uncontrollably, and can be diagnosed. There are several mechanisms which tumor cells use to escape the immune system. Briefly, cancer cells change the microenvironment around them, which suppresses the immune system and work as a barrier to protect them. Some enzymes, pathways, and suppressive cytokines are used to protect them from immune system attacks such as kynurenine pathway and Indoleamine 2,3-dioxygenase 1, IL6, IL10, and others. Staying parallel to this, cancer cells mutate to help them increase survival or develop resistance to apoptosis through overexpression of STAT3 and Bcl2, respectively. The third method is effecting on the immune checkpoint, leading to the inactivation of T cell-like PD1, CTLA4, BTLA, TIM3, and others. The last way is modulation a suppressive cell, Treg, and its formation. The tumor cells promote the differentiation of naive T cell to Treg, which acts as an off-switch on the immune system and turn off the response. IDO1, CCL2, CXCl2 are some ligands and enzymes are attributed to this mechanism [2-5].

Kynurenine Pathway and Indoleamine 2,3-Dioxygenase 1

As discussed above, one of the mechanisms to escape the immune system of cancer cells is the overexpression of IDO1 through the kynurenine pathway (KP). The pathway is a significant pathway of tryptophan metabolism which uses 95% of total free tryptophan, an essential amino acid. The remaining tryptophan is used for synthesis serotonin and melatonin through the methoxyindole pathway and protein synthesis (Figure 1) [6]. The kynurenine pathway induces the depletion of tryptophan with an oxidative ring-opening of the 2,3-carbon-carbon double bond of indole ring as the initial step. Through a processing sequence, some bioactive metabolites are generated such as kynurenic acid, 3-hydroxy kynurenic acid, 3-hydroxyanthranilic acid, and quinolinic acid. One of the end products of the KP is nicotinamide adenine dinucleotide (NAD^+) acting as a coenzyme in redox reactions of the body [7].

The first and rate-limiting step of the kynurenine pathway is carried out by one of three enzymes: indoleamine 2,3-dioxygenase 1 (IDO1), indoleamine 2,3-dioxygenase 2 (IDO2), and tryptophan 2,3- dioxygenase (TDO). Although they all catalyze the same biochemical reaction, they have entirely different functions and tissue expression. IDO2 is lastly characterized, and its functional role has not been understood fully yet. TDO is expressed mainly in the liver and responds to tryptophan levels throughout the intake tryptophan from food. Its substrate is specified to *L*-tryptophan. Recently, TDO is considered to tumor microenvironment [8, 9]. Meanwhile, IDO1 distributes in many tissues and organs. Its substrates including not only *L*-tryptophan but also *D*-tryptophan and several indol-type compounds, e.g., melatonin, are more numerous than those of TDO. Many studies demonstrated the central role of IDO1 in KP and immune modulation. Overexpression of IDO1 is related to immunosuppression and cancer diseases.

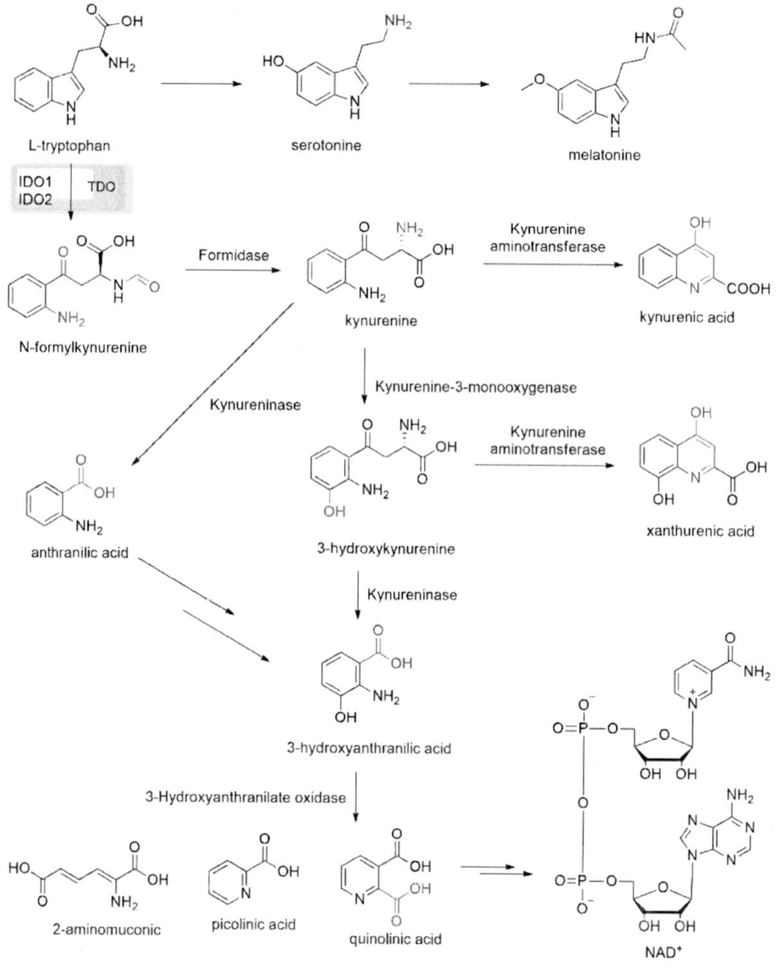

Figure 1. Metabolism of tryptophan through the kynurenine pathway and methoxyindole pathway. *Abbreviations: IDO: indoleamine 2,3-dioxygenase, TDO: tryptophan 2,3- dioxygenase.*

Catalytic Reaction Mechanism

IDO1 catalyzes the indole ring-opening process through forming its epoxide state, in which an oxygen molecule is inserted into tryptophan. Two atoms of oxygen molecule are added to Trp in a stepwise fashion. In the first step, an oxygen atom binds to the iron heme followed by Trp binding. The Trp binding is a radical addition into the double bond of indole ring to form Trp-epoxide intermediate, which is inserted the second oxygen atom from ferryl oxygen through nucleophilic addition-like at the C-2 position. The amino group of Trp acts as an intramolecular proton transfer system, which donates a proton to open epoxide ring, resulting in the formation of carbocation at 2-position and withdraws a proton from the hydroxy group resulted in transferring the remaining oxygen atom at the last step of the process (Figure 2) [10].

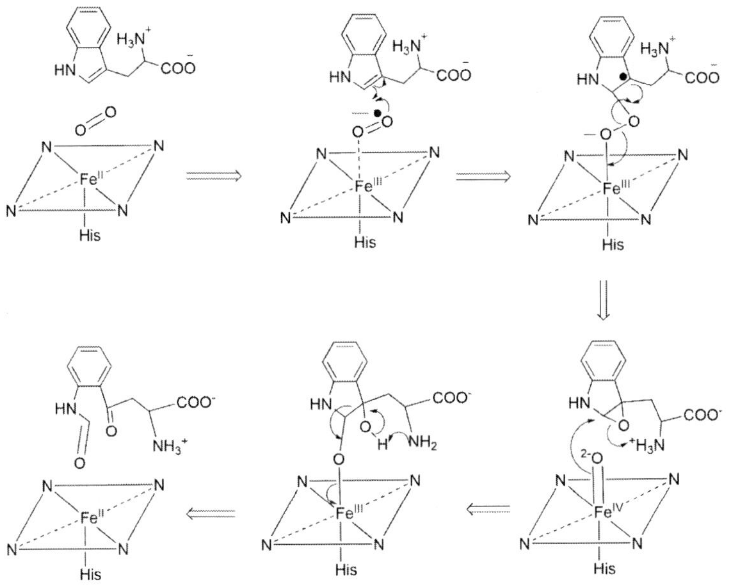

Figure 2. Reaction mechanism of oxidative ring-opening by IDO1.

Structure of IDO1

Since the first crystal structure of IDO1 in complex with 4-phenylimidazole (PI) was published by Sugimoto and Yoshitsugu [11], several other structure studies with other inhibitors have reported addition structures of enzyme: Arg-1, imidazothiazole [12], GDC-0919 [13], INCB14943 [14], and epacadostat [15]. Overall, IDO1 is a 45 kDa protein composed of a large catalytic domain holding the heme group, a small non-catalytic domain, and a loop connecting two domains. The larger domain is built by all helical structures which has 13 α-helices and two 3_{10}-helices, whereas the small domain is composed of two β-sheets, six α-helices, and three 3_{10}-helices. The active site locates in the small domain, and is composed of three regions: coenzyme heme, and two hydrophobic pockets: pocket A and pocket B. Pocket A is located above the sixth coordination site of the iron-heme; meanwhile pocket B is located at the binding site entrance (Figure 3a) [11].

In the heme environment, the mutagenesis studies demonstrated an essential role of His^{346} and Asp^{274} for IDO1 heme binding and enzyme catalyst. IDO1 will lose its catalytic activity without the presence of these two amino acids in the structure [16]. In the PI bound form, His^{346} provides the fifth coordinator of heme iron. It also contacts with heme 6-propionate to the water molecule in the proximal side. The other amino acid Asp^{274} interacts with heme 6-propionate through the formation of a salt bridge with Arg^{343} (Figure 3b) [11]. The sixth coordinator is nitrogen atom of PI suggesting to design new IDO1 inhibitors base on heme iron interaction.

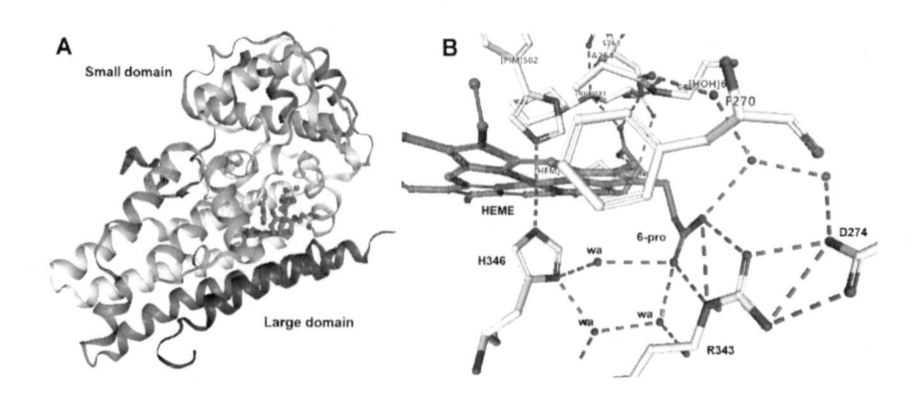

Figure 3. Structure of IDO-PI complex (PDB ID: 4E74). (a) The overall structure of human IDO. (b) Stereoview of the residues around the heme of IDO viewed from the side of heme plane. Abbreviations: C: Cys; S: Ser; F: Phe; R: Arg; D: Asp; H: His; L: Leu, wa: water, 6-pro: 6-propinoate.

Pocket A is composed of Cys^{129}, Ser^{167}, Tyr^{126}, Gly^{262}, Ala^{264}, and Phe^{163} which are differently positioned in each enzyme-inhibitor complex.

It suggests the flexible structure of pocket A. Pocket A is a small hydrophobic pocket which usually is occupied by aromatic rings like phenyl rings of phenylimidazole [12] or epacadostat [15], indole ring of GDC-0919 [13]. Also, Cys^{129}, Ser^{167}, and Ala^{264} play an essential role in the interaction with inhibitors by hydrogen bonding [15, 17]. Both hydrophobic and hydrophilic amino acids build pocket B. Among them, Phe^{226} and Arg^{231} are crucial for enzyme-inhibitor interactions and usually used to design IDO1 inhibitors. In particular, the π-interaction with Phe^{226} and electrostatic interactions with Arg^{231} are hypothesized for substrate/inhibitor binding in IDO1. Importantly, the crystal structure of IDO1 - GDC-0919 complex shows a hydrogen bonding with 7-propionate of heme structure, leading to the design of new inhibitors with extending interaction regions [13]. The scheme of IDO1 active site and its inhibitors were summarized in Figure 4. In general, the IDO1 active site is commonly divided into three regions: pocket A, pocket B, and a heme; potent inhibitors (IC_{50} < 100 nM) are consist of structures that can interact with all three regions.

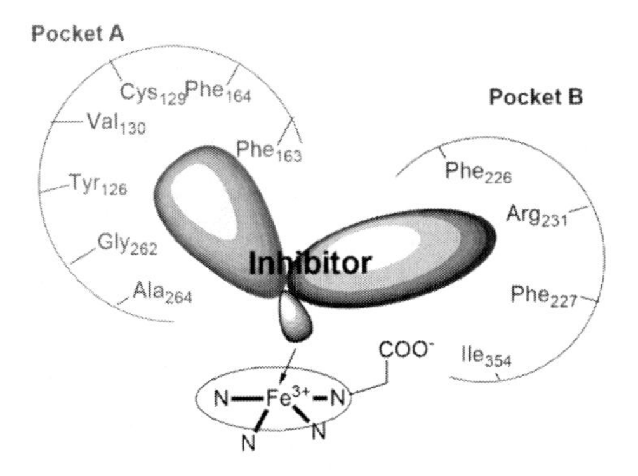

Figure 4. General scheme of IDO1 active site and its inhibitors.

IDO1 also contains an allosteric site (Si), which lies opposite to the active site through the heme phase. In the crystal structure of IDO1-3-indol ethanol (IDE) and tryptophan complexes, the allosteric site was based on Leu^{207}, Leu^{339}, Ala^{210}, Leu^{342}, Phe^{214}, Phe^{273}, Phe^{270}, and others. The OH group of IDE forms a hydrogen bond with 6-propionate of heme structure while its benzene ring interacts with the heme phase. In general, the binding changes slightly the enzymatic conformation; especially the branch chain of Phe[270] is pushed up due to the steric effects of the indole ring (Figure 5). By Phe[270] mutation and titration study, it is demonstrated that IDE acts as an inhibitor, and the Si site is an allosteric region [15].

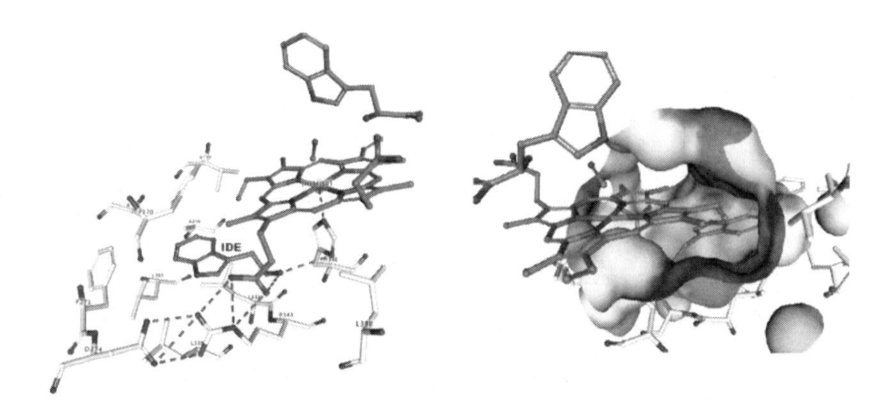

Figure 5. Crystal structure of the hIDO1-CN-Trp complex (PDB ID: 6E35) in a mixed ligand state and side view of the IDE-binding site. *Abbreviations: C: Cys; S: Ser; F: Phe; R: Arg; D: Asp; H: His; L: Leu; V: Val; A: Ala, IDE: 3-indol ethanol.*

Interestingly, the bonding of apo-IDO1 (non-heme IDO1 structure) and its heme cofactor is a dynamical process. IDO1 exists in the human body under two isoforms: full structure IDO1 and apo-IDO1. Compounds, which bind to apo-IDO1, also inhibit the activity of IDO1 and kynurenine pathway. It suggests that apo-IDO1 is an enzyme target to design a new class IDO1 inhibitor [18].

IDO1 and Cancer

As the first step enzyme (gatekeeper) of the kynurenine (KYN) pathway, IDO1, and KYN pathway share the same role in the regulation of immune responses. The expression of IDO1 in cells effects in microenvironments around them through two main theories: starving cells of tryptophan and toxicity of the metabolites in the KYN pathway (Figure 6) [19, 20].

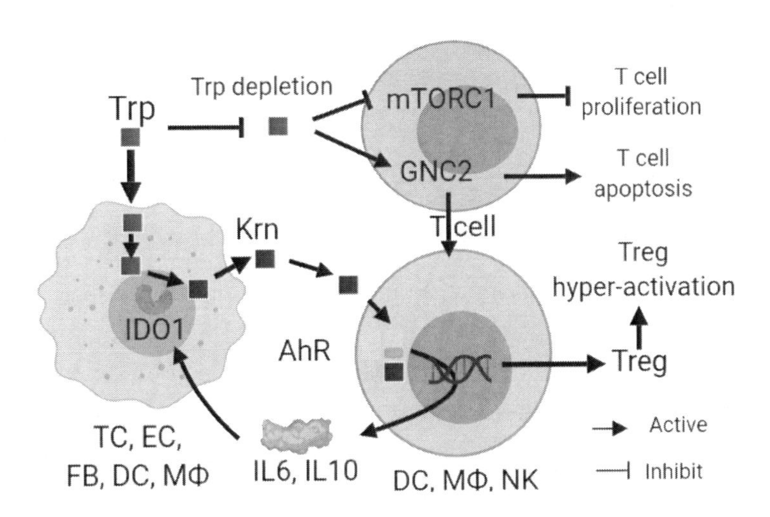

Figure 6. Mechanism of IDO pathway activity in immunosuppression. *TC: tumor cell, DC: dendritic cell, MΦ: macrophase, EC: endothelial cell, FB: fibroblast, IDO: indoleamine-2,3dioxegenase 1, Trp: tryptophan, Krn: kynurenine, mTORC1: mechanistic target of rapamycin (serine/threonine kinase) complex 1, GCN2: general control nonderepressible-2, ARH: aryl hydrocarbon receptor, Treg: regulatory T cell.*

In the tryptophan starvation theory, IDO1 activity directly reduces the concentration of local tryptophan which is an essential amino acid, thereby induces cell cycle arrest and increasing their susceptibility to apoptosis of T lymphocytes [22, 23]. The expression of IDO1 decreases tryptophan concentration leading to the accumulation of uncharged tryptophan transferring ribonucleic acid (tRNA) in cells. The elevation of tRNA actives the amino acid-sensitive general control nonderepressible 2 (GCN2) stress kinase pathway. GCN2 is a serine/threonine kinase that phosphorylates eukaryotic initiation factor 2α kinase (eIF2*α*) causing the downregulation of protein synthesis [24]. It also reduces the synthesis of fatty acid, which is essential for T cell proliferation and function [25]. As a result, it can be cytotoxic effects on various immune cells as $TCD8^+$, natural killer (NK), and invariant NK-T cells [26-28]. Besides, GCN2 promotes Treg differentiation and increases Treg activity de novo, resulting in immunosuppression [29].

Otherwise, depletion of tryptophan inhibits the mechanistic target of rapamycin complex 1 (mTORC1) and protein kinase C theta (PKCθ). Inhibition of mTORC1 leads to a response that includes activating autophagy, and energy in T cell [30], meanwhile, PKCθ is an essential enzyme for the activation of T cell [31].

In the tryptophan metabolite theory, the metabolites like kynurenine, 3-hydroxykynurenine, and 3-hydroxyanthreanilic acid can cause directly T cell cycle arrest and apoptosis [32], and indirectly suppress effector T cell by differentiation of regulatory T cell through aryl hydrocarbon receptor (AhR). AhR is a ligand-activated transcription receptor that ligates by tryptophan metabolites, translocates from the cytosol to the nucleus. Finally, AhR-ARNT (AhR nuclear translocator) complex forms with some superstructure and several co-transcription factors to promote the transcription of IL10 in DC and NK cells and IL6 in cancer cells and macrophages [33-35]. IL6 and cofactor PE2 have positive feedback on IDO1. The superstructure also increases IL10 by upregulating transcription factor c-Maf [36]. That all contribute to the conservation of naive $CD4^{+}T$ cells into FOXP3-expressing regulatory T cells, which inhibit the maturation and cytotoxicity of T cell [37].

IDO1 is overexpressed in many types of cancer cells and tumors such as prostate, colorectal, pancreatic, cervical, gastric, ovarian, and lung [19, 38-40]. Additionally, IDO1 can be induced by many agents such as inflammatory signals involving IFN-γ, lipopolysaccharides (LPS), and damage-associated molecular patterns (DAMP) or/and cytokines, including TNF-α, IL-6, IL-10 [41]. Some regulators also induct IDO1, including prostaglandin E2, inducible nitric oxide synthase (iNOS) and the tumor suppressor Bin 1 [42]. Among them, IFN-γ is one of the most IDO1 inductors. As the protection mechanism against cancer cells, cytokines, inflammatory agents is released to destroy cancer tumors and active the chemotaxis process. However, the released chemical inducts IDO1 of cancer cells, even cancer cells which do not constitutively over-express IDO1, resulting in an immune suppression system. The response of cancer to the immune system helps cancer survive and resist therapies. Several studies have reported that high IDO1 expression could be related to inadequate outcome chemotherapy and radiotherapy. The patient population, which is low in IDO1 expression, showed a good response to treatment. In clinical studies, IDO1 inhibitors showed promising results, leading IDO1 to become an attractive target in cancer treatment [43, 44].

DISCOVERY AND DEVELOPMENT OF IDO1 INHIBITORS

Historically, phenylimidazole was discovered as the first IDO1 inhibitor in 1989. It binds to heme structure by a nitrogen atom in imidazole ring, the structure of which is similar to indole ring of IDO1 substrate, tryptophan. Since the relationship of IDO1 and cancer has been reported in 2003, thousands of IDO1 inhibitors have been reported. Thanks to the advantage stemming from their small molecular structures, it is not surprising that IDO1 inhibitors have developed quickly. Although the FDA has no approved any IDO1 inhibitors until now, at least seven small molecule compounds are undergoing clinical trials. Among them, five compounds were disclosed chemical structure (Figure 7). Most of them were studied under combination with other cancer treatments such as chemotherapy, vaccine, and especially checkpoint inhibitors.

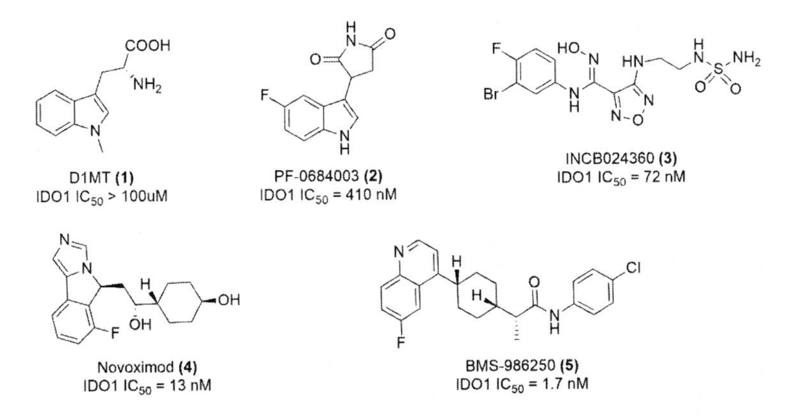

Figure 7. Structure of know IDO1 inhibitors under clinical trials.

Some pharmaceutical companies have joined IDO1 project in hope of bringing IDO1 drugs to the market soon. However, many compounds have not been published yet or published with a lack of information lead to difficulties in data summary and drug discovery and development. In general, IDO1 inhibitor can be classified into three generations:

- The first generation is inhibitors which bear indole-like rings as heme-binding groups
- The second generation is inhibitors which bond to heme structure but do not bear the indole-like motif.
- The third generation is inhibitors which fill in the apo-IDO1 structure.

THE FIRST GENERATION

Indole Motif Compounds

Most of the indole-containing analogs were the first known IDO1 inhibitors. They were designed based on tryptophan structure to form competitive inhibitors which replace of tryptophan and interact with heme structure. Normally, some small groups such as chlorine or fluorine were introduced on indole ring at 5 or 6 positions for pocket A filling and a linker at 2 or 3 positions for pocket B filling. However, they showed modest potencies and poor physical properties. Some strong potent inhibitors of group were published such as ketone-indole derivative **6** (IDO1 IC_{50} = 13 μM) [45, 46], arylthioindole derivative **7** (IDO1 IC_{50} = 7 μM) [47]. The low enzyme inhibition can be due to Trp itself shows only moderate affinity to IDO1. The *N*-methylation of indole ring yielded 1-methyl-Trp racemate, the IDO1 K_i of which reached 30 μM. Although the *S*-(*L*)-isomer (IDO1 K_i = 18 μM, IDO1 inhibition 63% at 100 μM) was found more active than D-isomer (IDO1 inhibition 12% at 100 μM), D-isomer showed better *in vitro* T cell effect and *in vivo* anticancer activity and become the first IDO1 inhibitor has been approved to clinical trial by

Newlink Genetics in 2008 under the name indoximod. Indoximod effects on the IDO pathway rather than directly on IDO1 enzyme (Figure 8) [48, 49].

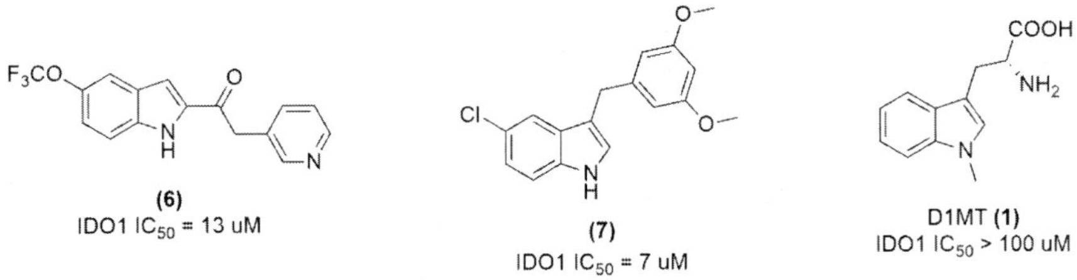

Figure 8. Structure of some indole-containing IDO1 inhibitors.

Indole structure became more attractive when iTeos Therapeutics and Pfizer brought one compound of the series, **PF-0684003**, to the clinical trial, [50]. The project started from HTS (high throughput screening) a library of 178 compounds. The compound **8** from HTS showed moderate potency on hIDO1 (IC_{50} = 3.0 μM), no activity on TDO and have excellent ligand efficiency (LE = 0.47 kcal/mol/HA). Studying on enantiomers showed that only one isomer (**8a**) had activity on IDO1 (IC_{50} = 1.8 μM). Subsequent studies on ADME showed attractive results, leading to the selection of **8a** as a lead compound for future optimization. A series of 5 or/and 6-substituted indole was synthesized and evaluated potency on IDO1. Only 5-chlorine, 5-bromine, and 5-fluorine have given higher potency than lead compound **8a,** suggesting that the position and size of substitutes are much important for potency. Docking study showed that the indole ring of **PF-0684003** lying a in narrow pocket A is crucial for bioactive results. The best compound (**2**) has been chosen for the pre-clinical trial. Compound **2** was reported to have a high oral bioavailability, high stability ($t_{1/2}$ = 16-19 hours) and tumor growth inhibition when combined with immune checkpoint inhibitors [51]. Therefore, **2** entered the phase I clinical trial for a brain tumor. Interestingly, two enantiomers of **2** have different potency (IC_{50} = 0.12 μM and IC_{50} = 54 μM) but, compound **2** is a racemic mixture due to repaid epimerization of chiral carbon [51, 52].

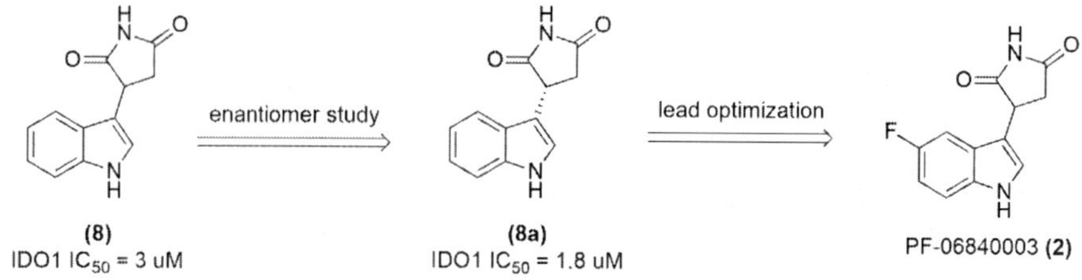

Figure 9. Process of PF-0684003 development.

Indazole Motif Compounds

Recently, some inhibitors based on the structure of indazole were published with micromolar potency. As an indole bioisostere, indazole structure also inhibits IDO1 and acts as the core skeleton. Substituent groups at 3-position were reported with a series of the aromatic ring. Linkers between aromatic ring and indazole were scanned such as amino, oxy, hydrazide; but they showed only moderate inhibitory activity.

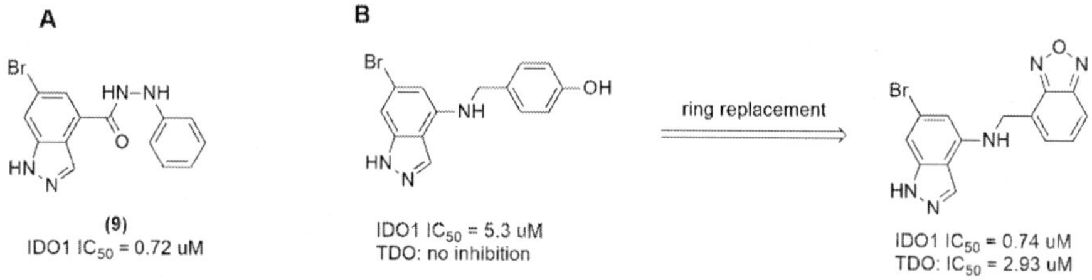

Figure 10. Some indazole derivatives as IDO1 inhibitors. *(A) 3-substituted indazole compound, (B) 4-substituted indazole development.*

The docking study revealed that despite the same scaffold, several 3-substitutes interact with pocket A, while others show interaction with pocket B. Also, because of restricted space in pocket A, the inhibitory efficacy of the 1*H*-indazole is sensitive to the 3-substitution. In this class, hydrazide derivatives showed the highest potency with some submicromolar inhibitory compounds (**9**, IC_{50} = 0.72 μM) [53]. For substituent groups at the 4-position, amine linker displays good potency.

By ring replacement, some inhibitors were found as submicromolar IC_{50} agents in IDO1, and TDO inhibitors (Figure 10) [54, 55].

Imidazole Compounds

Starting from 4-phenylimidazole (4PI), the heme binder identified as an IDO1 inhibitor (IDO1 IC_{50} = 48 μM) series of phenylimidazole derivatives was synthesized [56]. The X-ray structure of complex 4PI with IDO1 provides the way drug design in silico and synthesis of the new inhibitors. In the co-crystallization structure, imidazole plays as the sixth coordinate site of iron heme, and phenyl ring fills in pocket A by π-interaction Tyr[126] and Phe[163] [11]. Extension of the binding regions of 4PI to IDO1 was considered and followed three main pathways.

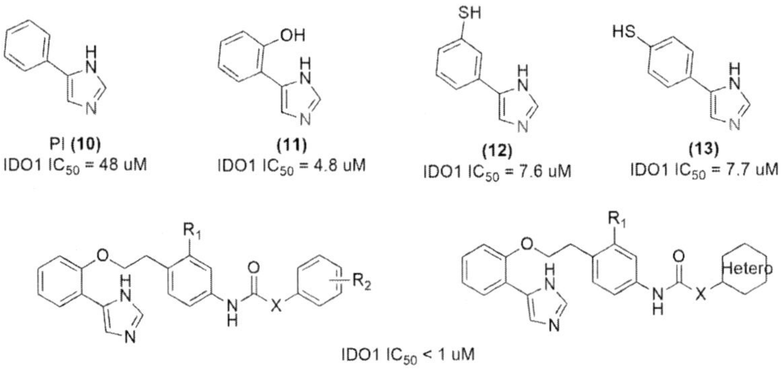

PI **(10)** IDO1 IC_{50} = 48 uM

(11) IDO1 IC_{50} = 4.8 uM

(12) IDO1 IC_{50} = 7.6 uM

(13) IDO1 IC_{50} = 7.7 uM

IDO1 IC_{50} < 1 uM

Figure 11. Structure of some active phenylidazole derivatives as IDO1 inhibitor.

The first way is introduction polar groups on phenyl ring or replacement phenyl ring by another bioisosteric ring to form hydrogen bonding with Cys[129] and Ser[167]. It was found that hydroxy group at *ortho*-position (IDO1 IC_{50} = 4.8 μM) and thio group at *meta*- (IDO1 IC_{50} = 7.6 μM) or *para*- (IDO1 IC_{50} = 7.7 μM) positions improve inhibitory potency [56]. A comparison between wild IDO1 and its Ser[167] mutated form confirmed the role of Ser[167] in interaction of enzyme with inhibitor (IDO1WT IC_{50} = 1.2 μM, IDO1S167A IC_{50} = 41 μM) [57]. However, when phenyl was replaced by its bioisosteres, the inhibitory potency was reduced or abolished [56]. In 2011, Newlink Genetics reported a series of inhibitors taking advantage of the 2-hydroxy group of phenyl ring; they led for nanomolar range inhibitors (Figure 11) [58].

On the other hand, extending inhibitor structure in pocket B was taken place as second structure-designed optimization. The extending structure focused on the hydrogen interaction with 7-propionate of heme structure, hydrophobic binding with Phe^{226}, and electrostatic interaction with Arg^{231}. A series of the bulky group was substituted at N^1, N^3, and C^2 of imidazole ring, but only substituents at N^3 proved higher activities than 4PI [56]. Recently, Brant and coauthors [17] found the *N*-indole-4-phenyl scaffold has inhibitory activity at a nanomolar concentration (**14**, IC_{50} = 33.8 nM). By using model docking, they found hydrogen bonding of NH indole with 7-propionate of heme structure at the entrance of B pocket (Figure 12).

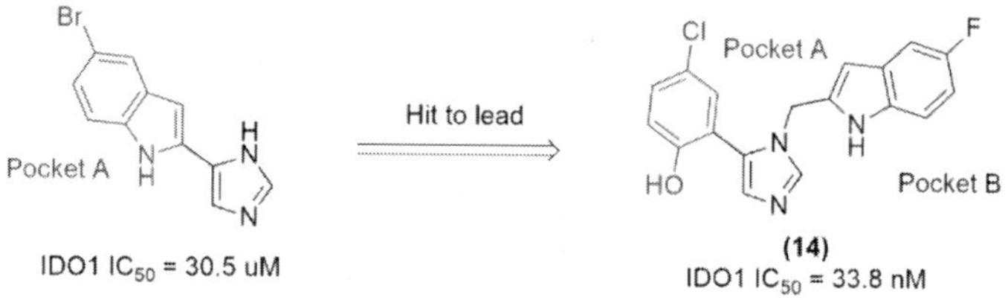

Figure 12. N-substituted of 4-PI development.

The third optimization route, imidazole ring, a heme binder, was replaced by its bioisosteric ring or fused imidazole scaffold. During replacement of imidazole ring by its bioisosteres (*N*-phenylimidazole, 4-phenylpyrazole, phenylthiazole, phenyltriazole, and phenyltetrazole), the new fragments lose or drop-down their potency [59]. The imidazole-containing fused tricyclic ring produced nanomolar range IDO1 inhibitors. Unluckily, most of them were patented with a little disclosed information. The first patent was disclosed about the tricyclic imidazoleisoindole ring, in which the cyclohexyl-ethanol-imidazoleisoindole motif showed the highest potency led the suitable property PK inhibitor, novoximod (IDO1 IC_{50} = 13 nM), entered in phase I clinical trial [60]. Molecular modeling studies revealed the interaction regions of inhibitor and enzyme: the nitrogen atom of imidazole ring is for heme binding, isoindole ring is into pocket A, and cyclohexylethanol structure occupies pocket B [13]. Based on the studies, hundreds of inhibitors were synthesized. There SAR was summarized in Figure 13.

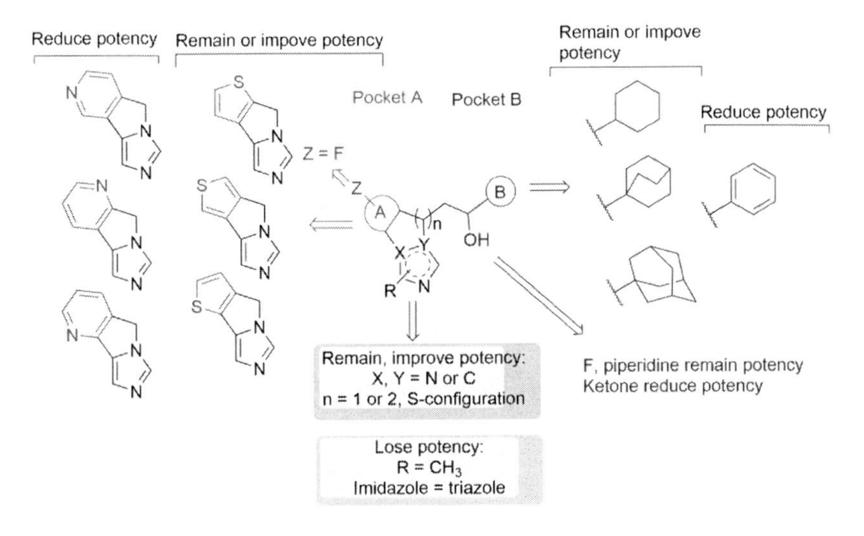

Figure 13. The SAR of fused imidazole scaffold.

Surprisingly, the replacement of fused-imidazole ring by another nitrogen-containing five-membered ring displayed no modification in inhibitory range, suggesting that the orientation of nitrogen of fused imidazole ring is less important [61]. However, methylation at 2-position of imidazole ring resulted in a complete loss in potency [13]. This result can be explained by the steric methyl group ability to prevent interaction between the imidazole ring and heme structure. On the other hand, the substitution of fluorine on the benzo group of imidazoleisoindole improves potency [13, 58, 60, 61] and five-membered heterocyclic rings replacement with benzo group keep potency [62]. Otherwise, six-membered heterocyclic rings cause to drop-down activity [63]. Further, the stereochemistry of CH in imidazoleisoindole seems to play an essential role in IDO1 activity, with *S*-configuration favors for higher potency [60]. In some cases, an extension of one more carbon in the middle five-membered ring of imidazoleisoindole results in also high IDO1 activity

[64]. In pocket B interacting region, modification of linker between imidazole and B-pocket interaction also yields a nanomolar range inhibitor, suggesting the way to improve PK and simplify chemistry. In addition, oxidation of the alcohol group to ketone reduced a potency cause by losing hydrogen bond [13, 58, 60, 61]; meanwhile, replacement the of hydroxy group with a fluorine or piperidine ring retained high potency [65]. Lately, compounds with the cyclohexyl ring replaced by bridged bi/tri-cyclic rings could maintain IDO1 activity, but this phenomenon is not correct for aromatic rings [66].

THE SECOND GENERATION

Carbonyl-Containing Compounds

Compounds with carbonyl group were mostly discovered by HTS from natural products. They usually bear quinone or iminoquinone motif, with the nanomolar IDO1 inhibition potency range [59, 67-69]. Recently, some good potent compounds in IDO1 were synthesized with the naphthoquinone core. Their SAR was studied and summarized in figure 14 [70].

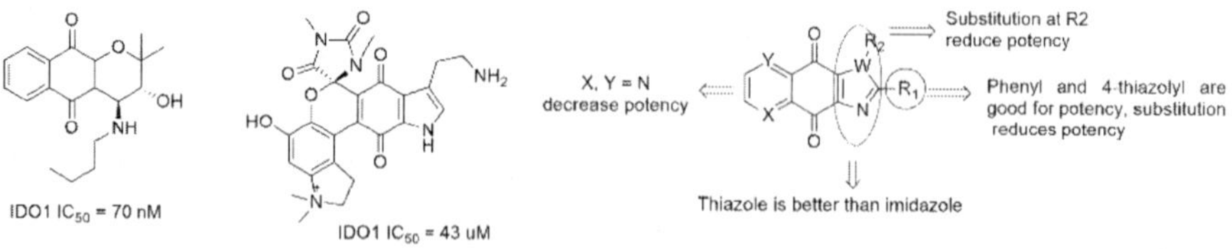

Figure 14. Structure of some natural IDO1 inhibitors and SAR of synthesized compounds.

Basing on the intermediate species structure in the catalytic process, some oxindole derivatives were designed and synthesized (Figure 15). Usually, they have a substitution at 3-position to fill in pocket B. Most of them showed moderate potency with some nanomolar range inhibitors [71].

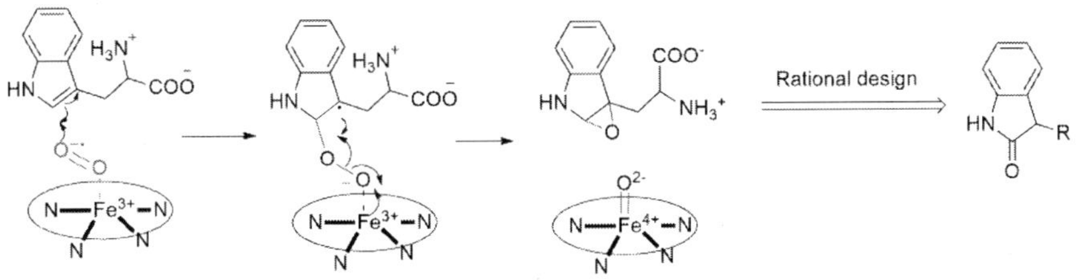

Figure 15. Rational design of oxidole derivative.

Sulfonylhydrazine Compounds

Sulfonylhydrazines derivatives are new non-indole motif IDO1 inhibitors, in which the oxygen of sulfonyl group plays as iron heme coordinator. Using HTS, the hit compound was identified as a potent IDO inhibitor (**15**, IDO1 IC_{50} = 167 nM). However, the compound did not show any IDO1 effect on the whole cell, suggesting its low permeability through the cell membrane. Introducing the acetamido group at the *para*-position of the benzo ring on phenyl sulfonyl resulted in an excellent cellular active molecule. The acetamido was fixed for various substitutes on phenylhydrazine ring, which lead to the discovery of compound **16**. The compound **16** showed the high potency in the enzyme assay and HeLa cell assay but did not inhibit tumor growth [72]. A re-optimization of the phenylsulfonyl ring brought compound **17**, which was a potent, selective, and orally bioavailability IDO1 inhibitor. Briefly, a good, submicromolar potency sulfonyl hydrazines derivative includes an iron coordinator, here O, two free NH group that can make hydrogen bond to surrounding heme structure, aromatic ring-containing bicyclic ring to fill pocket A and halide substituted on phenyl ring in pocket B (Figure 16) [73].

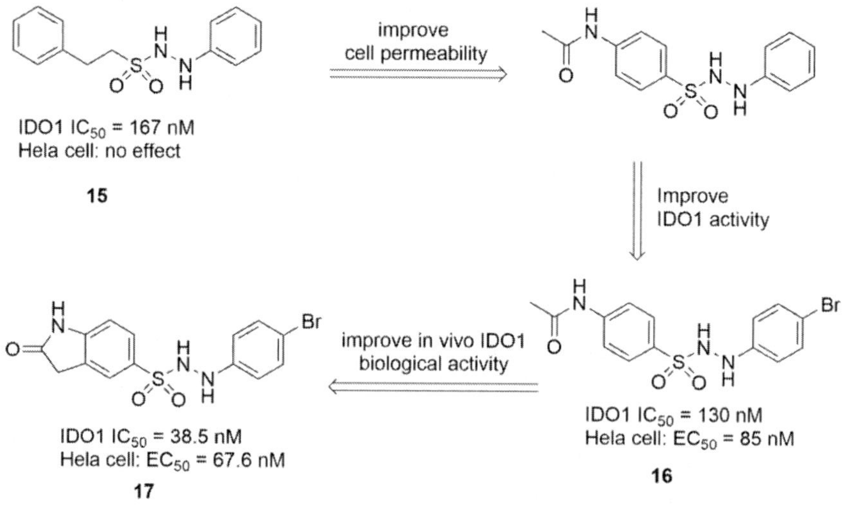

IDO1 IC_{50} = 167 nM
Hela cell: no effect

15

improve cell permeability

Improve IDO1 activity

IDO1 IC_{50} = 130 nM
Hela cell: EC_{50} = 85 nM

16

improve in vivo IDO1 biological activity

IDO1 IC_{50} = 38.5 nM
Hela cell: EC_{50} = 67.6 nM

17

Figure 16. Discovery and development of sulfonyl hydrazine derivatives as IDO1 inhibitors.

Single Aromatic Ring Compounds

Some single ring IDO1 inhibitors were rational design based on the structure of intermediate species in the catalytic process. They contain a group for iron-heme binding such as hydroxylamine, hydrazine, thiol.

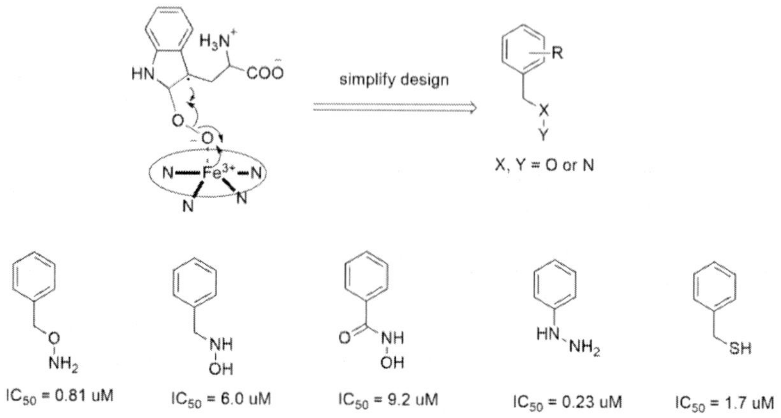

simplify design

X, Y = O or N

IC_{50} = 0.81 uM IC_{50} = 6.0 uM IC_{50} = 9.2 uM IC_{50} = 0.23 uM IC_{50} = 1.7 uM

Figure 17. Rational design of single aromatic derivatives.

Most of them give a moderate potency. In this group, although phenylhydrazine displayed the best activity on IDO1, it did not select as a hit because of its vulnerability to oxidation caused by heme. Therefore, the second potent hit compound, hydroxylamine was chosen for further development. The activity of some compounds and its design scheme was described in Figure 17 [59, 74].

Hydroxyamidine Compounds

N-hydroxyamidines are non-indole motif found as IDO1 inhibitors by Incyte company using HTS method. The project discovered the lead compound **18** with micromolar potency (IDO1 K_i = 1.5 μM). Compound **18** was tested its effects on HeLa cell, permeability, selection, and calculated ligand efficiency (LE) and confirmed binding by absorption spectroscopy. The results show a great hit for the discovery program. Further, a library of oxadiazole-carboximidamide was synthesized and evaluated activity in IDO1. A small group such as bromine, chlorine at 3-position on phenyl ring showed greater potency. Besides, 4-F on phenyl ring also slightly improved potency. The hydroxyamidine motif is the most important for potency, the replacement of which by other groups caused a complete loss in potency. A co-crystallization study of an N-hydroxyamidine derivative with IDO1 demonstrated the role of the hydroxy group through its coordinate bond with heme iron in the active site. The amino group also is important for enzyme affinity by hydrogen bonding with the heme structure. Finally, the lead compound **19** was discovery with great in vitro activity and selection (IDO1 IC50 = 59 nM, HeLa EC_{50} = 12 nM) [75]. However, **19** is poor bioavailability because of phase II glucuronidation reaction at the oxygen atom of the hydroxyamidines moiety. To tackle this problem, the compound was modified by adding a bulky group at amino C^3 in oxadiazole ring or replacement of oxadiazole ring by other heterocyclic hoping to hide the hydroxy group. However, the investigation of other heterocyclic ring indicates the important role of the oxadiazole ring for IDO1 potency.

Meanwhile, substitution at amino C^3 in oxadiazole ring brought very high protein binding inhibitors because of the lipophilicity. Therefore, polar groups were chosen for the next optimization. Only alkylsulfamide groups keep inhibitory potency and increase metabolic stability. All of the optimizations led to discover the great IDO1 inhibitor, epacadostat, which is undergoing clinical trial (Figure 18) [76, 77].

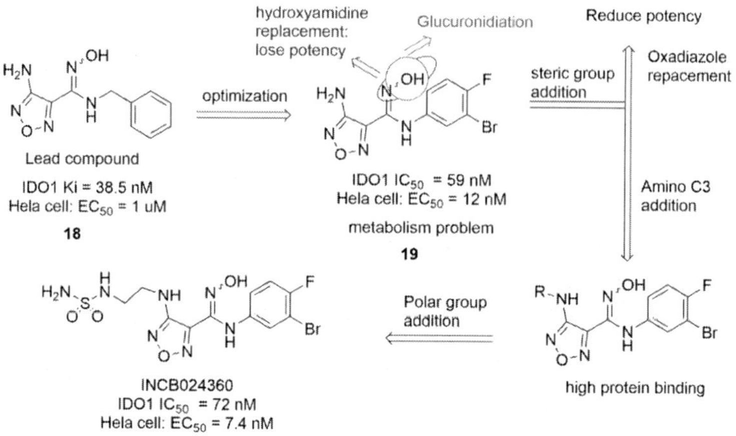

Figure 18. Hydroxyamidine derivatives discovery and development process.

THE THIRD GENERATION

The third generation of IDO1, which call BMS, was developed by Bristol-Myers Squibb Co. All of them are not heme coordinators; they interact directly with the active site of apo-IDO1. They bond with cellular apo-IDO1 and competing for the bond with heme of IDO1. They can be quinoline, pyridine or carboxylic derivatives, which nanomolar IC_{50} potency. Until now, BMS are the known strongest IDO1 inhibitors and kynurenine pathway inhibitors, of which BMS-986205 are undergoing phase I/II clinical [18]. However, their information has been not published frequently; most of them are kept under a secret. Recently, some patents and journals of the BMS series were published to help study their pharmacophore region [78-81]. In general, their pharmacophore has three parts: an aromatic ring interacts with Tyr[126], a hydrogen bonding region can interact with Ser[167], a π-interaction with Phe[270] and a hydrogen bonding donor for Arg[343]. All three regions are filled in A pocket and replaced the heme structure (Figure 19).

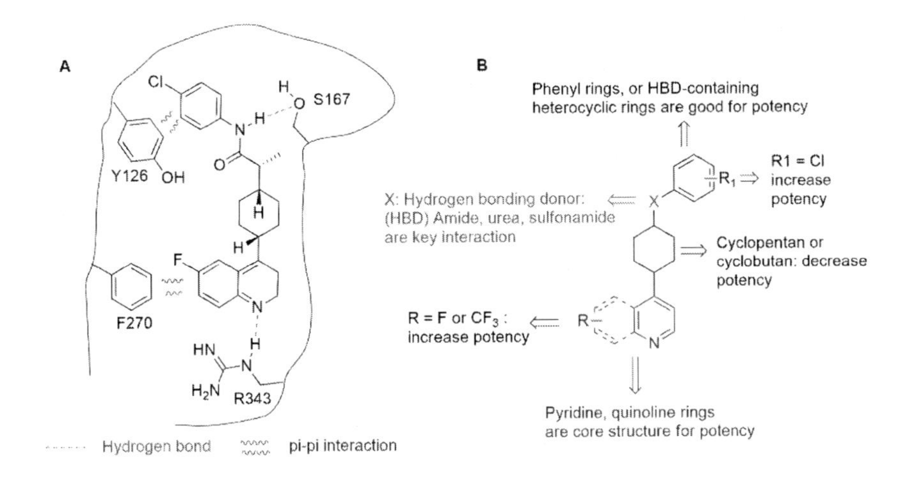

Figure 19. (A) Schematics of IDO1 and BMS-986205, (B) general SAR of BMS series.

FINAL CONSIDERATIONS

Thousands of IDO1 inhibitors were published, and some of them were entered rapidly in clinical trials promote that IDO1 is useful targets in cancer treatment. As small molecules, their preparation of library for bioactivity assay is not demanding and complicated. On the other hand, IDO1 inhibitors usually combine with immune checkpoint inhibitors to promote their effect and avoid resistance. Therefore, IDO1 is increasingly attractive to medicinal chemists. In only ten years, seven compounds have advanced to clinical trials. This breakthrough contributes to a promising possibility for commercialization in the near future. Besides, suitable IDO1 inhibitors also facilitate the elucidation of IDO1 active site. This brings about a positive potential for the design and development of new IDO1 inhibitors, helping to find other structure motifs. As one target in immunotherapy, IDO1 inhibitors were special attendance hoping to bring one of the first oral drugs in the immuno-oncology field. Epacadostat was entered rapidly to the clinical trial and is under the phase III trial now.

Recently, epacadostat failed on some phase III clinical trials and posed new challenges to IDO1 projects. Initially, TDO is considered as a modulation enzyme for tryptophan concentration in the liver. Therefore, most of the IDO1 clinical compounds were designed for IDO1 selectivity over TDO. However, nowadays, the connection between TDO and some cancer cell lines has been demonstrated. Because IDO1 and TDO both carry out the first step of the kynurenine pathway, the failure of epacadostat in clinical trials raises a question mark over the role of TDO: do cancer cells use TDO instead of IDO1 to overcome IDO1 inhibitors or not? The answer will affect on design strategy: selective IDO1 inhibition or dual IDO1/TDO inhibition.

REFERENCES

[1] Bray F, Ferlay J, Soerjomataram I, Siegel RL, Torre LA, Jemal A. Global cancer statistics 2018: GLOBOCAN estimates of incidence and mortality worldwide for 36 cancers in 185 countries. *CA: a cancer journal for clinicians.* 2018;68(6):394-424. doi: 10.3322/caac.21492.

[2] Kim R, Emi M, Tanabe K. Cancer immunoediting from immune surveillance to immune escape. *Immunol.* 2007;121(1):1-14. doi: 10.1111/j.1365-2567.2007.02587.x.

[3] Schreiber RD, Old LJ, Smyth MJ. Cancer Immunoediting: Integrating Immunity's Roles in Cancer Suppression and Promotion. *Science*. 2011;331(6024):1565-70. doi: 10.1126/science.1203486.

[4] Dunn GP, Bruce AT, Ikeda H, Old LJ, Schreiber RD. Cancer immunoediting: from immunosurveillance to tumor escape. *Nat Immunol.* 2002;3(11):991-8. doi: 10.1038/ni1102-991.

[5] Mittal D, Gubin MM, Schreiber RD, Smyth MJ. New insights into cancer immunoediting and its three component phases—elimination,

equilibrium and escape. *Curr Op Immunol.* 2014;27:16-25. doi: https://doi.org/10.1016/j.coi.2014.01.004.

[6] Oxenkrug GF. Metabolic syndrome, age-associated neuroendocrine disorders, and dysregulation of tryptophan—kynurenine metabolism. *Ann New York Ac Sci.* 2010;1199(1):1-14. doi: 10.1111/j.1749-6632.2009.05356.x.

[7] Dounay AB, Tuttle JB, Verhoest PR. Challenges and opportunities in the discovery of new therapeutics targeting the kynurenine pathway. *J Med Chem.* 2015;58(22):8762-82. doi: 10.1021/acs.jmedchem.5b00461.

[8] Pilotte L, Larrieu P, Stroobant V, Colau D, Dolušić E, Frédérick R, et al. Reversal of tumoral immune resistance by inhibition of tryptophan 2, 3-dioxygenase. *Proc Natio Ac Sci.* 2012;109(7):2497-502. doi: 10.1073/pnas.1113873109.

[9] Opitz CA, Litzenburger UM, Sahm F, Ott M, Tritschler I, Trump S, et al. An endogenous tumour-promoting ligand of the human aryl hydrocarbon receptor. *Nature.* 2011;478:197. doi: 10.1038/nature 1049.

[10] Yeung AW, Terentis AC, King NJ, Thomas SR. Role of indoleamine 2, 3-dioxygenase in health and disease. *Clin Sci.* 2015;129(7):601-72. doi: 10.1042/CS20140392.

[11] Sugimoto H, Oda S-i, Otsuki T, Hino T, Yoshida T, Shiro Y. Crystal structure of human indoleamine 2, 3-dioxygenase: catalytic mechanism of O2 incorporation by a heme-containing dioxygenase. *Proc Nati Aca Sci.* 2006;103(8):2611-6. doi: 10.1073/pnas.0508996 103.

[12] Tojo S, Kohno T, Tanaka T, Kamioka S, Ota Y, Ishii T, et al. Crystal structures and structure–activity relationships of imidazothiazole derivatives as IDO1 inhibitors. *ACS Med Chem Lett.* 2014;5(10): 1119-23. doi: 10.1021/ml500247w.

[13] Peng YH, Ueng SH, Tseng CT, Hung MS, Song JS, Wu JS, et al. Important hydrogen bond networks in indoleamine 2, 3-dioxygenase 1 (IDO1) inhibitor design revealed by crystal structures of

imidazoleisoindole derivatives with IDO1. *J Med Chem.* 2015;59(1):282-93. doi: 10.1021/acs.jmedchem.5b01390.

[14] Wu Y, Xu T, Liu J, Ding K, Xu J. Structural insights into the binding mechanism of IDO1 with hydroxylamidine based inhibitor INCB14943. *Biochem Biophys Res Commun.* 2017;487(2):339-43. doi: 10.1016/j.bbrc.2017.04.061.

[15] Lewis-Ballester A, Pham KN, Batabyal D, Karkashon S, Bonanno JB, Poulos TL, et al. Structural insights into substrate and inhibitor binding sites in human indoleamine 2,3-dioxygenase 1. *Nat Commun.* 2017;8(1):1693. doi: 10.1038/s41467-017-01725-8.

[16] Littlejohn TK, Takikawa O, Truscott RJ, Walker MJ. Asp274 and his346 are essential for heme binding and catalytic function of human indoleamine 2,3-dioxygenase. *J Biol Chem.* 2003;278(32):29525-31. doi: 10.1074/jbc.M301700200.

[17] Brant MG, Goodwin-Tindall J, Stover KR, Stafford PM, Wu F, Meek AR, et al. Identification of Potent Indoleamine 2,3-Dioxygenase 1 (IDO1) Inhibitors Based on a Phenylimidazole Scaffold. *ACS Med Chem Lett.* 2018;9(2):131-6. doi: 10.1021/acsmedchemlett.7b00488.

[18] Nelp MT, Kates PA, Hunt JT, Newitt JA, Balog A, Maley D, et al. Immune-modulating enzyme indoleamine 2,3-dioxygenase is effectively inhibited by targeting its apo-form. *Proc Natio Aca Sci.* 2018;115(13):3249-54. doi: 10.1073/pnas.1719190115.

[19] Uyttenhove C, Pilotte L, Théate I, Stroobant V, Colau D, Parmentier N, et al. Evidence for a tumoral immune resistance mechanism based on tryptophan degradation by indoleamine 2,3-dioxygenase. *Nat Med.* 2003;9(10):1269-74. doi: 10.1038/nm934.

[20] Löb S, Königsrainer A, Rammensee H-G, Opelz G, Terness P. Inhibitors of indoleamine-2,3-dioxygenase for cancer therapy: can we see the wood for the trees? *Nat Rev Cancer.* 2009;9:445. doi: 10.1038/nrc2639.

[21] Moon YW, Hajjar J, Hwu P, Naing A. Targeting the indoleamine 2,3-dioxygenase pathway in cancer. *J Immunother Cancer.* 2015;3:51. doi: 10.1186/s40425-015-0094-9.

[22] Munn DH, Shafizadeh E, Attwood JT, Bondarev I, Pashine A, Mellor AL. Inhibition of T Cell Proliferation by Macrophage Tryptophan Catabolism. *J Exp Med.* 1999;189(9):1363-72. doi: 10.1084/jem.189.9.1363.

[23] Geon Kook Lee HJP, Megan Macleod, Phillip Chandler, David H Munn, and Andrew L Mellor. Tryptophan deprivation sensitizes activated T cells to apoptosis prior to cell division. *Immunol.* 2002;107(4):452-60. doi: 10.1046/j.1365-2567.2002.01526.x.

[24] Munn DH, Sharma MD, Baban B, Harding HP, Zhang Y, Ron D, et al. GCN2 kinase in T cells mediates proliferative arrest and anergy induction in response to indoleamine 2,3-dioxygenase. *Immunity.* 2005;22(5):633-42. doi: 10.1016/j.immuni.2005.03.013.

[25] Eleftheriadis T, Pissas G, Antoniadi G, Liakopoulos V, Stefanidis I. Indoleamine 2,3-dioxygenase depletes tryptophan, activates general control non-derepressible 2 kinase and down-regulates key enzymes involved in fatty acid synthesis in primary human CD4+ T cells. *Immunol.* 2015;146(2):292-300. doi: 10.1111/imm.12502.

[26] Frumento G, Rotondo R, Tonetti M, Damonte G, Benatti U, Ferrara GB. Tryptophan-derived Catabolites Are Responsible for Inhibition of T and Natural Killer Cell Proliferation Induced by Indoleamine 2,3-Dioxygenase. *J Exp Med.* 2002;196(4):459-68. doi: 10.1084/jem.20020121.

[27] Molano A, Illarionov PA, Besra GS, Putterman C, Porcelli SA. Modulation of invariant natural killer T cell cytokine responses by indoleamine 2,3-dioxygenase. *Immunol Lett.* 2008;117(1):81-90. doi: 10.1016/j.imlet.2007.12.013.

[28] Wang D, Saga Y, Mizukami H, Sato N, Nonaka H, Fujiwara H, et al. Indoleamine-2,3-dioxygenase, an immunosuppressive enzyme that inhibits natural killer cell function, as a useful target for ovarian cancer therapy. *Int J Oncol.* 2012;40(4):929-34. doi: 10.3892/ijo.2011.1295.

[29] Sharma MD, Baban B, Chandler P, Hou DY, Singh N, Yagita H, et al. Plasmacytoid dendritic cells from mouse tumor-draining lymph nodes directly activate mature Tregs via indoleamine 2,3-

dioxygenase. *J Clin Invest.* 2007;117(9):2570-82. doi: 10.1172/JCI31911.

[30] Metz R, Rust S, Duhadaway JB, Mautino MR, Munn DH, Vahanian NN, et al. IDO inhibits a tryptophan sufficiency signal that stimulates mTOR: A novel IDO effector pathway targeted by D-1-methyl-tryptophan. *Oncoimmunol.* 2012;1(9):1460-8. doi: 10.4161/onci.21716.

[31] Altman A, Villalba M. Protein Kinase Cθ (PKC&theta): A Key Enzyme in T Cell Life and Death. *J Biochem.* 2002;132(6):841-6. doi: 10.1093/oxfordjournals.jbchem.a003295.

[32] Grohmann U, Fallarino F, Puccetti P. Tolerance, DCs and tryptophan: much ado about IDO. *Trend Immunol.* 2003;24(5):242-8. doi: 10.1016/S1471-4906(03)00072-3.

[33] Wang C, Ye Z, Kijlstra A, Zhou Y, Yang P. Activation of the aryl hydrocarbon receptor affects activation and function of human monocyte-derived dendritic cells. *Clin Exp Immunol.* 2014;177(2):521-30. doi: 10.1111/cei.12352.

[34] Wagage S, John B, Krock BL, Hall AOH, Randall LM, Karp CL, et al. The Aryl Hydrocarbon Receptor Promotes IL-10 Production by NK Cells. *J Immunol.* 2014;192(4):1661-70. doi: 10.4049/jimmunol.1300497.

[35] Ulrike M. Litzenburger CAO, Felix Sahm, Katharina J. Rauschenbach, Saskia Trump, Marcus Winter, Martina Ott, Katharina Ochs, Christian Lutz, Xiangdong Liu, Natasa Anastasov, Irina Lehmann, Thomas Höfer, Andreas von Deimling, Wolfgang Wick, and Michael Platten. Constitutive IDO expression in human cancer is sustained by an autocrine signaling loop involving IL-6, STAT3 and the AHR. *Oncotarget.* 2014;5:1038-51. doi:

[36] Apetoh L, Quintana FJ, Pot C, Joller N, Xiao S, Kumar D, et al. The aryl hydrocarbon receptor interacts with c-Maf to promote the differentiation of type 1 regulatory T cells induced by IL-27. *Nat Immunol*. 2010;11:854. doi: 10.1038/ni.1912.

[37] Wainwright DA, Dey M, Chang A, Lesniak MS. Targeting Tregs in Malignant Brain Cancer: Overcoming IDO. *Front Immunol.* 2013;4:116. doi: 10.3389/fimmu.2013.00116.

[38] Brandacher G, Perathoner A, Ladurner R, Schneeberger S, Obrist P, Winkler C, et al. Prognostic value of indoleamine 2,3-dioxygenase expression in colorectal cancer: effect on tumor-infiltrating T cells. *Clin Cancer Res.* 2006;12(4):1144-51. doi: 10.1158/1078-0432.Ccr-05-1966.

[39] Ben-Haj-Ayed A, Moussa A, Ghedira R, Gabbouj S, Miled S, Bouzid N, et al. Prognostic value of indoleamine 2,3-dioxygenase activity and expression in nasopharyngeal carcinoma. *Immunol Lett.* 2016;169:23-32. doi: 10.1016/j.imlet.2015.11.012.

[40] Godin-Ethier J, Hanafi L-A, Piccirillo CA, Lapointe R. Indoleamine 2,3-Dioxygenase Expression in Human Cancers: Clinical and Immunologic Perspectives. *Clin Cancer Res.* 2011;17(22):6985-91. doi: 10.1158/1078-0432.Ccr-11-1331.

[41] Taylor MW FG. Relationship between interferon-gamma, indoleamine 2,3-dioxygenase, and tryptophan catabolism. *FASEB J.* 1991;5(11):2516-22. doi:

[42] Balachandran VP, Cavnar MJ, Zeng S, Bamboat ZM, Ocuin LM, Obaid H, et al. Imatinib potentiates antitumor T cell responses in gastrointestinal stromal tumor through the inhibition of Ido. *Nat Med.* 2011;17:1094. doi: 10.1038/nm.2438.

[43] Creelan BC, Antonia S, Bepler G, Garrett TJ, Simon GR, Soliman HH. Indoleamine 2,3-dioxygenase activity and clinical outcome following induction chemotherapy and concurrent chemoradiation in Stage III non-small cell lung cancer. *Oncoimmunol.* 2013;2(3): e23428. doi: 10.4161/onci.23428.

[44] Liu H, Shen Z, Wang Z, Wang X, Zhang H, Qin J, et al. Increased expression of IDO associates with poor postoperative clinical outcome of patients with gastric adenocarcinoma. *Scientific Reports.* 2016;6:21319. doi: 10.1038/srep21319.

[45] Dolušić E, Larrieu P, Blanc S, Sapunaric F, Pouyez J, Moineaux L, et al. Discovery and preliminary SARs of keto-indoles as novel

indoleamine 2,3-dioxygenase (IDO) inhibitors. *Eur J Med Chem.* 2011;46(7):3058-65. doi: 10.1016/j.ejmech.2011.02.049.

[46] Dolušić E, Larrieu P, Blanc S, Sapunaric F, Norberg B, Moineaux L, et al. Indol-2-yl ethanones as novel indoleamine 2,3-dioxygenase (IDO) inhibitors. *Bioorg Med Chem.* 2011;19(4):1550-61. doi: 10.1016/j.bmc.2010.12.032.

[47] Coluccia A, Passacantilli S, Famiglini V, Sabatino M, Patsilinakos A, Ragno R, et al. New Inhibitors of Indoleamine 2,3-Dioxygenase 1: Molecular Modeling Studies, Synthesis, and Biological Evaluation. *J Med Chem.* 2016;59(21):9760-73. doi: 10.1021/acs.jmedchem.6b00718.

[48] Lob S, Konigsrainer A, Schafer R, Rammensee H-G, Opelz G, Terness P. Levo- but not dextro-1-methyl tryptophan abrogates the IDO activity of human dendritic cells. *Blood.* 2008;111(4):2152-4. doi: 10.1182/blood-2007-10-116111.

[49] Hou D-Y, Muller AJ, Sharma MD, DuHadaway J, Banerjee T, Johnson M, et al. Inhibition of Indoleamine 2,3-Dioxygenase in Dendritic Cells by Stereoisomers of 1-Methyl-Tryptophan Correlates with Antitumor Responses. *Cancer Res.* 2007;67(2):792-801. doi: 10.1158/0008-5472.Can-06-2925.

[50] iTeos, Pyrrolidine-2,5-dion derivatives, pharmaceutical compositions and methods for use as IDO1 inhibitors. WO2015173764A1. 2015.

[51] Crosignani S, Bingham P, Bottemanne P, Cannelle H, Cauwenberghs S, Cordonnier M, et al. Discovery of a Novel and Selective Indoleamine 2,3-Dioxygenase (IDO-1) Inhibitor 3-(5-Fluoro-1H-indol-3-yl)pyrrolidine-2,5-dione (EOS200271/PF-06840003) and Its Characterization as a Potential Clinical Candidate. *J Med Chem.* 2017;60(23):9617-29. doi: 10.1021/acs.jmedchem.7b00974.

[52] Pfizer i, Combinations comprising a pyrrolidine-2,5-dion IDO1 inhibito and an anti-body. WO2016181349A1. 2016.

[53] Pradhan N, Paul S, Deka SJ, Roy A, Trivedi V, Manna D. Identification of Substituted 1H-Indazoles as Potent Inhibitors for Immunosuppressive Enzyme Indoleamine 2,3-Dioxygenase 1. *ChemistrySelect.* 2017;2(20):5511-7. doi: 10.1002/slct.201700906.

[54] Qian S, He T, Wang W, He Y, Zhang M, Yang L, et al. Discovery and preliminary structure–activity relationship of 1H-indazoles with promising indoleamine-2,3-dioxygenase 1 (IDO1) inhibition properties. *Bioorg Med Chem.* 2016;24(23):6194-205. doi: 10.1016/j.bmc.2016.10.003.

[55] Yang L, Chen Y, He J, Njoya EM, Chen J, Liu S, et al. 4,6-Substituted-1H-Indazoles as potent IDO1/TDO dual inhibitors. *Bioorg Med Chem.* 2019;27(6):1087-98. doi: 10.1016/j.bmc.2019.02.014.

[56] Kumar S, Jaller D, Patel B, LaLonde JM, DuHadaway JB, Malachowski WP, et al. Structure Based Development of Phenylimidazole-Derived Inhibitors of Indoleamine 2,3-Dioxygenase. *J Med Chem.* 2008;51(16):4968-77. doi: 10.1021/jm800512z.

[57] Tomek P, Palmer BD, Flanagan JU, Sun C, Raven EL, Ching L-M. Discovery and evaluation of inhibitors to the immunosuppressive enzyme indoleamine 2,3-dioxygenase 1 (IDO1): Probing the active site-inhibitor interactions. *Eur J Med Chem.* 2017;126:983-96. doi: 10.1016/j.ejmech.2016.12.029.

[58] Genetics N, Imidazole derivatives as IDO inhibitors. WO2011056652A1. 2011.

[59] Rohrig UF, Majjigapu SR, Vogel P, Zoete V, Michielin O. Challenges in the Discovery of Indoleamine 2,3-Dioxygenase 1 (IDO1) Inhibitors. *J Med Chem.* 2015;58(24):9421-37. doi: 10.1021/acs.jmedchem.5b00326.

[60] Geneticks N, *Fused imidazole derivatives useful as IDO inhibitors.* WO2012142237A. 2012.

[61] Genetics N, *Tricyclic compounds as inhibitos of immunosuppression mediated by tryptophan metabolization.* WO2014159248A1. 2014.

[62] Group CTTp, *Tricyclic compound serving as immunomodulator.* WO2017140274A1. 2017.

[63] PLC RP, *6,7-heterocyclic fused 5H-pyrrolo[1,2-C]imidazole derivatives and their use as indoleamine 2,3-diexygenase (IDO)*

and/or tryptophan 2,3-dioxygenase (TDO2) modulators. WO2016059412A1. 2016.
[64] Sciences Sl, Fused imidazole derivatives as IDO/TDO inhibitors. WO2017075341A1. 2017.
[65] Merck, Cyclohexyl-ethyl substituted diaza-and triaza-tricyclic compounds as indole-amine-2,3-dioxygenase (IDO) antagonists for the treatment of cancer. WO2016037026A1. 2016.
[66] Pharma Hi, Heterocycles useful as IDO and TDO inhibitors. WO2016165613A1. 2016.
[67] Kumar S, Malachowski WP, DuHadaway JB, LaLonde JM, Carroll PJ, Jaller D, et al. Indoleamine 2,3-Dioxygenase Is the Anticancer Target for a Novel Series of Potent Naphthoquinone-Based Inhibitors. *J Med Chem.* 2008;51(6):1706-18. doi: 10.1021/jm7014155.
[68] Research Lifm, Novel IDO inhibitors and method of use thereof. WO2008115804A1. 2008.
[69] Colubia TuoB, Substituted quinone indoleamine 2,3-dioxygenase (IDO) inhibitors and syntheses and uses thereof. WO2008052352A1. 2008.
[70] Pan L, Zheng Q, Chen Y, Yang R, Yang Y, Li Z, et al. Design, synthesis and biological evaluation of novel naphthoquinone derivatives as IDO1 inhibitors. *Eur J Med Chem.* 2018;157:423-36. doi: 10.1016/j.ejmech.2018.08.013.
[71] Paul S, Roy A, Deka SJ, Panda S, Srivastava GN, Trivedi V, et al. Synthesis and evaluation of oxindoles as promising inhibitors of the immunosuppressive enzyme indoleamine 2,3-dioxygenase 1. *Medchemcomm.* 2017;8(8):1640-54. doi: 10.1039/c7md00226b.
[72] Cheng MF, Hung MS, Song JS, Lin SY, Liao FY, Wu MH, et al. Discovery and structure-activity relationships of phenyl benzenesulfonylhydrazides as novel indoleamine 2,3-dioxygenase inhibitors. *Bioorg Med Chem Lett.* 2014;24(15):3403-6. doi: 10.1016/j.bmcl.2014.05.084.
[73] Lin SY, Yeh TK, Kuo CC, Song JS, Cheng MF, Liao FY, et al. Phenyl Benzenesulfonylhydrazides Exhibit Selective Indoleamine

2,3-Dioxygenase Inhibition with Potent in Vivo Pharmacodynamic Activity and Antitumor Efficacy. *J Med Chem.* 2016;59(1):419-30. doi: 10.1021/acs.jmedchem.5b01640.

[74] Malachowski WP, Winters M, DuHadaway JB, Lewis-Ballester A, Badir S, Wai J, et al. O-alkylhydroxylamines as rationally-designed mechanism-based inhibitors of indoleamine 2,3-dioxygenase-1. *Eur J Med Chem.* 2016;108:564-76. doi: 10.1016/j.ejmech.2015.12.028.

[75] Yue EW, Douty B, Wayland B, Bower M, Liu X, Leffet L, et al. Discovery of Potent Competitive Inhibitors of Indoleamine 2,3-Dioxygenase with in Vivo Pharmacodynamic Activity and Efficacy in a Mouse Melanoma Model. *J Med Chem.* 2009;52(23):7364-7. doi: 10.1021/jm900518f.

[76] Yue EW, Sparks R, Polam P, Modi D, Douty B, Wayland B, et al. INCB24360 (Epacadostat), a Highly Potent and Selective Indoleamine-2,3-dioxygenase 1 (IDO1) Inhibitor for Immuno-oncology. *ACS Med Chem Lett.* 2017;8(5):486-91. doi: 10.1021/acsmedchemlett.6b00391.

[77] Co. I, *1,2,5-oxadiazoles as inhibitors of indoleamine 2,3-dioxygenase.* WO2010005958A2. 2010.

[78] Bioscience F, *Immunoregulatory agents.* WO2016073770A1. 2016.

[79] Biosciences F, *Immunoregulatory agents.* WO2016073738A2. 2016.

[80] Bioseciences F, *Immunoregulatory agents.* WO2016073774A2. 2016.

[81] BMS, *Inhibitors of indoleamine 2,3-dioxygenase and methods of their use.* WO2018039512A1. 2018.

In: Advances in Medicinal Chemistry … ISBN: 978-1-53616-368-1
Editor: E. Ferreira da Silva-Júnior

Chapter 7

AN UPDATE ON EG5 KINESIN INHIBITORS FOR THE TREATMENT OF CANCER

Paolo Guglielmi[1], PhD, Daniela Secci[1], PhD, Giulia Rotondi[1] and Simone Carradori[2], PhD

[1]Department of Drug Chemistry and Technology, Sapienza University of Rome, Rome, Italy
[2]Department of Pharmacy, "G. D'Annunzio" University of Chieti-Pescara, Chieti, Italy

ABSTRACT

Cancer is one of the most crucial worldwide health problems, affecting millions of people every year. Therapy against this multi-etiological illness takes advantage from compounds able to attack targets playing a pivotal role in the tumor cell organization. Among the drugs used in the clinical treatment of cancer, there are the antimitotic agents, which induce the mitotic arrest and subsequent cell death by binding to tubulin and limiting microtubule dynamics (microtubule-targeting agents). However, the direct interaction with microtubules, which are crucial for cell division, motility, and several other functions, also accounts for their toxicity affecting their therapeutic window negatively. Nevertheless, another opportunity can be pursued to obtain anticancer drugs, taking advantage from proteins which directly interact with microtubules.

Eg5 (also known as kinesin spindle protein – KSP) is one of these proteins and belongs to kinesins, a large family of motor proteins involved in cytokinesis, intracellular vesicle and organelle transport, and mitosis, by unidirectionally moving along microtubules using ATP hydrolysis to provoke motile force for the movement. The inhibition of Eg5 causes a characteristic monoastral spindle phenotype and mitotic arrest. Eg5 is overexpressed in a large plethora of malignancies such as leukemia, breast, lung, ovarian, bladder, and pancreatic cancer while being almost not detected in non-proliferative tissues. Taking into account all these issues, novel heterocyclic compounds able to target this enzyme could affect mitosis without disrupting microtubule dynamics. Nowadays, some inhibitors targeting Eg5 have entered clinical trials either as monotherapies or in combination with other drugs. In this chapter, the researches published in the years 2013-2019 regarding kinesin spindle protein inhibitors will be discussed.

Keywords: Eg5 kinesin, KSP, inhibitors, clinical development, cancer

INTRODUCTION

Biological Properties and Functions of Kinesin Eg5

Antimitotic agents represent an important class of drugs for the treatment of cancer [1]. To date the antimitotic drugs available in the clinic (taxanes, vinca alkaloids, and colchicine) target tubulin [2], the structural constituent of microtubules that take part in cell growth and division, motility and intracellular trafficking [3]. These drugs inhibit microtubule dynamics at low concentrations, causing an aberrant mitotic spindle formation, cell cycle arrest, and cell death. Nevertheless, their clinical use has been limited by various side effects, such as nausea, vomiting, diarrhea, myelotoxicity, mostly neurotoxicity, and drug resistance [1]. Alternative strategies to block the cell division include targets that are involved in mitosis-specific roles. The kinesin class (microtubule-dependent motor proteins) plays critical roles in the cell division, including the establishment of spindle bipolarity [4]. The kinesin superfamily consists of over 650 distinct proteins divided into 14 subfamilies according to their sequence similarity and functions. Among these, the most studied is Eg5, also known as Kiff11 or kinesin spindle protein, KSP [5].

Eg5 is a bipolar homotetrameric protein with pairs of *N*-terminal plus-end directed motor domains on either end/at the opposite end [6, 7]. Briefly, every of this subunit contains a kinesin motor domain of 350 aa [8] consisting of two binding domains [5], the ATP and the microtubule binding pocket, an internal stalk domain, that forms a coiled coil [9], and a tail domain. The kinesin motor domain ensures ATPase activity, through that generates motile force for the movement [1, 8] and the microtubule binding property [8], whereas the stalk domain assists in the formation of the tetrameric structure [6]. This structural organization, different from other kinesins that function as dimers [9], allows Eg5 to crosslink and push spindle poles apart via the sliding of antiparallel microtubules [8]. Indeed, during mitotic prophase, Eg5 motors move microtubules that extend from each of the duplicated centrosomes, driving bipolar spindle formation. This Eg5 capability contributes to the formation, maintenance, and elongation of a bipolar spindle. Therefore, blocking Eg5 leads to an efficacious arrest of the cell cycle [9].

Eg5 has obtained much interest in cancer topic since its overexpression has been observed in a large variety of tumors, including those of the breast, colon, lung, ovary and its depletion or gene silencing, using different methods, leads to apoptosis in several tumor cell lines [4]. Also, Eg5 has been correlated to valuable prognosis in certain types of tumors, as demonstrated by recent studies [10, 11]. Therefore, Eg5 has been identified as a promising target for cancer therapy.

The main classes of Eg5 inhibitors are usually ATP-uncompetitive inhibitors that bind to an allosteric site located in the motor domain and formed by helix $\alpha2$, loop L5 and helix $\alpha3$ [2, 7] trapping the motor in a state with no ADP release and a weak affinity for microtubules [12]. Interestingly Loop L5, although it is present in all kinesins, is longer in KSP than in the other classes, contributing to the high selectivity of these drugs [13]. Recently, new binding sites have been discovered with two new candidates: FCTP and BI8. The molecule FCPT seems to bind to a site near Loop L5 but with a competitive mechanism to ATP and the BI8 agent appears to bind to a novel pocket in Eg5 formed by helix α4 and α6 [7].

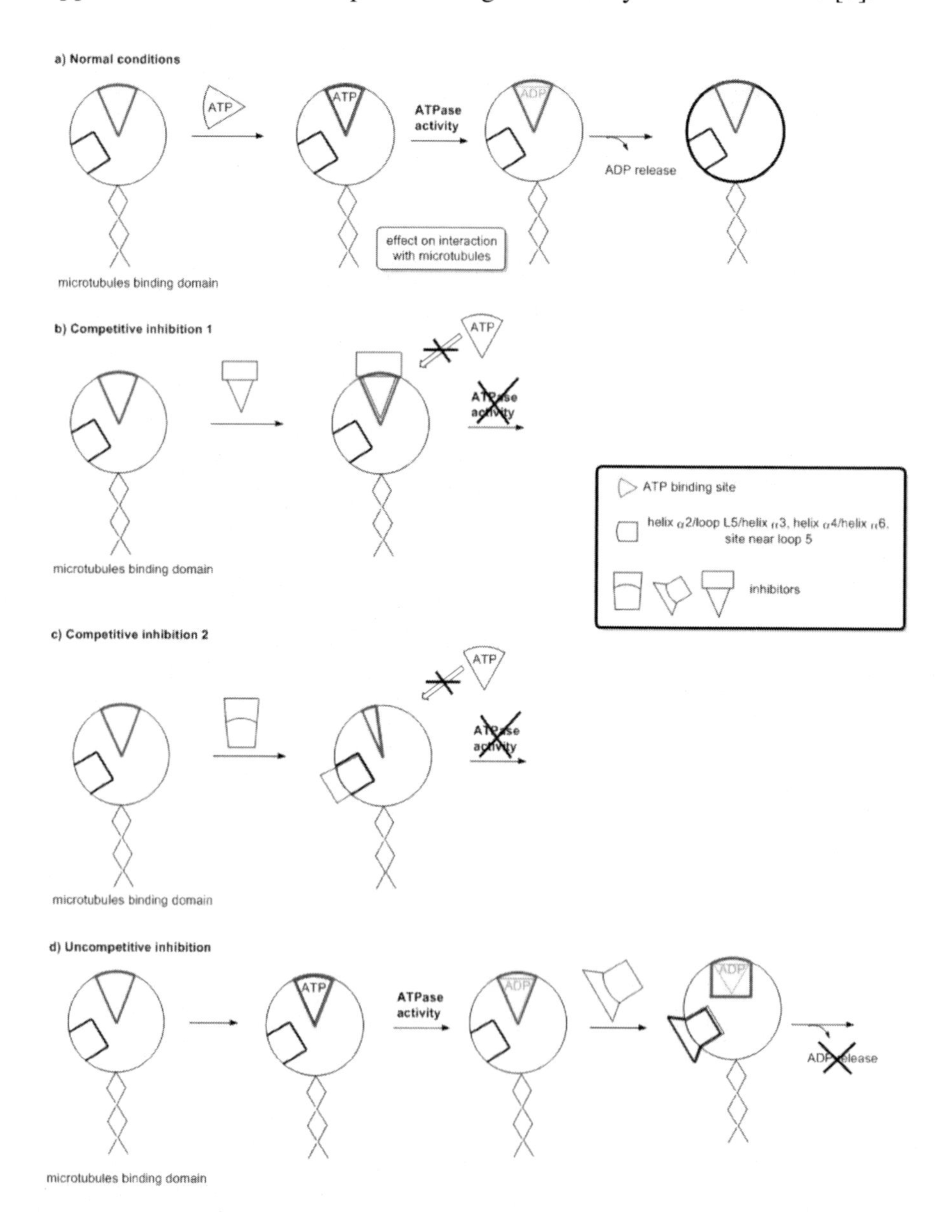

Figure 1. Type of Eg5 kinesin inhibition involving ATPase active site.

The most significant development of inhibitors, able to bind to allosteric sites (often, but not always, obtaining uncompetitive inhibition), instead of the active site endowed with ATPase activity (giving ATP-competitive inhibition, Figure 1, a-d), is related to the presence of this catalytic domain in other proteins, which could negatively affect the selectivity.

To date, several Eg5 inhibitors have entered in clinical phase, but they have not prompted a significant clinical response in cancer patients [14]. Indeed, although nine inhibitors reached the clinic, only filanesib (ARRY-520, Figure 2) exhibited activity in patients with relapsed or refractory multiple myeloma [5]. This unfortunate and disappointing result has been correlated to numerous reasons: a poor overlap between preclinical and clinical data, the upregulation of compensatory pathways (e.g., KIF15, kinesin-12) and the activation of resistance mechanisms [5, 14].

Eg5 Inhibitors

Eg5 Inhibitors Entered in Clinical Trials

In the attempt to evaluate how small molecules could perturb specific protein functions, Mayer and colleagues [15] performed a high throughput phenotypic screen, observing that one among the 139 tested compounds produced cells containing monoastral microtubule array surrounded by a ring of chromosomes. Due to this monoastral phenotype of the spindle, the 1,4-dihydropyrimidine-based molecule eliciting this effect was named Monastrol (Figure 2, a), the first selective inhibitor of mitotic kinesin discovered.

Monastrol (Figure 2, a) induced reversible mitotic arrest, acting as an allosteric inhibitor of Eg5 and blocking ADP release through the establishment of an Eg5-ADP-monastrol ternary complex [16–18]. From this initial discovery, many scaffolds have been investigated for their ability to bind to Eg5 [19], leading to highly potent and selective inhibitors also endowed with significant antitumor activity, that in some instances allowed their entrance in clinical phase trials (Figure 2) [5].

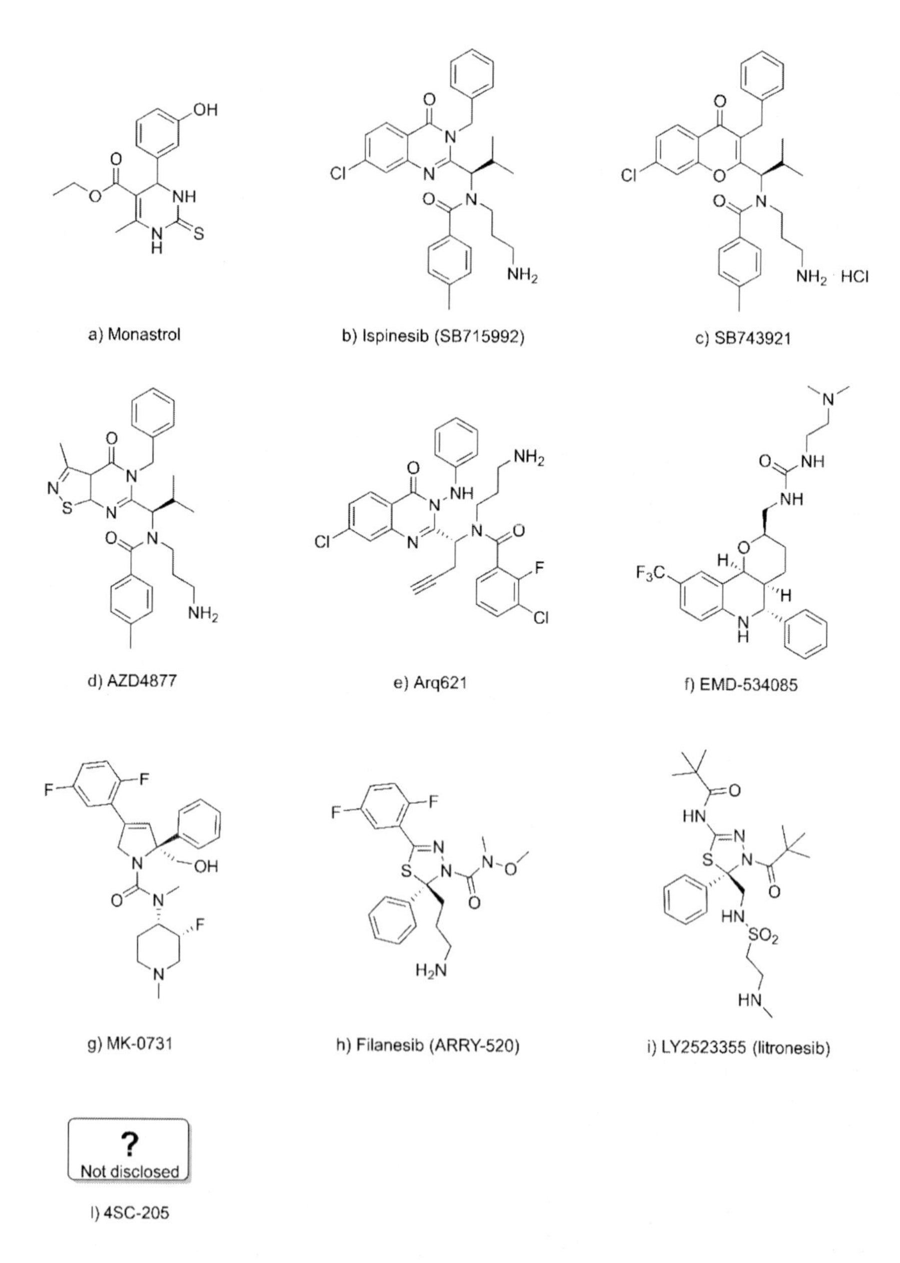

Figure 2. Structures of compounds entered in clinical trials.

Ispinesib (also known as SB-715992, Figure 2, b), the first Eg5 inhibitor exhibiting *in vivo* anti-tumor activity, was evaluated in 16 different clinical trials as single drug or in combination with other compounds for the treatment of different tumor types [20]. Its discovery started from the optimization of the close analog CK0106023 (Figure 3, a), that in tumor-bearing mice showed antitumor activity comparable to or exceeding that of paclitaxel [21, 22]. Similar to Monastrol, ispinesib works as a selective and allosteric inhibitor of Eg5 binding to the α2/L5/α3 region [23]. Ispinesib demonstrated broad *in vivo* antitumor activity along with reasonable overall drug-like properties; however, it moderately inhibits hERG, cytochrome P450 (CYP) 3A4 and exhibited high toxicity rates at the doses used in a large scale animal study, reducing the therapeutic window [24].

The results showed by ispinesib encouraged the development of further Eg5 inhibitors based on the similar scaffold, leading to SB-743921, AZD-4877, and Arq621. SB-743921 (Figure 2, c) was obtained through the replacement of quinazolinone core with a chromen-4-one one, exhibiting improvement of activity in enzymatic and cellular assays, although retaining binding ability against hERG channel and CYP3A4 inhibition [21, 25].

Taking into account the inhibitors developed at that time, AstraZeneca® researchers designed and synthesized a series of Eg5 inhibitors containing bicyclic scaffold, where the chloro-phenyl ring of ispinesib and SB-743921 was replaced with 5- and 6-membered heterocyclic rings [26]. All the obtained compounds were evaluated as Eg5 inhibitors, and most of them showed sub-micromolar/nanomolar inhibitory activity. AZD4877 (Figure 2, d) displayed low nanomolar inhibitory activity in enzymatic (Eg5) as well as cellular assays (Colo205 cell line), along with favorable pharmacokinetic profile and appreciable *in vivo* efficacy. These features led this inhibitor to be considered a good candidate for clinical evaluation, participating in 6 different clinical trials for different tumor types [27–29].

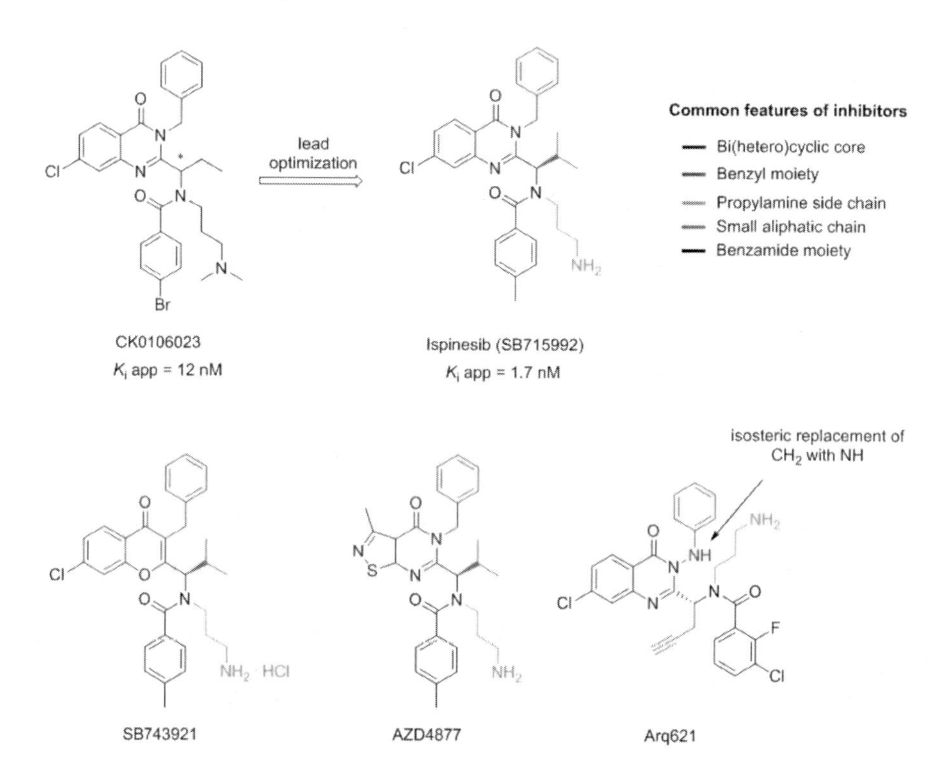

Figure 3. Lead optimization of CK0106023 to obtain Ispinesib and common features of Ispinesib with analog compounds.

Few details are available about the molecular development of inhibitor Arq621 (Figure 2, e). From the structural point of view, this compound shares the quinazolinone bicyclic core with Ispinesib and the propylamine side chain with SB743921 and AZD4877 (Figure 3). However, few differences consisting in the isosteric replacement of the methylene benzyl moiety with an amine one (aniline) together with the exchange of aliphatic isopropyl chain with the one containing propargyl group, can be observed. Arq621 showed anti-tumor activity with potencies in the low nanomolar range against colon, NSCLC, gastric, and hematologic cancer cell lines. In a clinical trial enrolling 48 patients with a solid tumor, this drug was well-tolerated without evidence of bone marrow toxicity and DNA damage [30].

EMD 534085 (Figure 2, f) is an Eg5 directed inhibitor developed by Merck Serono® Laboratories and obtained from the optimization of hexahydropyranoquinoline (HHPQs) core (for more information about

SAR studies see references [21, 31]). EMD 534085 exhibited very low nanomolar affinity towards Eg5 (IC_{50} = 8 nM), also limiting the proliferation of HCT116 cells effectively (IC_{50} = 30 nM). In a phase I study involving 44 patients with advanced solid tumors or lymphoma, EMD 53085 exhibited well tolerability but, at the same time, scarce antitumor activity in monotherapy, leading to the end of its development [32]. MK-0731 is a 2,4-diaryl-2,5-dihydropyrrole-based molecule discovered at the Merck® Research Laboratories [33–40] through the optimization involving four different core scaffolds (Figure 2, g and Figure 4). The hit-compound, found through high-throughput screening, underwent different modifications and changes with the purpose to ameliorate enzymatic and cellular activity. At the same time, attempts were performed to obtain the reduction of hERG binding and solubility improvement [21]. In a phase I trial involving patients with solid tumors, MK-0731 was well tolerated with a maximum tolerated dose of 17 mg/m^2, after which neutropenia became predominant. At this dose, the treatment resulted in prolonged disease stabilization rather than tumor remission [41].

Among the changes attempted in the development of the MK-0731 analogs, the insertion of propylamine chain was useful in the enhancement of solubility and potency, although augmented affinity for hERG and susceptibility to Pgp-efflux were detected. Nevertheless, this structural feature was kept in the 1,3,4-thiadiazole derivative ARRY-520 (Figure 2, h; and Figure 4), another Eg5 inhibitor, whose development started from a pyrrolinone hit by way of oxadiazoline optimization, up to the discovery of thiadiazoline core [42]. This drug displayed high selectivity against other kinesin motor proteins, good drug-like properties, along with the absence of CYP3A4 inhibition. ARRY-520 was effective against 13 of the 16 different histological tumor xenograft models used to evaluate its *in vivo* activity, exhibiting a preference for hematological cancers (seven of the eight clinical trials involving ARRY-520 were on hematological malignancies) [43].

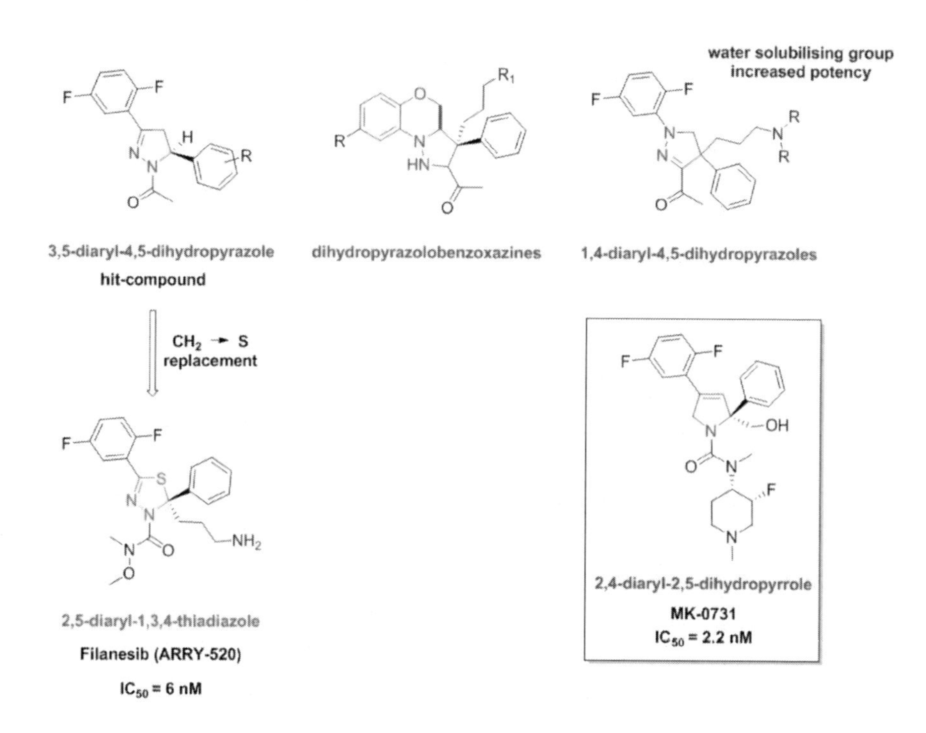

Figure 4. Development of MK-0731 and similarities with ARRY-520.

Litronesib (also called LY2523355, Figure 2, i) shares the thiadiazoline core with ARRY-520, although some differences have been introduced in this inhibitor whose discovery derived from an extensive structure-activity relationship (SAR) program based on K858, a precursor compound identified by a phenotype-based screening for mitotic arrest of HCT116 cells [44]. LY2523355 exhibited broad-spectrum both in antiproliferative activity against cancer cell lines and in preclinical xenograft tumor models [45]. 4SC-205 is an Eg5 inhibitor entered in the clinical trial whose structure has not been disclosed yet. In the clinical trial performed to investigate its safety and tolerability, ascending doses of the drug were orally administered, exhibiting safety up to 200 mg (ow), 100 mg (tw) and 30 mg (con) [46].

Below, the researches regarding kinesin spindle protein (KSP, Eg5) inhibitors divided into structural categories will be discussed, taking into account the works published in the years 2013-2019.

Dihydropyrimidine-Based (Monastrol Analogs) Eg5 Inhibitors

Dihydropyrimidine is one of the most important and explored motif for the development of Eg5 directed drugs, being the scaffold of monastrol, the first KSP inhibitor discovered (Figure 2, a); however, since the antimitotic activity of monastrol is quite low, efforts to ameliorate this feature have been attempted using different strategies.

In this regard, taking advantage by the insertion of a heterocyclic moiety, a series of monastrol analogues bearing 1,3,4-oxadiazole ring have been synthesized [47] and tested (at fixed concentration of 10 μM) against 60 different cell lines derived from nine types of cancer (leukemia, melanoma, lung, colon, CNS, ovarian, renal, prostate and breast) (Figure 5).

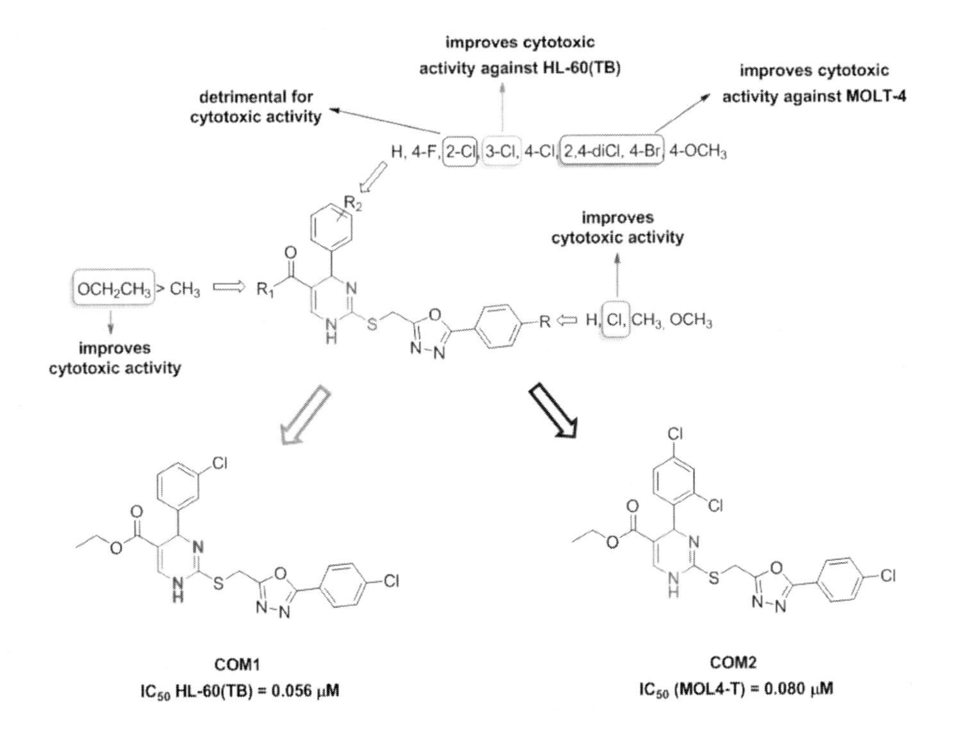

Figure 5. Monastrol analogs bearing 1,3,4-oxadiazole moiety.

This initial screening displayed the presence of compounds possessing a broad spectrum of antimitotic activity, albeit the highest percentages of growth inhibition (GI%) were observed against leukemia cell lines HL-60(TB) and MOLT-4. Focusing the attention on 6 compounds with the best GI%, the median growth inhibitory concentration (IC_{50}) against the most sensitive cell lines (HL-60(TB) and MOLT-4) have been evaluated using MTT assay. These compounds shared some features as (*i*) the ethoxyl group constituting the R_1 moiety (similar to monastrol), and (*ii*) the presence of chlorine atom at the *para* position of the isoxazole-bound phenyl ring, whose substitution with a methyl group or hydrogen atom, impaired the activity against both the cell lines. However, differences in the activity towards the two tumor cell lines were elicited mainly by the R_3 substituents nature. 3-Cl increased the cytotoxic activity against HL-60(TB), while reduced drastically the efficacy towards MOL-4T which, on the contrary, was preferentially inhibited by derivatives bearing 4-Br or 2,4-diCl as R3 substituent. The compounds COM1 and COM2 (Figure 5) tested on HL-60(TB) and MOLT-4, respectively, induced the cell cycle arrest at the G2/M phase, which in turn led to apoptosis. Despite the absence of tests performed directly towards Eg5, the behavior of these molecules resembled the one of monastrol, used as a reference drug in the experiments. Furthermore, docking studies showed that COM1/2 can allocate inside the allosteric binding site of Eg5 with energy scores higher than monastrol.

Exploiting the optimization of Biginelli synthesis, 37 novel dihydropirimidin-2-(thio)one derivatives were synthesized (Figure 6, a) [48] and evaluated as Eg5 inhibitors to determine the effect towards different breast cancer cells MCF-7 and MDA-MB-231 cell lines were selected as models of breast cancer cells [49]. Initial screening was accomplished in order to determine which compounds were able to elicit optimal antitumor activity, obtaining ten candidates highly active against MCF-7 and 30 against MDA-MB-231, the latter resulting very sensitive to these DHPM derivatives. However, only five candidates, exhibiting the highest decrease in cell viability against both cell lines, were further investigated (COM3-COM7, Figure 6, a).

a)

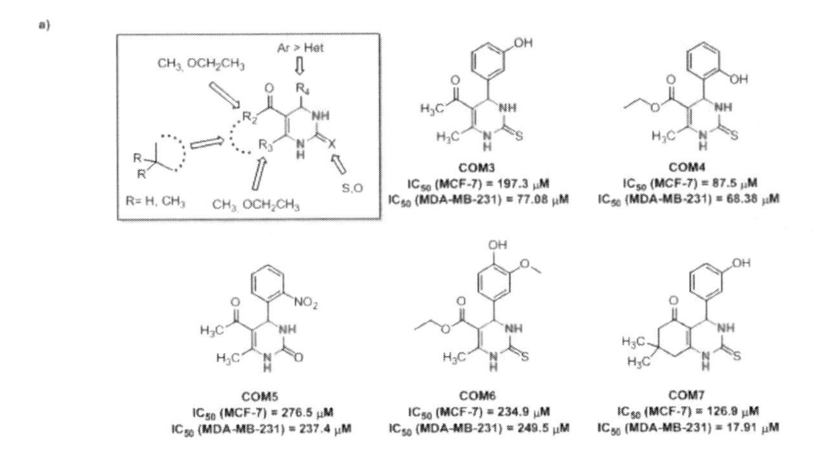

b)

Figure 6. a) Dihydropirimidin-2-(thio)one derivatives and b) N^1-substituted dihydropyrimidin-2-thione derivatives.

The compounds COM3-COM7 were tested, at first, against human fibroblasts, displaying (except for COM4) maximum non-cytotoxic concentration towards normal cell between 0.8 and 1 mM, almost 10-fold more than IC_{50} found for cancer cells. Therefore, the cytotoxic effect was exclusive for the tumor cells, with IC_{50} values in the micromolar range, although some structural features of molecules negatively affected it (see COM5 and COM6). The inhibitory activity against Eg5 was for COM6 and COM3 higher than Monastrol (while was comparable for the others) inducing the formation of the monopolar spindle in MCF-7 cells during mitosis and leading, ultimately, to the apoptosis response.

Further insights into the Monastrol scaffold came from studies focused on two sets of N^1-substituted dihydropyrimidin-2-thione derivatives with different aryl moieties [50]. Initial molecular docking analysis was accomplished to select putative Eg5 inhibitors, finding four compounds which potentially bound to Eg5 better than Monastrol (Figure 6, b). Their antitumor activity was determined against two glioma cell lines, U138 and C6. The cell viability results underlined the importance of 3-hydroxyl group resembling that of Monastrol and leading to higher cytotoxicity (Figure 6, b, green square). Nevertheless, C6 cell line was more sensitive than U138 in the cell viability assays and the two most potent derivatives caused the arrest of the cell cycle in the G2/M phase, mainly in U138 compared to C6. These outcomes could be related to faster cell death of C6 cells or to the presence of more than one mechanism of action along with Eg5 inhibition, whose occurrence was demonstrated by immunocytochemistry assays, showing the formation of the typical monopolar spindles.

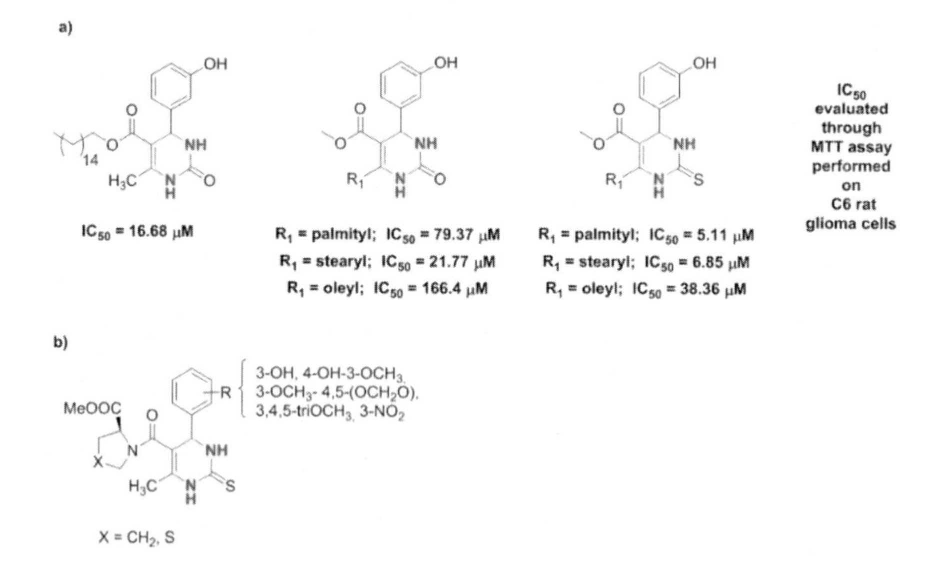

Figure 7. Putative Eg5 inhibitors, lacking cellular/enzymatic assays.

Finally, other compounds designed based on Monastrol scaffold are reported in Figure 7, a. However, even if structurally compatible with other Eg5 inhibitors, these derivatives did not display a clear interaction with Eg5. Therefore, further deepening of their SAR and properties cannot be done at this time [51]. Similarly, novel proline/cyclized cysteine derivatives connected with Monastrol moiety (Figure 7, b) have been explored only with docking studies, lacking enzymatic or cellular assays [52].

Pyrimidine-Based Eg5 Inhibitors

Pyridine derivatives are of great interest for the development of biologically active compounds. In the attempts to develop new inhibitors of HIV-1/2, Al-Masoudi, and coworkers [53] designed a series of 27 pyridine-2-one derivatives (Figure 8, a) that were tested against Eg5 enzyme using malachite green ATPase assay. Only two compounds (blue rectangle in Figure 8, a) exhibited very weak inhibition of KSP, which was insufficient to proceed with further studies. The same authors also developed a series of ruthenium complexes using Monastrol analogs as linkers of the central metal ion, obtaining moderate results with only one of the three complexes able to inhibit Eg5 with an IC_{50} = 30 μM [54]. Using structure-based virtual screening Zhang and collaborators [55] analyzed 500000 compounds as putative inhibitors of Eg5, paying attention to 50 of them with the best scores.

Among these, 37 commercially available compounds were tested in the Eg5 ATPase assay, obtaining one molecule, bearing a triazolo-pyrimidine core, with moderate inhibitory activity (SRI35566, IC_{50} = 65 μM; Figure 8, a). Successive efforts aiming at identifying structural analogs of SRI35566 led to the discovery of the other two compounds endowed with pyrazolo-pyrimidine (SRI35565) and triazolo-pyrimidine core (SRI35564), respectively (Figure 8, b). The selected candidates displayed a lower affinity for Eg5 than parent drug (SRI35565, IC_{50} = 78.9 μM; SRI35564, IC_{50} = 118.3 μM); however, all three molecules enhanced the percentage of

HCT116 cells containing monopolar spindle, which is directly connected to Eg5 inhibition. Using the same concentration of inhibitors useful to obtain the monopolar spindle formation on HTC116 cells (80 μM), the authors also evaluated their effectiveness in suppressing colony formation. From the tests, SRI35564 resulted as the most active compound, exhibiting the highest cytotoxicity. However, no potent Eg5 inhibitory activity of SRI35564 was observed. Therefore, a potential off-target effect should be taken into account.

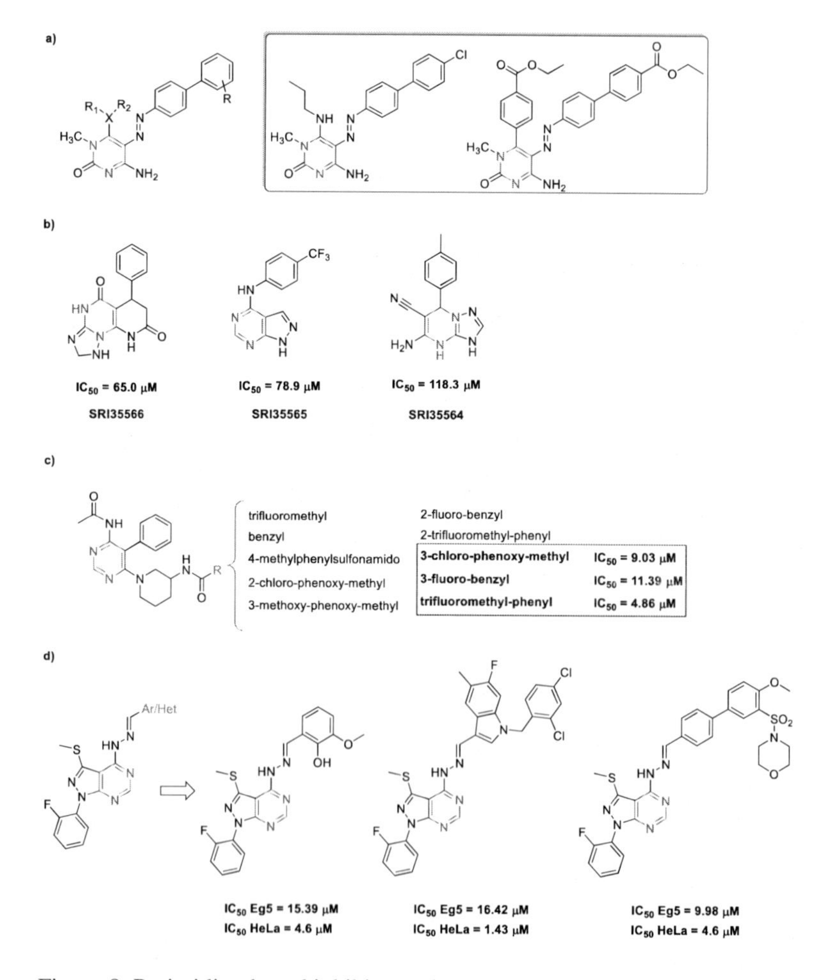

Figure 8. Pyrimidine-based inhibitors of Eg5.

A series of pyridine-pyrimidine derivatives (Figure 8, c) were synthesized and evaluated as inhibitors of KSP [56]. Among the ten tested compounds, only three showed inhibitory activity in the micromolar range. The docking studies, performed in order to understand if these molecules bound to the site formed by helix α2/helix α3/loop L5, or to the one constituted by helix α4/helix α6, were not resolutive. The same research group further explored the pyrimidine core, producing pyrazolo-pyrimidine derivatives [57] also obtaining in this case only 3 compounds, able to inhibit Eg5 in the micromolar range (Figure 8, d). Even then, docking studies were not resolutive in order to understand the binding site of the inhibitors. Performing MTT assay, these three compounds were screened for their antiproliferative activity against HeLa cells, finding low micromolar IC_{50}'s (Figure 8, d) which did not fit with Eg5 inhibition results, paving the way to other off-target mechanisms of cytotoxicity.

Benzimidazole-Based Eg5 Inhibitors

Benzimidazole is a valid scaffold for the development of Eg5 inhibitors, and its importance was further increased by the discovery of a novel binding-pocket where benzimidazoles easily bind also in the presence of classical inhibitors in the L5/α2/α3 site [58]. One of the compounds implicated in this discovery (BI8, Figure 9, a), was in-depth investigated through X-ray crystallography [59], showing as the high-affinity binding pocket is formed by helix α4 and helix α6. The interaction of BI8 inside this binding site induces a conformational change of Eg5, producing a series of effect which, ultimately, inhibited the ADP release from the active site.

Carbajales and coworkers [60] combined structure-based drug design with multicomponent reaction synthetic approaches, in order to realize 18 novel benzimidazole-based derivatives evaluated for their *in vitro* inhibitory activity against Eg5 (at the fixed concentration of 5 μM) and their activity (cytotoxicity and G2/M blockage) towards HeLa cells. The six most active candidates showed low micromolar inhibitory activity towards Eg5 with IC_{50} spanning from 1.37 to 2.52 μM. Two bromines on the benzimidazole phenyl ring were more effective than chlorines in improving the inhibitory activity, while halogen removal produced ineffective compounds. The presence of a carbonyl moiety (Y = O, Figure 9, c) also ameliorated the activity. The stereochemistry was an essential issue for the inhibition; the two most active derivatives were optically resolved in the single enantiomers (although the absolute configuration was not assigned) and tested singularly showing differences in Eg5 inhibition. Cytotoxicity tests performed on HeLa cells, displayed antiproliferative effects in the low micromolar range, with EC_{50} values lying between 3.63 and 49.8 μM, and the presence of monopolar spindle. The differences in cellular activity among the enantiomers were much-amplified respect to the enzymatic ones, being the (supposed) *S*-enantiomers 6/7-times more potent than *R*-ones.

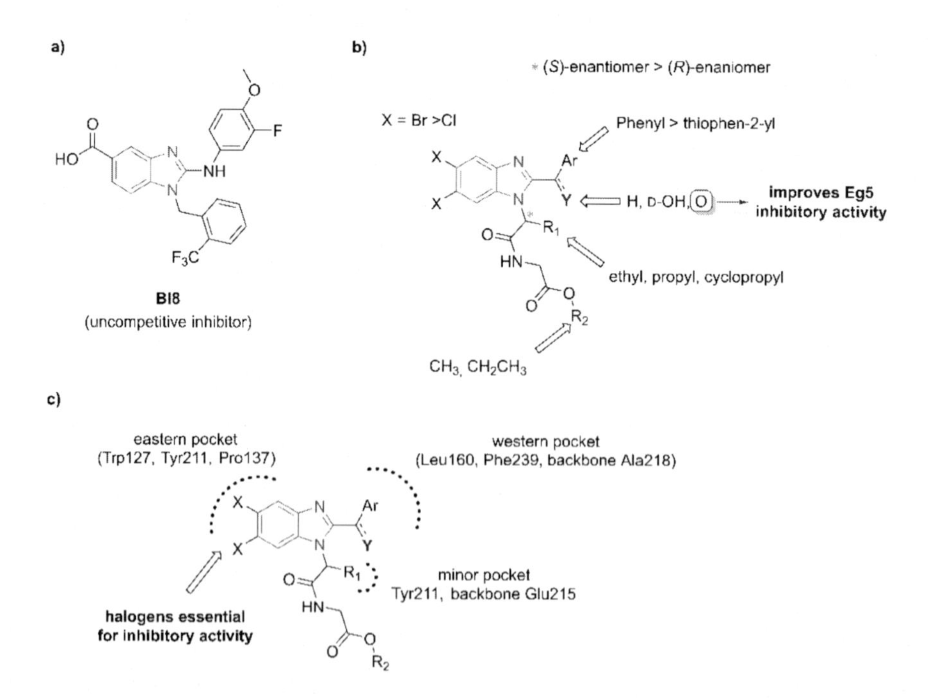

Figure 9. a) Structure of BI8; b) benzimidazole derivatives; c) pocket occupancy inhibitors inside the allosteric binding site of KSP.

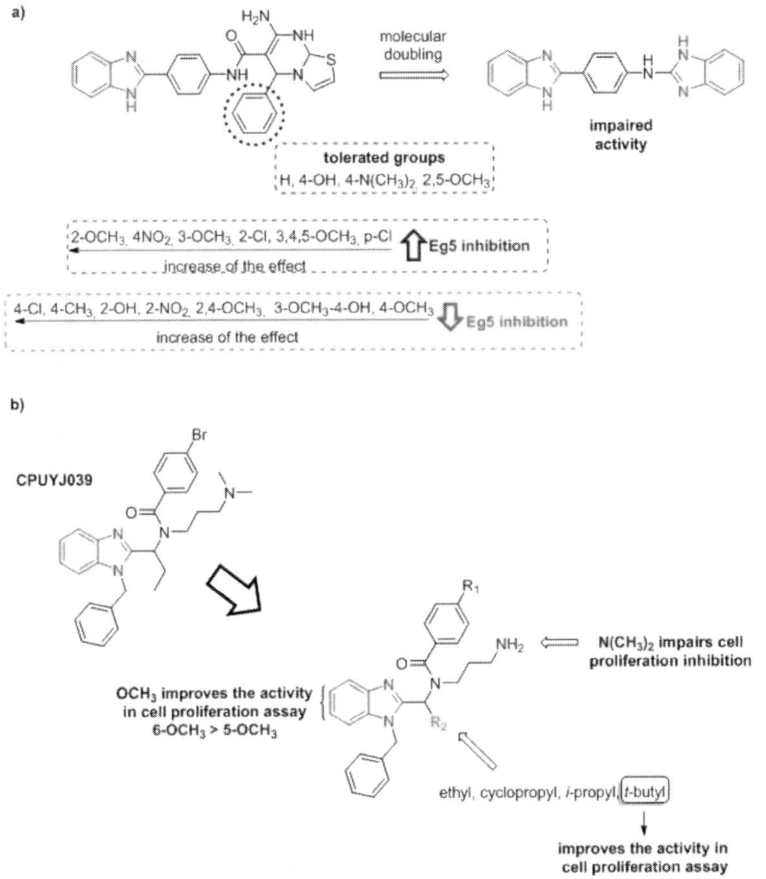

Figure 10. a) Structures of benzimidazole-thiazolo[3,2-*a*]pyrimidine-6-carboxamide derivatives; b) analogs of CPUYJ039.

Further insights on this scaffold have been accomplished through the synthesis of molecules designed with benzimidazole connected by phenyl ring to thiazolo[3,2-*a*]pyrimidine-6-carboxamide moiety (Figure 10, a) [61]. An initial screening was carried out against colorectal cancer cell line HCT116, human liver cancer cell line HepG2, and human ovarian cancer cell line A2780, displaying that HCT116 cell line was the more sensitive to the tested compounds. In particular, six molecules exhibited a concentration required for 50% inhibition of cell viability lower than 2 μM ($1.20 < IC_{50}$ (μM) HCT116 < 1.93), while the molecular doubling did not elicit the hoped results, producing weak cytotoxic compounds. Successive studies focused on the inhibitory activity assessment of these candidates against Eg5, through MT-activated ATPase assay, obtaining low nanomolar IC_{50} values ($0.0009 < IC_{50}$ (μM) < 0.0055). Importantly, the two compounds bearing respectively 2-OCH_3 (IC_{50} HCT116 = 1.20 μM) and 4-NO_2 (IC_{50} HCT116 = 1.32 μM), which exhibited higher cytotoxicity against HCT116 cells, showed the best inhibitory activity towards Eg5 (IC_{50} 2-OCH_3 = 0.0009 μM; IC_{50} 4-NO_2 = 0.0010 μM), with a good correlation between enzymatic and cellular data.

Using efficient synthetic procedures based on multicomponent reaction approach, a series of analogs of the potent Eg5 inhibitor CPUYJ039 (Figure 10, b), have been explored [62]. The synthesized compounds were tested on two human cancer cell lines (HCT116 and HeLa) and some of them showed cell proliferation inhibition in the micromolar range ($1.0 < IC_{50}$ (μM) HCT116 < 18.0; $1.3 < IC_{50}$ (μM) HeLa < 19.0). In general, the *N*-dimethyl moiety at the end of the propylamino chain was detrimental for the activity which was, on the contrary, ameliorated by the presence of methoxyl group bound to the phenyl ring of benzimidazole core. The CPUYJ039 analogs exhibiting the best inhibition values in cellular assays furnished the presence of monoastral spindles (detected with immunofluorescence assay) which were in accordance with their inhibitory activity of KSP in the low micromolar range ($1.8 < IC_{50}$ (μM) Eg5 < 10.0).

Thiadiazoline-Based Eg5 Inhibitors

Thiadiazoline heterocycle is another fundamental scaffold for the development of new Eg5 inhibitors, being the core of the two compounds entered in clinical trial filanesib and litronesib. Along with these two molecules, a third one (K858) endowed with thiadiazoline scaffold, was discovered using high-throughput morphology-based screening [44].

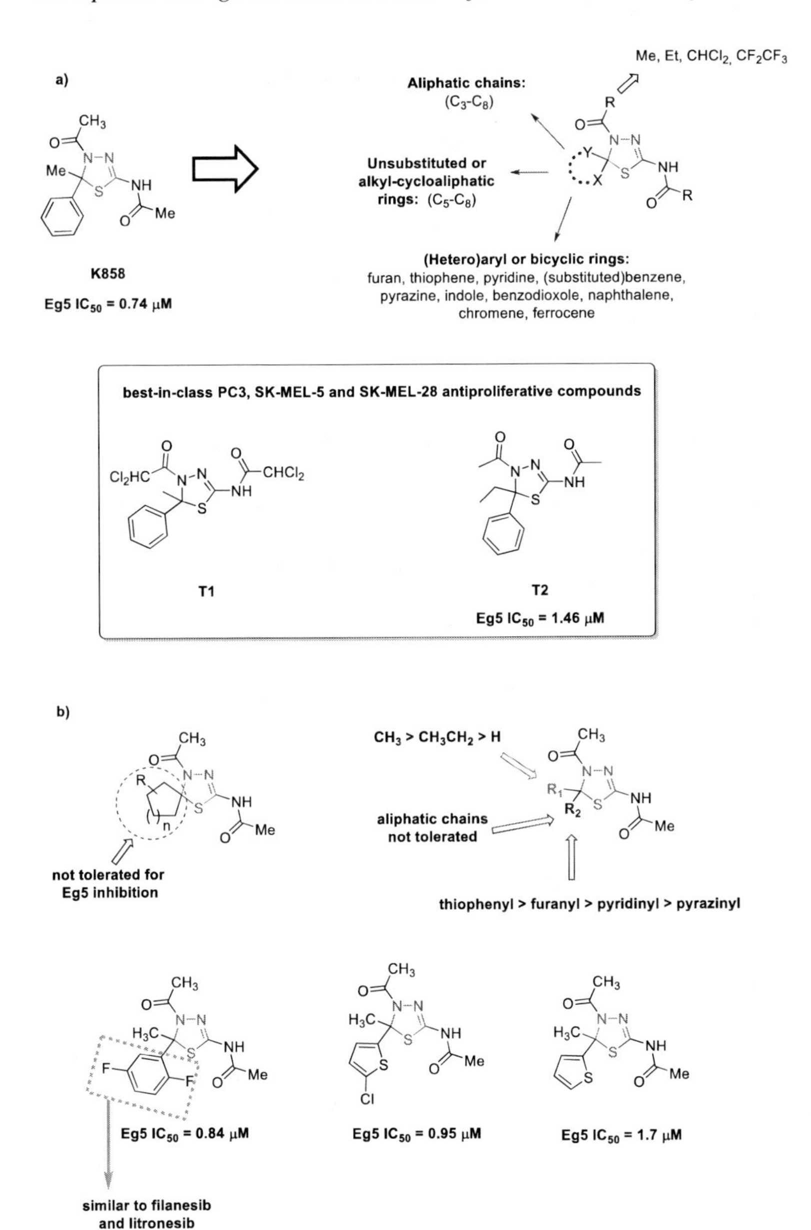

Figure 11. K858 and inhibitors based on K858 scaffold.

K858 selectively inhibited Eg5 and was able to induce mitotic arrest, inhibiting cell growth in human colorectal cancer HCT116 cells, without affecting the microtubule polymerization. In successive studies, K858 exhibited antiproliferative activity against four different breast cancer cell lines, albeit also inducing the up-regulation of survivin, an anti-apoptotic molecule [63]. This effect has also been observed when K858 was tested against two human glioblastoma cell lines (U-251 and U-87), displaying antitumor activity and the over-expression of survivin [64]. Recently K858 has been deeply investigated by Kozielsky's group [65], obtaining the co-crystallization of the inhibitor within Eg5 enzyme. Molecular insights into the K858 binding pocket displayed that the inhibitor bound in the well-known site formed by helix $\alpha 2$, loop L5 and helix $\alpha 3$, where it established a very complex interaction network which almost clashes with the not-sophisticated structure of the inhibitor. Furthermore, only the (*R*)-enantiomer was found inside the enzyme, indicating enantio-selectivity for K858. In the same work the author explored, from the enzymatic point of view, some derivatives published by De Monte and coworkers [1], who realized a vast library (103 compounds) of 1,3,4-thiadiazoles that were tested against PC3 prostate cancer cells, at first, and then evaluated also against SK-MEL-5 and SK-MEL-28 melanoma cell lines. From the intensive screening done on the PC3 cells, only 9 compounds exhibited adequate antiproliferative effects, while two were endowed with antitumor activity for SK-MEL-5 and SK-MEL-28 melanoma cell lines (T1 and T2, Figure 11, a). These inhibitors displayed inhibitory activity against Eg5 comparable with K858. With the aim to unraveling the real ability of some of these 103 thiadiazolines to inhibit Eg5, Kozielski's group evaluated the effect of 50 of them on the basal ATPase activity of Eg5, taking into account that results of cell-based assays can suffer from insufficient drug-like properties [65]. All the compounds distinct from the substituents bound at position 5 (Figure 11, b), showing as its proper substitution was essential for the Eg5 inhibitory activity. For example, to obtain a higher activity, one of the two substituents must be the methyl group, albeit the ethyl one can be tolerated, while the presence of a single hydrogen atom provoked the drop of activity.

The presence of alkyl chains was not tolerated as well as their cyclic form, inducing the complete loss of affinity against Eg5. (Un)substituted phenyl and heterocyclic rings seemed to be more tolerated when the other group is methyl. The compound showing the best inhibitory activity, albeit slightly lower than K858, was the one bearing the 2,5-difluorophenyl ring that is the moiety shared with filanesib and litronesib. Moreover, the 5-chloro-thiophen-2-yl moiety was also well-tolerated, showing $IC_{50} < 1$ μM. The substitution of the methyl group with hydrogen keeping the 1-naphthyl ring induced the loss of inhibitory activity ($IC_{50} > 50$ μM), underlying the importance of the methyl group at position 5. Previously, the K858 scaffold was investigated by Yamamoto and colleagues [66], in order to optimize its structure and increase the mitotic phase accumulation activity. Using different synthetic approaches, the authors obtained a series of K858 derivatives useful to evaluate the effect of the substitutions performed on the scaffold (Figure 12, a-d).

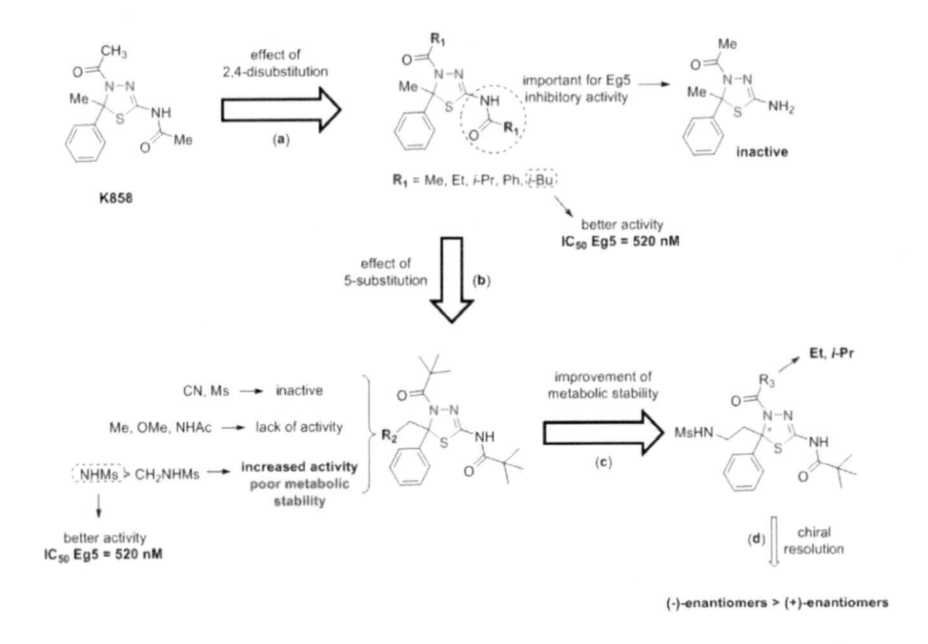

Figure 12. Development of new thiadiazoline inhibitors of Eg5, based on K858.

In the firsts attempts, the effects of substituents at positions 2 and 4 were determined (TIA2, a). The first screening was done evaluating the ability to induce mitotic accumulation (MI_{20}: concentration required for mitotic accumulation against 20% of the cells tested) followed, for the compounds that "passed" the MI_{20} assay, by antiproliferative activity (GI_{50}) and IC_{50} assessment of ATPase activity of Eg5. Among the attempted substituents, the bulky pivaloyl one elicited the best inhibitory activity (Eg5 IC_{50} = 520 nM), while the excessive increase of dimension (e.g., phenyl ring) negatively affected the activity reducing or eliminating the ability to induce mitotic accumulation. After the selection of the best-tolerated group at the positions 2 and 4, the attention was focused on the group at position 5, introducing polar groups able to increase microsomal stability (Figure 12, b). While some moieties negatively affected the antimitotic activity (e.g., CN or methylsulfonyl represented as Ms), the two groups bearing nitrogen atom bound to methylsulfonyl group increased the ability to induce mitotic accumulation, cellular growth, and Eg5 inhibition. Nevertheless, this increment in antitumor activity went together with an insufficient mouse liver microsomal stability. Finally, changes at position 3 were performed (Figure 12, c) using ethyl or *i*-propyl group despite *i*-butyl. These substitutions were effective in ameliorating the microsomal stability; furthermore, their optical resolution displayed that the chiral properties affected the activity with the *(-)*-enantiomers that exhibited a more potent mitotic accumulation activity and Eg5 inhibitory activity than *(+)*-enantiomers and the corresponding racemates (Figure 12, d).

Other derivatives based on this scaffold were designed and synthesized starting from MK-0731, albeit this compound contains dihydropyrrole moiety, designing a series of constrained analogs containing spirocyclic moiety [67] (Figure 13). These compounds were evaluated on HCT-116 colon cancer cells, checking the levels of phos-Histone H3, a marker of mitosis, to determine the effect of the molecules. Among the tested bicyclic system containing the spiro bond, the chromane was the most effective, showing better results than thiochromane or tetrahydroquinoline, while changing the ring dimensions had detrimental effects. Keeping constant the chromane moiety, the nature of the R substituents was evaluated, obtaining the best results for the methyl group. Bulky substituents, or the ones endowed with hydroxyl or aminic head, impaired the activity against HCT-116. Intending to introduce polar groups without reducing the activity, the authors tried to append it to the six-membered spiro ring system. The insertion of an ethylamino group dramatically improved the activity leading to nanomolar activity (Phos HH3, EC_{50} = 5 nM), further improved by chiral resolution and selection of the eutomer; this compound (SP1, Figure 12) potently induced mitotic arrest, with monoastral spindle formation, characteristic of KSP inhibition.

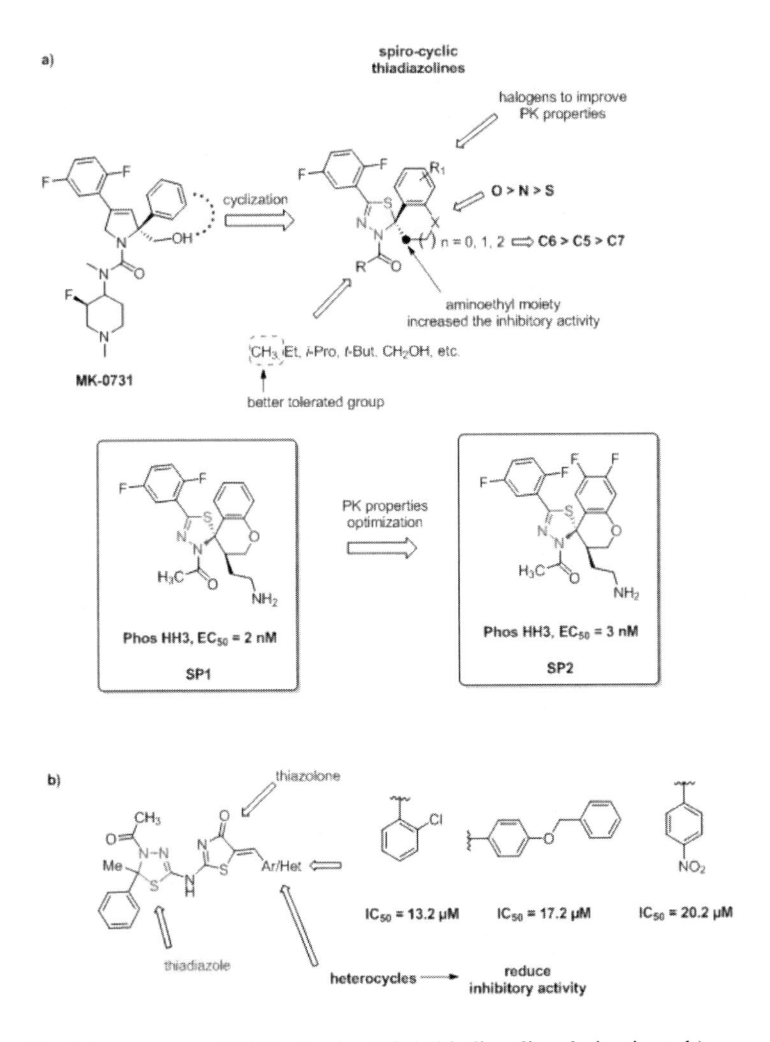

Figure 13. a) Structures and SAR of spiro 1,3,4-thiadiazoline derivatives; b) thiadiazole-thiazolone hybrids.

In order to improve pharmacokinetic properties substituents were placed on the chromane phenyl ring, obtaining the compound SP2 (Figure 12), discovering that the presence of the 6,7-difluoro substitution positively affected the pharmacokinetic properties. This compound was dosed orally and tested in the treatment of Colo-205 colon tumor xenograft mouse model, showing antitumor activity.

Taking advantage from molecular hybridization, the 1,3,4-thiadiazole scaffold and the thiazolone one were linked together to obtain a series of thiadiazole-thiazolone hybrids, as potential inhibitors of KSP (Figure 12, b) [68]. These derivatives were evaluated for their inhibitory activity against the basal and the MT-stimulated ATPase of Eg5. While in the basal condition these compounds did not show Eg5 inhibition, micromolar inhibitory activity was obtained in the MT-stimulated one. The best inhibitor of this series containing a 2-chlorophenyl ring exhibited IC_{50} = 13.2 μM, that is quite far from the one of K858 used as the reference drug (IC_{50} = 0.80 μM). Other substituents were ineffective in ameliorating the inhibitory activity towards Eg5, with Br, OMe, Me, F that did not allow to reduce the IC_{50} values. Furthermore, the presence of heterocyclic systems as furan, thiophene, or pyridine produced compounds with inhibitory activity in the high micromolar range (IC_{50} > 100 μM).

Natural Compounds Endowed with Eg5 Inhibitory Activity

After the discovery of Terpendole-E (TerE), the first natural-product able to inhibit Eg5 [69], other natural occurring analogs of TerE or containing completely different scaffolds have been explored. Terpendole-E is a compound containing a fused indole-diterpene system produced by the fungus *Chaunopycis alba*. The selection of a fungus strain able to over-produce TerE allowed performing prolonged culturing, overcoming the issue of instability in the culture media and the difficult to isolate possible metabolites [70]. Indeed, the prolonged exposition to enzymatic equipment led to the shunt metabolite production of TerE obtaining ten indole-diterpene compounds (Figure 14) which were isolated and evaluated at the enzymatic and cellular level [71]. The first discovered was the 11-ketopaspaline (11-keto), the TerE oxidized analogs containing 11-carbonyl moiety despite hydroxyl one. This molecule displayed enhanced inhibitory activity against Eg5 (microtubule-stimulated ATPase activity assay, IC_{50} TerE = 19.0 μM; IC_{50} 11-keto = 3.0 μM) as well as its analogue deprived of 11-carbonyl moiety, paspaline (pas), albeit in a lesser extent (IC_{50} 11-keto = 7.1 μM) [71].

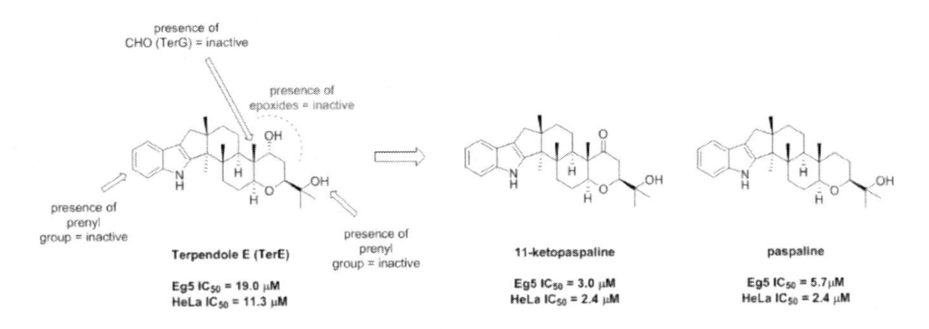

Figure 14. Terpendole-E and related metabolites.

All the other metabolites were ineffective against Eg5 being endowed with IC_{50} > 100 μM, although some of them showed cytotoxicity against HeLa cells. However, from the analysis of G2/M-phase population was clear that only TerE, 11-keto and pas, carried out their effect through Eg5 inhibition, while other mechanisms must be taken into account for the other metabolites.

The presence of carbonyl moiety of 11-keto also ameliorated the cytotoxicity against HeLa cells, that was four times more effective than TerE. This analog arrested cell-cycle progression in the M phase, inducing the monoastral spindle formation [70]. Further insights were done in order to unravel the inhibitory mechanism of TerE and 11-keto against Eg5 mutants that showed resistance to *S*-trityl-L-cysteine (STLC) and GSK-1, two inhibitors which bind to the L5/α2/α3 portion of Eg5 motor domain. TerE and 11-keto were effective against all the Eg5 mutants, accounting for a different binding site inside Eg5 or different inhibitory mechanisms concerning STLC and GSK-1.

Using an *in silico* analysis, Ognunwa and coauthors [72] selected *(+)*-Morelloflavone (MF) form *Garcinia dulcis* (Figure 15), among the 40 plant-derived bioflavonoids screened, as a potential inhibitor of Eg5, being able to interact with L5/α2/α3 site potentially. Its inhibitory activity was determined against both basal and microtubule-activated ATPase activities of Eg5, displaying moderate potency with IC_{50} of 100 and 96 μM respectively, suggesting that the presence of microtubules did not affect the enzyme-inhibitor binding. Varying the amount of ATP and morelloflavone used in the inhibitory activity tests, the authors did not observe competitive inhibition, supporting the postulated interaction at the Eg5 allosteric site. Additional in-depth analysis performed by the same research group corroborated the presence of a stable Eg5-MF complex that elicited a series of changes in Eg5 structure, affecting ATPase substrate binding and conversion, as well as the interaction with microtubules, ultimately, inducing Eg5 inhibition (73).

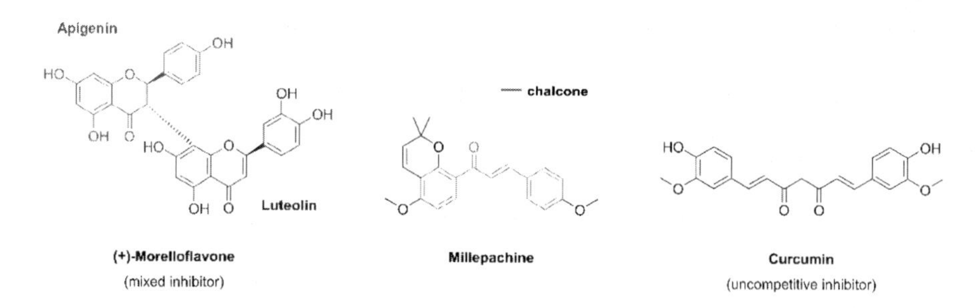

Figure 15. Structures of naturally occurring compounds: *(+)*-Morelloflavone, Millepachine, and curcumin.

An extensive study on curcumin, obtained from the rhizome of the plant *Curcuma longa*, revealed the ability of this natural compound to inhibit the basal and MT-stimulated ATPase of Eg5 [74], inducing the cell cycle arrest at G2/M phase, along with the formation of the monopolar spindle. Further analysis, performed on the Eg5-437H (containing the motor domain, the neck linker and the stalk region of the enzyme) was the basis to postulate the existence of a different mechanism of action concerning Monastrol. Competitive binding assay displayed that both the ligands can simultaneously bind to the protein accounting for different binding sites of curcumin on Eg5-437H. Docking analysis addressed this new binding site of curcumin into a pocket delimited by loop L8, helix $\alpha5$ and sheets $\beta4$, $\beta5$, and $\beta6$, where the molecules established hydrophobic and electrostatic interactions. This site is far away from the nucleotide and microtubule binding sites; therefore, the influence on the ATPase activity and microtubule binding activity of Eg5 could be related to conformational change induced by this molecule.

Another natural-compound able to interfere with Eg5 activity is the chalcone Millepachine (MIL), produced by *Millettia pachycarpa.* This molecule also acted as topoisomerase II inhibitor, inducing G2/M arrest and apoptosis through DNA damage. Focusing the research on the effects on the spindle assembly, Wu and colleagues [75] proved that MIL induced G2/M arrest inhibiting the normal spindle assembly, without affecting microtubule polymerization and exhibiting this outcome against different cancer cell lines (human hepatocarcinoma cell line HepG2, human ovarian cancer cell line A2780S, human non-small-cell lung cancer cell line H460, human anaplastic thyroid cancer cell line C643).

STLC Derivatives

Since the discovery of the *S*-trityl-*L*-cysteine (STLC) ability to effectively inhibit Eg5, some attempts have been focused on the improvement and development of this scaffold [76]. Kozielsky's research group [77] reported new analogs based on the STLC scaffold to improve the potency and *in vivo* efficacy of previously reported compounds (Figure16, a). These novel derivatives were tested to assess their *in vitro* and *in vivo* efficacy evaluating K_i^{app} (basal ATPase inhibitory activity) and GI_{50} (inhibition growth) against K562 human leukemia cells.

Previous studies reported that small alkyl electron-donating and hydrophobic substituents, placed at the *meta*-position of one of the phenyl rings, strengthened the interactions in the P3 pocket. In order to evaluate whether the repositioning of substituents could affect the interactions with the binding site, a small number of thioethanamines with substituents in *para* position have been synthesized. The *p*-methyl, *p*-ethyl, and *p*-methoxyl (R group in Figure 16, a) analogs exhibited a comparable activity to the previous related compound. Instead, the insertion of the *p*-ethoxyl and *p*-trifluoromethoxyl led to less activity, probably due to inappropriate physicochemical properties. At this stage, analogs based on the butanamine scaffold, containing a single substituent in *meta* or *para*-position as *m*-ethyl, *m*-isopropyl, and *p*-methoxyl, were synthesized. These novel compounds were found to be more active than the thioethanamine ones, exhibiting a comparable activity to their previous related leads. To improve metabolic and pharmacokinetic properties of the mentioned above compounds, a series of new analogs based on the thioethanamine scaffold with disubstituted phenyl rings using, in turn, a fluorine atom or alkyl substituents were synthesized. The *o*-fluorine (R3), when combined with a *meta*- or *para*-methyl (R_1), decreased the inhibitory activity, while if combined with *p*-methoxy, led to an improvement.

The dialkylphenyl and tetralene derivatives proved comparable Eg5 basal activity to the monosubstituted analogs, but when the carboxylic acid was introduced at the α-position of the tail, the improvement of activity was evident. The most potent K_i^{app} was exhibited by the 3,4-dimethyl triphenyl thioethanamine compound S1 (K_i^{app} = 1.2 nM) instead, the most potent growth inhibition activity was displayed by 3-ethyl-4-methyltriphenylthioethanamine derivative S2 (GI_{50} = 34 nM). Notably, these compounds were more active than their counterparts lacking the α-carboxylate, stressing the central role of the physicochemical balance for cellular activity. The modifications made on S1 were translated on the butanamine scaffold. The resulting compound *rac*-S3 was found to be less potent in the *in vitro* assay but reported 3-fold more active in the cellular growth inhibition assay (GI_{50} = 23.4 nM). All the modifications described to date led to an increase of hydrophobicity. To develop compounds with an optimal physicochemical balance, further modifications based on the insertion of the *m*-hydroxy or primary/secondary amides (R_4) combined with a lipophilic substituent (R_1) were done. The *m*-phenol analog with *m*-chlorophenyl led to an activity decrease. Instead, the *m*-phenol analogs with *m*-ethyl, and *m*-methylphenyl substituents were more active than the related compounds with one phenyl ring substitution. Notably, the compound with *m*-phenol and *p*-methylphenyl substituent was the most active thioethanamine α-carboxylic-free derivative. After identifying the best leads of these series, Kozielsky and colleagues developed several tests including the previous related described compounds, the clinical candidate ispinesib and its more potent second-generation analog SB-743921 as benchmarks.

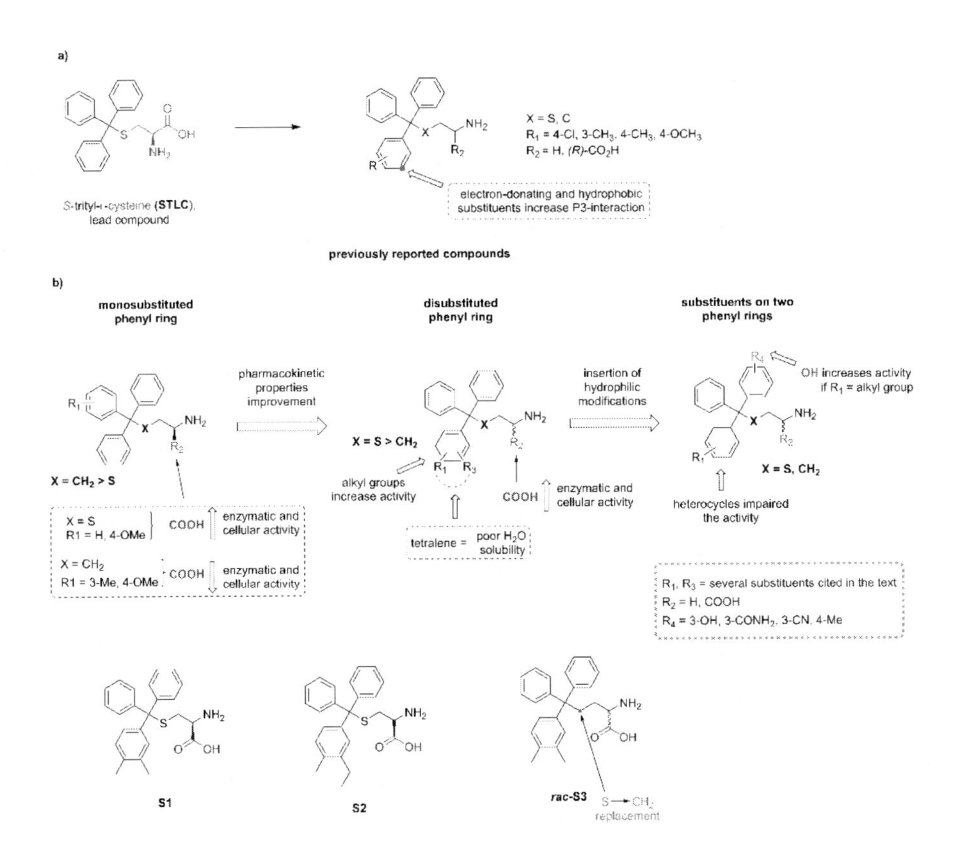

Figure 16. *S*-trityl-*L*-cysteine and relative analogs.

We decided to focus the descriptions of the biological tests on the three most potent analogs S1, S2, and *rac*-S3 (Figure 16, b). Their activity in an MT-stimulated Eg5 ATPase assay was evaluated; S1 and *rac*-S3 possessed a value of K_i^{app} inferior to 10 nM. Furthermore, the selectivity of *rac*-S3, the most potent compound of the entire series, was assessed towards several human kinesins as Kif3B (kinesin-2; organelle transport), neuron-specific Kif5A and Kif5B/conventional kinesin (both kinesin-1; cargo transport), Kif7 (kinesin-4, involved in Hedgehog signaling), Kif9 (kinesin-9; regulation of matrix degradation), and Kif20A/MKLP-2 and Kif20B/MPP1(both kinesin-6; required for cytokinesis), showing no affinity at a maximum inhibitor concentration of 200 μM.

Then, these compounds were tested against a panel of cell lines derived from several types of cancer. They exhibited to be more active than the original lead compounds, whereas S1 and *rac*-S3 were more potent than the clinical candidate ispinesib and equal to SB-743921. The *in vitro* and *in vivo* physicochemical properties and ADMET of these three compounds and the most active carboxylate-free derivative were evaluated to make a direct comparison of their properties. The result of the tests confirmed the trend that carboxylate-free compounds have more liabilities than those possessing it. In the effort to ameliorate these data, they proposed other modifications based on the modulation of the primary amine basicity (Figure 17). Since the alkylation of primary amines produced a loss of activity, they attempted to modulate the pK_a of the amine group by β-fluorination with one or two fluorine atoms (R_1). Nevertheless, these modifications did not provide any improvement.

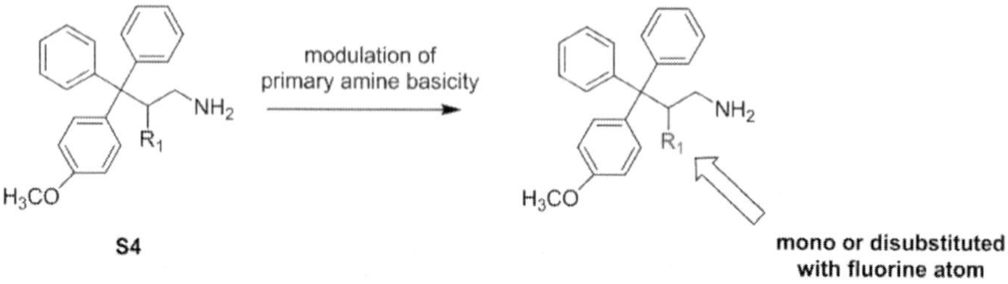

Figure 17. Improvement of physicochemical properties of STLC derivatives.

They tested the S1 and S2 derivatives that differed in term of cell activity. Although its high potency in cell assay, S2 displayed higher clearance in human hepatocytes inducing a less availability in human plasma, and it demonstrated to be a weak inhibitor of CYP259. These data could explain the poor *in vivo* activity, tested on lung cancer xenograft models, compared to the *in vitro* efficacy. The results are different for *rac*-S3 that showed higher turbidimetric solubility, increased metabolic profile in human hepatocytes, no inhibition of the common CYP isoforms and hERG channel. These results, compared to those of the clinical candidate SB-743921, stressed that this analog possesses an optimal ADMET profile. The xenograft model confirmed these promising data since the treatment led to total tumor regression by day 17, a result that is comparable to that obtained with the clinical candidate ispinesib. Through this approach of lead optimization, *rac*-S3 was identified as the most potent of the STLC derivatives to date, with properties that make it an optimal candidate for further clinical investigations.

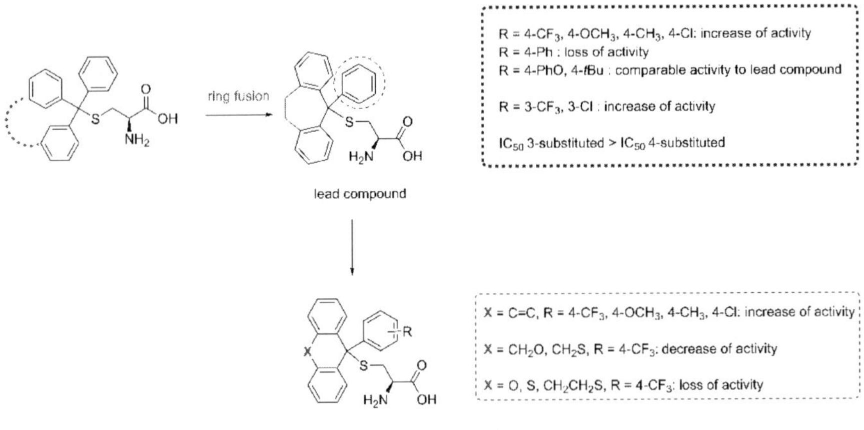

Figure 18. Optimization of fused trityl derivatives.

Based on the studies relative to the crystal structure of KSP in complex with STLC, Ogo and collaborators [78] developed a series of STLC derivatives blocking two phenyl rings of the trityl group with several alkyl chains (Figure 18). The insertion of this structural element could increase the hydrophobic interactions with the binding site. They synthesized a derivative with an ethylene linker between two phenyl rings of the trityl group and evaluated its ATPase inhibitory activity. Since it proved to be 9-fold active than STLC (KSP ATPase IC_{50} = 209 nM), they developed modifications based on the optimization of linker chain-type and size and the role of substituents on the non-linked phenyl ring. These novel derivatives were tested to assess their KSP ATPase inhibition and cytotoxicity against HCT116 cells (IC_{50}), including the previous related described compounds and the clinical candidates ispinesib and filanesib as benchmarks. In the first part of their study, the authors evaluated the substituent effects on the nonlinked phenyl ring of the lead compound using several substituents that differ for size and position. The obtained data indicate that compounds endowed with small and lipophilic substituents at the *para*-position are the most potent in terms of ATPase inhibition and cytotoxicity. Although their ATPase activity is comparable to ispinesib and filanesib, their cytotoxicity is less potent.

After determining the most potent substituents, they investigated the chain-type and size of the linker. Concerning the chain, the replacement of the C–C single bond with the C=C double bond led to an improvement of the cytotoxic activity probably due to their more tolerated pharmacokinetic profile. Instead, the replacement with CH_2O and CH_2S produced a decrease in activity. About the size, the substitution of the C–C single bond with O, S, and CH_2CH_2S induced a dramatic loss of activity. These results suggest that the optimal linker-type and size in combination with a small and lipophilic *p*-substituent in the non-linked phenyl ring ensured a potent inhibition of KSP with IC_{50} values comparable to those of the clinical candidates used as benchmarks. After determining the selectivity against Eg5, to understand whether these derivatives affect the formation of the bipolar spindle, Ogo and coworkers investigated their effects on cell progression using the MI_{50} value (50% mitotic index induction concentration). Microscopic analysis revealed the formation of monopolar spindles, the typical feature of KSP inhibition. These compounds were tested in the HCT116 xenograft model demonstrating their potent antitumor activity *in vivo* even though the amount, schedule, and roof of administration should be optimized.

Novel Approaches for Eg5 Inhibition

A novel molecular approach, different from the ones so far shown, concerned the use of photochromic compounds, able to reversibly change their configurations and physicochemical characteristics depending on ultraviolet (UV) and visible (vis) light irradiation. Maruta and coauthors [79] deeply investigated this kind of approach to inhibit Eg5. In one of the first studies, they evaluated the opportunity to photocontrol Eg5 ATPase activity covalently incorporating two types of photochromic molecules (azobenzene and spiropyran) into the loop L5. For this purpose, five Eg5 mutants differing for the position of the single reactive cysteine in this loop were produced. Only one among the five mutants produced and bound to P1 exhibited changes in ATPase activity upon UV-vis transition, showing a moderate (24%) and completely reversible alteration after the *cis-trans* isomerization (Figure 19, a).

The second photochromic compound P2 (Figure 19, b), endowed with STLC trityl group, did not elicit activity regardless of the Eg5 mutant type. P3 (Figure 19, c) elicited the same Eg5 mutant that was able to incorporate P1 compound, also inducing a decrease of 30% and 21% of activity if it is bound to the spiropyran or merocyanine, respectively. After this first observation, further insights were obtained using compound P4 (Figure 20, a), containing the trityl group of STLC, along with *N*-acetyl-*L*-cysteine and azobenzene moiety [80]. In this case, the photochromic molecule was evaluated as an inhibitor and was not included inside loop L5. The effect of *cis/trans* isomerization was evaluated for the ATPase activity of Eg5 in basal, and microtubule activated conditions, exhibiting a significant inhibitory activity of the *trans*-isomer than the *cis* one; similarly, *trans*-isomer was more effective in reducing microtubule gliding velocity respect to *cis* one.

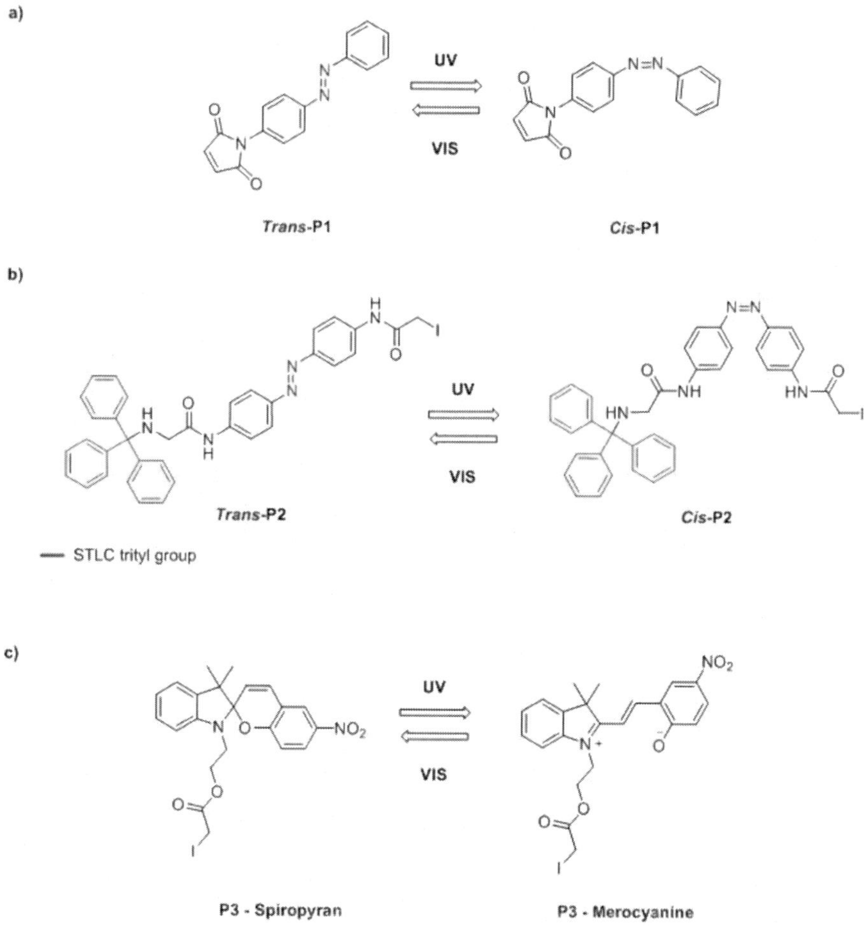

a)

UV

VIS

Trans-P1 *Cis*-P1

b)

UV

VIS

Trans-P2 *Cis*-P2

— STLC trityl group

c)

UV

VIS

P3 - Spiropyran P3 - Merocyanine

Figure 19. Loop L5-incorporated photochromic compounds.

Albeit these inhibitors were endowed with inhibitory activity in the low micromolar range, they did not possess a real "On/Off" effect. In this regard, a further attempt was done with the compound P5, (Figure 20, b), containing spiropyran-merocyanin system; in this case, two photochromic spiropyran moieties were bound through a carboxylic acidic linker [81].

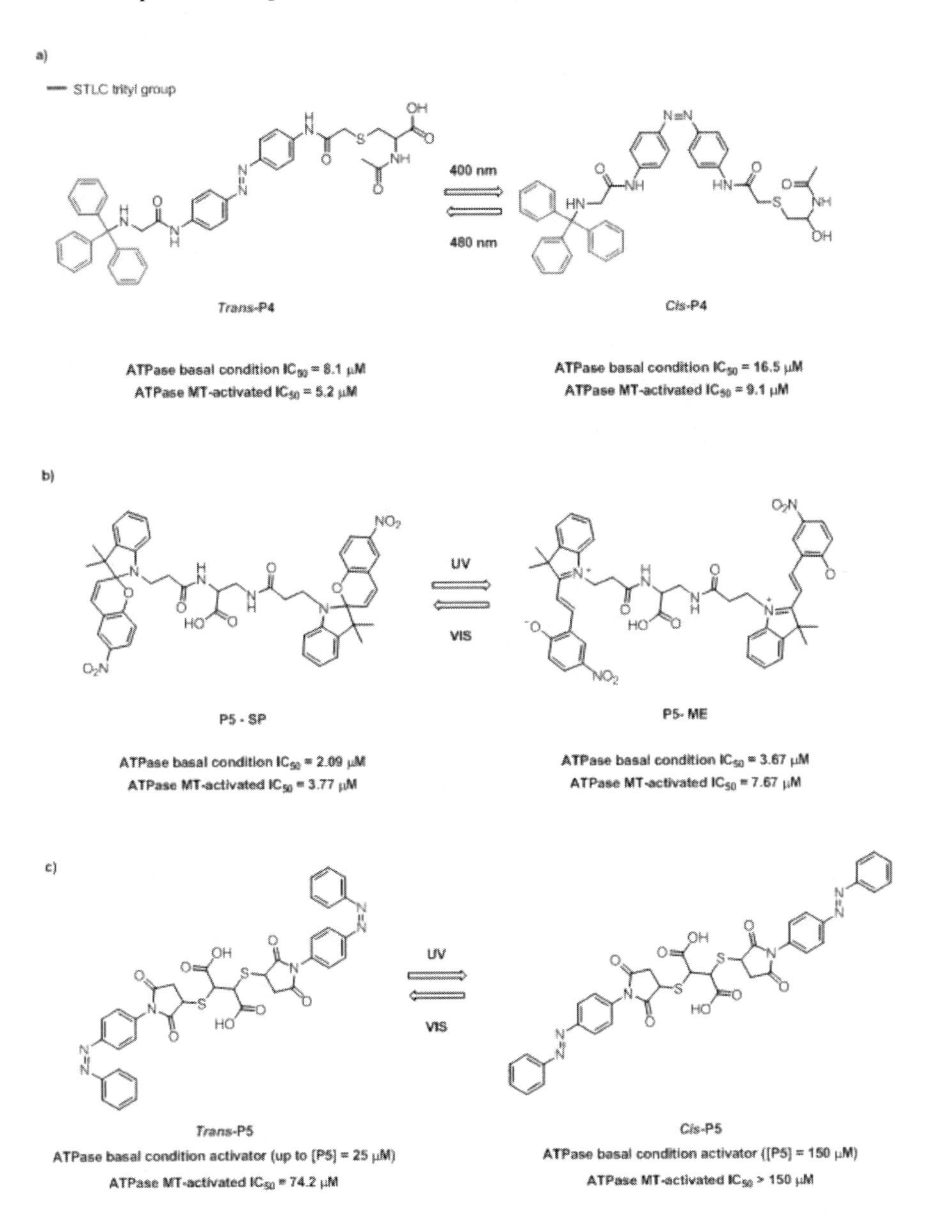

Figure 20. Photochromic compounds endowed with Eg5 inhibitory activity.

The transition from single-headed *trans/cis*-azobenzene system to two-headed spiropyran-merocyanin, positively affected the inhibitory activity against Eg5 in both basal and MT-activated activity. Moreover, while for the *trans/cis*-azobenzene, the IC_{50} was higher in the MT-activated ATPase assay of Eg5 activity, the spiropyran-merocyanin system displayed better inhibitory activity in the basal ATPase assay. The significant inhibitory activity of spiropyran form respect to the merocyanine one could be addressed to the hydrophobic closed ring of the first, resembling the ispinesib one, which was converted in the zwitterionic open form of the second (a similar behavior was also observed for spiropyran-based compound P3, Figure 19, c). Pursuing the studies on the two-headed structure, the compound P5 containing two azobenzene moieties was synthesized [82] (Figure 20, c). Astoundingly, the *trans* form evaluated in a basal condition dose-dependent ATPase assay exhibited biphasic alteration of activity: at lower concentrations it was an activator, increasing the activity by 400% at 25 μM, while further increasing of P5 concentration caused the reduction of ATPase activity (ATPase activity inhibited at 150 μM). The *cis*-isomer was an activator at higher concentration, increasing the ATPase activity by 400% at 125 μM. Conversely, the MT-stimulated ATPase activity was inhibited by *trans*-P5 with IC_{50} = 74.2 μM, while the inhibitory activity of the cis isomer was very low, obtaining in this way an almost "On/Off" photoswitching system.

Miscellanea

In this section, we collected different chemotypes which have been proposed as Eg5 inhibitors without the possibility to be categorized in the previous classes (Table 1). When possible, the presence of some SAR hints is reported in the text associated with the figure.

Table 1. Miscellanea of compounds tested as Eg5 inhibitors

Structure	Features of the compounds	Ref.
a b For Eg5 IC_{50} a>b; a b For Eg5 IC_{50} b>a; X, Y, Z = CH_2 > N (competitive inhibitors)	To increase solubility, a series of diarylamine derivatives were synthesized formally removing the (blue) bond of carbazole core, leading to nonplanar analogs. The new derivatives inhibited Eg5 in an ATP-competitive manner, as the lead compound, and in a similar extent. However, the introduced molecular changes elicited the hoped results with improved solubility. Furthermore, the selectivity of the best inhibitor was evaluated against a panel of proteins endowed with ATPase activity (centromere-associated protein E (CENP-E), Kid, mitotic kinesin-like protein 1 (MKLP-1), KIF4A, KIFC3, KIF14, and KIF2A) showing the absence of inhibition of these off-targets.	[83]
Eg5 EC_{50} = 1.18 μM	The tetrazole-based Eg5 inhibitor, YL001, was discovered through shape similarity-based virtual screening. It showed inhibitory activity in the low micromolar range against Eg5 and was effective (5.94 ≤ IC_{50} (μM) ≤ 76.25) against thirteen human cancer cell lines.	[84]
better tolerated moiety for Eg5 inhibition	A series of *β*-carboline derivatives were evaluated for their cytotoxic activity against two cancer cell lines. The most active compounds in cell assays were also evaluated against Eg5 enzyme, showing micromolar inhibitory activity. Furthermore, inhibitory activity against topoisomerase-I was also detected, accounting for the multitarget anticancer activity of these compounds. The selectivity of the inhibitors was assessed against a panel of 468 typical, atypical and mutant human kinases at a concentration of 1 μM, displaying no interaction with other enzymes.	[85]

FINAL CONSIDERATIONS

Taking into account all these papers, new heterocyclic molecules able to target mitosis without disrupting microtubule dynamics were considered as an essential topic of study in Medicinal Chemistry. Nowadays, several inhibitors targeting Eg5 have entered clinical trials either as monotherapies or in combination with other drugs. In this chapter, we have exhaustively collected the researches regarding kinesin spindle protein (KSP, Eg5) inhibitors, taking into account the works published in the years 2013-2019, and we have extrapolated robust SARs within each scaffold.

FUNDING

None.

DECLARATION OF INTEREST

The author has no relevant affiliations or financial involvement with any organization or entity with a financial interest in or financial conflict with the subject matter or materials discussed in the manuscript. This includes employment, consultancies, honoraria, stock ownership or options, expert testimony, grants or patents received or pending, or royalties.

REFERENCES

[1] De Monte C, Carradori S, Secci D, D'Ascenzio M, Guglielmi P, Mollica A, et al. Synthesis and pharmacological screening of a large library of 1,3,4-thiadiazolines as innovative therapeutic tools for the treatment of prostate cancer and melanoma. *Eur J Med Chem* 2015;105:245–62. doi:10.1016/j.ejmech.2015.10.023.

[2] Perez-Melero C. KSP inhibitors as antimitotic agents. *Curr Top Med Chem* 2014;14:2286–311. doi:10.2174/156802661466614113009553 2.

[3] Mukhtar E, Adhami VM, Mukhtar H. Targeting microtubules by natural agents for cancer therapy. *Mol Cancer Ther* 2014;13:275–84. doi:10.1158/1535-7163.mct-13-0791.

[4] Liu M, Ran J, Zhou J. Non-canonical functions of the mitotic kinesin Eg5. Thorac Cancer 2018;9:904–10. doi:10.1111/1759-7714.12792.

[5] Burger A. Recent findings and future directions for interpolar mitotic kinesin inhibitors in cancer therapy. *J Med Chem* 2005;6:829–829. doi:10.1021/jm00342a066.

[6] Makala H, Ulaganathan V. Identification of novel scaffolds to inhibit human mitotic kinesin Eg5 targeting the second allosteric binding site using in silico methods. *J Recept Signal Transduct* 2018;38:12–9. doi:10.1080/10799893.2017.1387922.

[7] Waitzman JS, Rice SE. Mechanism and regulation of kinesin-5, an essential motor for the mitotic spindle. *Biol Cell* 2014;106:1–12. doi:10.1111/boc.201300054.

[8] Verhey KJ, Cochran JC, Walczak CE. The Kinesin Superfamily. In: Kozielski, FSB F, editor. *Kinesins and Cancer,* vol. 12, Dordrecht: Springer Netherlands; 2015, p. 1–26. doi:10.1007/978-94-017-9732-0_1.

[9] Mann BJ, Wadsworth P. Kinesin-5 regulation and function in mitosis. *Trends Cell Biol* 2019;29:66–79. doi:10.1016/j.tcb.2018.08.004.

[10] Liu C, Zhou N, Li J, Kong J, Guan X, Wang X. Eg5 overexpression is predictive of poor prognosis in hepatocellular carcinoma patients. *Dis Markers* 2017;2017:1–9. doi:10.1155/2017/2176460.

[11] Jin Q, Huang F, Wang X, Zhu H, Xian Y, Li J, et al. High Eg5 expression predicts poor prognosis in breast cancer. *Oncotarget* 2017;8:62208–16. doi:10.18632/oncotarget.19215.

[12] Groen AC, Needleman D, Brangwynne C, Gradinaru C, Fowler B, Mazitschek R, et al. A novel small-molecule inhibitor reveals a

possible role of kinesin-5 in anastral spindle-pole assembly. *J Cell Sci* 2008;121:2293–300. doi:10.1242/jcs.024018.

[13] Chen GY, Kang YJ, Gayek AS, Youyen W, Tüzel E, Ohi R, et al. Eg5 inhibitors have contrasting effects on microtubule stability and metaphase spindle integrity. *ACS Chem Biol* 2017;12:1038–46. doi:10.1021/acschembio.6b01040.

[14] Dumas ME, Sturgill EG, Ohi R. Resistance is not futile: Surviving Eg5 inhibition. *Cell Cycle* 2016;15:2845–7. doi:10.1080/15384101.2016.1204864.

[15] Mayer TU. Small molecule inhibitor of mitotic spindle bipolarity identified in a phenotype-based screen. *Science* 1999;286:971–4. doi:10.1126/science.286.5441.971.

[16] Kapoor TM, Mayer TU, Coughlin ML, Mitchison TJ. Probing spindle assembly mechanisms with monastrol, a small molecule inhibitor of the mitotic kinesin, Eg5. *J Cell Biol* 2000;150:975–88. doi:10.1083/jcb.150.5.975.

[17] DeBonis S, Simorre JP, Crevel I, Lebeau L, Skoufias DA, Blanzy A, et al. Interaction of the mitotic inhibitor monastrol with human kinesin Eg5. *Biochemistry* 2003;42:338–49. doi:10.1021/bi026716j.

[18] Yan Y, Sardana V, Xu B, Homnick C, Halczenko W, Buser CA, et al. Inhibition of a mitotic motor protein: where, how, and conformational consequences. *J Mol Biol* 2004;335:547–54. doi:10.1016/j.jmb.2003.10.074.

[19] El-Nassan HB. Advances in the discovery of kinesin spindle protein (Eg5) inhibitors as antitumor agents. *Eur J Med Chem* 2013;62:614–31. doi:10.1016/j.ejmech.2013.01.031.

[20] (ClinicalTrials.gov, https://clinicaltrials.gov/ct2/results?cond=ispinesib&term=&cntry=&state=&city=&dist=).

[21] Good JAD, Berretta G, Anthony NG, Mackay SP. The Discovery and Development of Eg5 Inhibitors for the Clinic. *Kinesins and Cancer, Dordrecht:* Springer Netherlands; 2015, p. 27–52. doi:10.1007/978-94-017-9732-0_2.

[22] Beraud C, Wood KW, Crompton A, Roth S, Sakowicz R, Finer JT, et al. Antitumor activity of a kinesin inhibitor. *Cancer Res* 2005:3276–80. doi:10.1158/0008-5472.can-03-3839.

[23] Talapatra SK, Schüttelkopf AW, Kozielski F. The structure of the ternary Eg5-ADP-ispinesib complex. *Acta Crystallogr Sect D Biol Crystallogr* 2012;68:1311–9. doi:10.1107/S0907444912027965.

[24] Carol H, Lock R, Houghton PJ, Morton CL, Kolb EA, Gorlick R, et al. Initial testing (stage 1) of the kinesin spindle protein inhibitor ispinesib by the pediatric preclinical testing program. *Pediatr Blood Cancer* 2009;53:1255–63. doi:10.1002/pbc.22056.

[25] Jackson JR, Gilmartin A, Dhanak D, Knight S, Parrish C, Luo L, et al. A second generation KSP inhibitor, SB-743921, is a highly potent and active therapeutic in preclinical models of cancer. *Clin Cancer Res* 2006;12:B11 LP-B11.

[26] Theoclitou ME, Aquila B, Block MH, Brassil PJ, Castriotta L, Code E, et al. Discovery of (+)-N-(3-aminopropyl)-N-[1-(5-benzyl-3-methyl-4-oxo-[1,2] thiazolo[5,4-*d*]pyrimidin-6-yl)-2-methylpropyl]-4-methylbenzamide (AZD4877), a kinesin spindle protein inhibitor and potential anticancer agent. *J Med Chem* 2011;54:6734–50. doi:10.1021/jm200629m.

[27] (ClinicalTrials.gov, https://clinicaltrials.gov/ct2/results?cond=&term=AZD4877&cntry=&state=&city=&dist=)ht.

[28] Kantarjian HM, Padmanabhan S, Stock W, Tallman MS, Curt GA, Li J, et al. Phase I/II multicenter study to assess the safety, tolerability, pharmacokinetics and pharmacodynamics of AZD4877 in patients with refractory acute myeloid leukemia. Invest New Drugs 2012;30:1107–15. doi:10.1007/s10637-011-9660-2.

[29] Infante JR, Kurzrock R, Spratlin J, Burris HA, Eckhardt SG, Li J, et al. A Phase i study to assess the safety, tolerability, and pharmacokinetics of AZD4877, an intravenous Eg5 inhibitor in patients with advanced solid tumors. *Cancer Chemother Pharmacol* 2012;69:165–72. doi:10.1007/s00280-011-1667-z.

[30] Chen LC, Rosen LS, Iyengar T, Goldman JW, Savage R, Kazakin J, et al. First-in-human study with ARQ 621, a novel inhibitor of Eg5:

Final results from the solid tumors cohort. *J Clin Oncol* 2011;29:3076. doi:10.1200/jco.2011.29.15_suppl.3076.

[31] Schiemann K, Finsinger D, Zenke F, Amendt C, Knöchel T, Bruge D, et al. The discovery and optimization of hexahydro-2H-pyrano[3,2-c]quinolines (HHPQs) as potent and selective inhibitors of the mitotic kinesin-5. *Bioorganic Med Chem Lett* 2010;20:1491–5. doi:10.1016/j.bmcl.2010.01.110.

[32] Hollebecque A, Deutsch E, Massard C, Gomez-Roca C, Bahleda R, Ribrag V, et al. A phase I, dose-escalation study of the Eg5-inhibitor EMD 534085 in patients with advanced solid tumors or lymphoma. *Invest New Drugs* 2013;31:1530–8. doi:10.1007/s10637-013-0026-9.

[33] Cox CD, Breslin MJ, Mariano BJ, Coleman PJ, Buser CA, Walsh ES, et al. Kinesin spindle protein (KSP) inhibitors. Part 1: The discovery of 3,5-diaryl-4,5-dihydropyrazoles as potent and selective inhibitors of the mitotic kinesin KSP. *Bioorg Med Chem Lett* 2005;15:2041–5. doi:10.1016/j.bmcl.2005.02.055.

[34] Fraley ME, Garbaccio RM, Arrington KL, Hoffman WF, Tasber ES, Coleman PJ, et al. Kinesin spindle protein (KSP) inhibitors. Part 2: The design, synthesis, and characterization of 2,4-diaryl-2,5-dihydropyrrole inhibitors of the mitotic kinesin KSP. *Bioorganic Med Chem Lett* 2006;16:1775–9. doi:10.1016/j.bmcl.2006.01.030.

[35] Garbaccio RM, Fraley ME, Tasber ES, Olson CM, Hoffman WF, Arrington KL, et al. Kinesin spindle protein (KSP) inhibitors. Part 3: Synthesis and evaluation of phenolic 2,4-diaryl-2,5-dihydropyrroles with reduced hERG binding and employment of a phosphate prodrug strategy for aqueous solubility. *Bioorganic Med Chem Lett* 2006;16:1780–3. doi:10.1016/j.bmcl.2005.12.094.

[36] Cox CD, Torrent M, Breslin MJ, Mariano BJ, Whitman DB, Coleman PJ, et al. Kinesin spindle protein (KSP) inhibitors. Part 4:1 Structure-based design of 5-alkylamino-3,5-diaryl-4,5-dihydropyrazoles as potent, water-soluble inhibitors of the mitotic kinesin KSP. *Bioorganic Med Chem Lett* 2006;16:3175–9. doi:10.1016/j.bmcl.2006.03.040.

[37] Coleman PJ, Schreier JD, Cox CD, Fraley ME, Garbaccio RM, Buser CA, et al. Kinesin spindle protein (KSP) inhibitors. Part 6: Design and synthesis of 3,5-diaryl-4,5-dihydropyrazole amides as potent inhibitors of the mitotic kinesin KSP. B*ioorg Med Chem Lett* 2007;17:5390–5. doi:10.1016/j.bmcl.2007.07.046.

[38] Garbaccio RM, Tasber ES, Neilson LA, Coleman PJ, Fraley ME, Olson C, et al. Kinesin spindle protein (KSP) inhibitors. Part 7: Design and synthesis of 3,3-disubstituted dihydropyrazolobenzoxazines as potent inhibitors of the mitotic kinesin KSP. Bioorganic *Med Chem Lett* 2007;17:5671–6. doi:10.1016/j.bmcl.2007.07.067.

[39] Roecker AJ, Coleman PJ, Mercer SP, Schreier JD, Buser CA, Walsh ES, et al. Kinesin spindle protein (KSP) inhibitors. Part 8: Design and synthesis of 1,4-diaryl-4,5-dihydropyrazoles as potent inhibitors of the mitotic kinesin KSP. *Bioorganic Med Chem Lett* 2007;17:5677–82. doi:10.1016/j.bmcl.2007.07.074.

[40] Cox CD, Breslin MJ, Whitman DB, Coleman PJ, Garbaccio RM, Fraley ME, et al. Kinesin spindle protein (KSP) inhibitors. Part V: Discovery of 2-propylamino-2,4-diaryl-2,5-dihydropyrroles as potent, water-soluble KSP inhibitors, and modulation of their basicity by β-fluorination to overcome cellular efflux by P-glycoprotein. *Bioorg Med Chem Lett* 2007;17:2697–702. doi:10.1016/j.bmcl.2007.03.006.

[41] Holen K, DiPaola R, Liu G, Tan AR, Wilding G, Hsu K, et al. A phase i trial of MK-0731, a Kinesin Spindle Protein (KSP) inhibitor, in patients with solid tumors. *Invest New Drugs* 2012;30:1088–95. doi:10.1007/s10637-011-9653-1.

[42] Allen S, Robinson JE, Zhao Q, Hans J, Lyssikatos J, Aicher T, et al. The Discovery and Optimization of Kinesin Spindle Protein (KSP) Inhibitors: Path to ARRY-520 n.d.:426205.

[43] (Clinicaltrials.gov, https://clinicaltrials.gov/ct2/results?cond=&term=ARRy-520&cntry=&state=&city=&dist=).

[44] Nakai R, Iida S, Takahashi T, Tsujita T, Okamoto S, Takada C, et al. K858, a Novel Inhibitor of Mitotic Kinesin Eg5 and Antitumor

Agent, Induces Cell Death in Cancer Cells. *Cancer Res* 2009;69:3901–9. doi:10.1158/0008-5472.CAN-08-4373.

[45] Ye XS, Fan L, Van Horn RD, Nakai R, Ohta Y, Akinaga S, et al. A Novel Eg5 Inhibitor (LY2523355) Causes Mitotic Arrest and Apoptosis in Cancer Cells and Shows Potent Antitumor Activity in Xenograft Tumor Models. *Mol Cancer Ther* 2015;14:2463–72. doi:10.1158/1535-7163.mct-15-0241.

[46] Mross KB, Scharr D, Richly H, Frost A, Bauer S, Krauss B, et al. First-in-human study of 4SC-205 (AEGIS), a novel oral inhibitor of Eg5 kinesin spindle protein. J Clin Oncol 2014;32:2564. doi:10.1200/jco.2014.32.15_suppl.2564.

[47] Ragab FAF, Abou-Seri SM, Abdel-Aziz SA, Alfayomy AM, Aboelmagd M. Design, synthesis and anticancer activity of new monastrol analogues bearing 1,3,4-oxadiazole moiety. *Eur J Med Chem* 2017;138:140–51. doi:10.1016/j.ejmech.2017.06.026.

[48] Ramos LM, Guido BC, Nobrega CC, Corrêa JR, Silva RG, De Oliveira HCB, et al. The biginelli reaction with an imidazolium-tagged recyclable iron catalyst: Kinetics, mechanism, and antitumoral activity. *Chem - A Eur J* 2013;19:4156–68. doi:10.1002/chem. 201204314.

[49] Guido BC, Ramos LM, Nolasco DO, Nobrega CC, Andrade BYG, Pic-Taylor A, et al. Impact of kinesin Eg5 inhibition by 3,4-dihydropyrimidin-2(1H)-one derivatives on various breast cancer cell features. *BMC Cancer* 2015;15:1–15. doi:10.1186/s12885-015-1274-1.

[50] Gonçalves IL, Rockenbach L, das Neves GM, Göethel G, Nascimento F, Porto Kagami L, et al. Effect of N-1 arylation of monastrol on kinesin Eg5 inhibition in glioma cell lines. *Medchemcomm* 2018;9:995–1010. doi:10.1039/c8md00095f.

[51] De Oliveira FS, De Oliveira PM, Farias LM, Brinkerhoff RC, Sobrinho RCMA, Treptow TM, et al. Synthesis and antitumoral activity of novel analogues monastrol–fatty acids against glioma cells. *MedChemComm* 2018;9:1282–8. doi:10.1039/C8MD00169C.

[52] Shaheer Malik M, Seddigi ZS, Bajee S, Azeeza S, Riyaz S, Ahmed SA, et al. Multicomponent access to novel proline/cyclized cysteine tethered monastrol conjugates as potential anticancer agents. *J Saudi Chem Soc* 2019:1–11. doi:10.1016/j.jscs.2019.01.003.

[53] Al-Masoudi NA, Kassim AG, Abdul-Reda NA. Synthesis of potential pyrimidine derivatives via suzuki cross-coupling reaction as HIV and kinesin Eg5 Inhibitors. *Nucleosides, Nucleotides and Nucleic Acids* 2014;33:141–61. doi:10.1080/15257770.2014.880475.

[54] Al-Masoudi WA, Al-Masoudi NA, Weibert B, Winter R. Synthesis, X-ray structure, in vitro HIV and kinesin Eg5 inhibition activities of new arene ruthenium complexes of pyrimidine analogs. *J Coord Chem* 2017;70:2061–73. doi:10.1080/00958972.2017.1334259.

[55] Zhang W, Zhai L, Lu W, Boohaker RJ, Padmalayam I, Li Y. Discovery of Novel Allosteric Eg5 Inhibitors Through Structure-Based Virtual Screening. *Chem Biol Drug Des* 2016;88:178–87. doi:10.1111/cbdd.12744.

[56] Muthuraja P, Himesh M, Prakash S, Venkatasubramanian U, Manisankar P. Synthesis of N-(1-(6-acetamido-5-phenylpyrimidin-4-yl) piperidin-3-yl) amide derivatives as potential inhibitors for mitotic kinesin spindle protein. *Eur J Med Chem* 2018;148:106–15. doi:10.1016/j.ejmech.2018.02.010.

[57] Muthuraja P, Veeramani V, Prakash S, Himesh M, Venkatasubramanian U, Manisankar P. Structure-activity relationship of pyrazolo pyrimidine derivatives as inhibitors of mitotic kinesin Eg5 and anticancer agents. *Bioorg Chem* 2019;84:493–504. doi:10.1016/j.bioorg.2018.12.014.

[58] Sheth PR, Shipps GW, Seghezzi W, Smith CK, Chuang CC, Sanden D, et al. Novel benzimidazole inhibitors bind to a unique site in the kinesin spindle protein motor domain. *Biochemistry* 2010;49:8350–8. doi:10.1021/bi1005283.

[59] Ulaganathan V, Talapatra SK, Rath O, Pannifer A, Hackney DD, Kozielski F. Structural Insights into a Unique Inhibitor Binding Pocket in Kinesin Spindle Protein. *J Am Chem Soc* 2013;135:2263–72. doi:10.1021/ja310377d.

[60] Carbajales C, Prado MÁ, Gutiérrez-De-Terán H, Cores Á, Azuaje J, Novio S, et al. Structure-based design of new KSP-Eg5 inhibitors assisted by a targeted multicomponent reaction. *ChemBioChem* 2014;15:1471–80. doi:10.1002/cbic.201402089.

[61] Abd El-All AS, Magd-El-Din AA, Ragab FAF, Elhefnawi M, Abdalla MM, Galal SA, et al. New Benzimidazoles and Their Antitumor Effects with Aurora A Kinase and KSP Inhibitory Activities. *Arch Pharm* (Weinheim) 2015;348:475–86. doi:10.1002/ardp.201400441.

[62] Carbajales C, Sawada J ichi, Marzaro G, Sotelo E, Escalante L, Sánchez-Díaz Marta A, et al. Multicomponent Assembly of the Kinesin Spindle Protein Inhibitor CPUYJ039 and Analogues as Antimitotic Agents. *ACS Comb Sci* 2017;19:153–60. doi:10.1021/acscombsci.6b00166.

[63] De Iuliis F, Taglieri L, Salerno G, Giuffrida A, Milana B, Giantulli S, et al. The kinesin Eg5 inhibitor K858 induces apoptosis but also survivin-related chemoresistance in breast cancer cells. *Invest New Drugs* 2016;34:399–406. doi:10.1007/s10637-016-0345-8.

[64] Taglieri L, Rubinacci G, Giuffrida A, Carradori S, Scarpa S. The kinesin Eg5 inhibitor K858 induces apoptosis and reverses the malignant invasive phenotype in human glioblastoma cells. *Invest New Drugs* 2018;36:28–35. doi:10.1007/s10637-017-0517-1.

[65] Talapatra SK, Tham CL, Guglielmi P, Cirilli R, Chandrasekaran B, Karpoormath R, et al. Crystal structure of the Eg5 - K858 complex and implications for structure-based design of thiadiazole-containing inhibitors. *Eur J Med Chem* 2018;156:641–51. doi:10.1016/j.ejmech.2018.07.006.

[66] Yamamoto J, Amishiro N, Kato K, Ohta Y, Ino Y, Araki M, et al. Synthetic studies on mitotic kinesin Eg5 inhibitors: Synthesis and structure-activity relationships of novel 2,4,5-substituted-1,3,4-thiadiazoline derivatives. *Bioorg Med Chem Lett* 2014;24:3961–3. doi:10.1016/j.bmcl.2014.06.034.

[67] Mansoor UF, Angeles AR, Dai C, Yang L, Vitharana D, Basso AD, et al. Discovery of novel spiro 1,3,4-thiadiazolines as potent, orally

bioavailable and brain penetrant KSP inhibitors. *Bioorg Med Chem* 2015;23:2424–34. doi:10.1016/j.bmc.2015.03.052.

[68] Khathi SP, Chandrasekaran B, Karunanidhi S, Tham CL, Kozielski F, Sayyad N, et al. Design and synthesis of novel thiadiazole-thiazolone hybrids as potential inhibitors of the human mitotic kinesin Eg5. *Bioorg Med Chem Lett* 2018;28:2930–8. doi:10.1016/j.bmcl.2018.07.007.

[69] Nakazawa J, Yajima J, Usui T, Ueki M, Takatsuki A, Imoto M, et al. A novel action of terpendole E on the motor activity of mitotic kinesin Eg5. *Chem Biol* 2003;10:131–7. doi:10.1016/S1074-5521(03)00020-6.

[70] Tarui Y, Chinen T, Nagumo Y, Motoyama T, Hayashi T, Hirota H, et al. Terpendole e and its derivative inhibit STLC- and GSK-1-resistant Eg5. *ChemBioChem* 2014;15:934–8. doi:10.1002/cbic.201300808.

[71] Nagumo Y, Motoyama T, Hayashi T, Hirota H, Aono H, Kawatani M, et al. Structure-Activity Relationships of Terpendole E and its natural derivatives. *Chemistry Select* 2017;2:1533–6. doi:10.1002/slct.201602015.

[72] Ogunwa TH, Taii K, Sadakane K, Kawata Y, Maruta S, Miyanishi T. Morelloflavone as a novel inhibitor of mitotic kinesin Eg5. *J Biochem* 2019;0:1–9. doi:10.1093/jb/mvz015.

[73] Ogunwa TH, Laudadio E, Galeazzi R, Miyanishi T. Insights into the molecular mechanisms of Eg5 inhibition by (+)-Morelloflavone 2019:1–19.

[74] Raghav D, Sebastian J, Rathinasamy K. Biochemical and Biophysical characterization of curcumin binding to human mitotic kinesin Eg5: Insights into the inhibitory mechanism of curcumin on Eg5. *Int J Biol Macromol* 2018;109:1189–208. doi:10.1016/j.ijbiomac.2017.11.115.

[75] Wu W, Liu F, Su A, Gong Y, Zhao W, Liu Y, et al. The effect and mechanism of millepachine-disrupted spindle assembly in tumor cells. *Anticancer Drugs* 2018;29:449–56. doi:10.1097/CAD.0000000000000618.

[76] DeBonis S, Skoufias DA, Lebeau L, Lopez R, Robin G, Margolis RL, et al. In vitro screening for inhibitors of the human mitotic kinesin Eg5 with antimitotic and antitumor activities. *Mol Cancer Ther* 2004;3:1079–90.

[77] Good JAD, Wang F, Rath O, Kaan HYK, Talapatra SK, Podgórski D, et al. Optimized S-trityl-L-cysteine-based inhibitors of Kinesin Spindle Protein with potent in vivo antitumor activity in lung cancer xenograft models. *J Med Chem* 2013;56:1878–93. doi:10.1021/jm3014597.

[78] Ogo N, Ishikawa Y, Sawada JI, Matsuno K, Hashimoto A, Asai A. Structure-guided design of novel l-Cysteine derivatives as potent KSP inhibitors. *ACS Med Chem Lett* 2015;6:1004–9. doi:10.1021/acsmedchemlett.5b00221.

[79] Ishikawa K, Tamura Y, Maruta S. Photocontrol of mitotic kinesin Eg5 facilitated by thiol-reactive photochromic molecules incorporated into the loop L5 functional loop. *J Biochem* 2014;155:195–206. doi:10.1093/jb/mvt111.

[80] Ishikawa K, Tohyama K, Mitsuhashi S, Maruta S. Photocontrol of the mitotic kinesin Eg5 using a novel S-trityl-l-cysteine analogue as a photochromic inhibitor. *J Biochem* 2014;155:257–63. doi:10.1093/jb/mvu004.

[81] Sadakane K, Takaichi M, Maruta S. Photo-control of the mitotic kinesin Eg5 using a novel photochromic inhibitor composed of a spiropyran derivative. *J Biochem* 2018;164:239–46. doi:10.1093/jb/mvy046.

[82] Sadakane K, Alrazi IMD, Maruta S. Highly efficient photocontrol of mitotic kinesin Eg5 ATPase activity using a novel photochromic compound composed of two azobenzene derivatives. *J Biochem* 2018;164:295–301. doi:10.1093/jb/mvy051.

[83] Takeuchi T, Oishi S, Kaneda M, Ohno H, Nakamura S, Nakanishi I, et al. Kinesin spindle protein inhibitors with diaryl amine scaffolds: Crystal packing analysis for improved aqueous solubility. *ACS Med Chem Lett* 2014;5:566–71. doi:10.1021/ml500016j.

[84] Wang Y, Wu X, Du M, Chen X, Ning X, Chen H, et al. Eg5 inhibitor YL001 induces mitotic arrest and inhibits tumor proliferation. *Oncotarget* 2017;8:42510–24. doi:10.18632/oncotarget.17207.

[85] Abdelsalam MA, Aboulwafa OM, M Badawey ESA, El-Shoukrofy MS, El-Miligy MM, Gouda N, et al. Design, synthesis, anticancer screening, docking studies and in silico ADME prediction of some β-carboline derivatives. *Future Med Chem* 2018;10:1159–75. doi:10.4155/fmc-2017-0206.

About the Editor

Edeildo F. da Silva-Júnior (PhD)
Pharmaceutical Sciences Institute
Federal University of Alagoas, Maceió - Brazil

Edeildo Ferreira da Silva-Júnior studied Pharmacy at the Federal University of Alagoas, Maceió – Brazil. Subsequently, he received his Master's degree in Pharmaceutical Sciences by the Federal University of Alagoas, Brazil, as well. Since his Master degree, he is working in the medicinal chemistry of Neglected Tropical Diseases (NTD's), collecting valuable experiences on the researches addressed to the development of novel synthetic compounds with activity against *Leishmania spp.* and *Trypanosoma cruzi*, in association with collaborators from the University of Strasbourg – France. He has Ph.D. degree in Chemistry and Biotechnology, with focus on the Medicinal and Biological Chemistry, obtained at the Federal University of

Alagoas. During his Ph.D., he developed part of his project and thesis at the Institute of Pharmacy and Biochemistry from the Johannes Gutenberg University of Mainz, Germany. Current, he has a post-doctoral position at the Federal University of Alagoas, where he acts as Project-Leader and Laboratory-Head at the Institute of Pharmaceutical Sciences. Furthermore, he develops projects involving collaborators from the Johannes Gutenberg University of Mainz and Wurzburg University (Germany). He has patents and high-qualified publications in organic and medicinal chemistry fields, such as *Bioorganic Chemistry, Molecules, Current Topics in Medicinal Chemistry, Biosensors & Bioelectronics, Bioorganic & Medicinal Chemistry, Current Drug Metabolism, Biomedicine & Pharmacotherapy,* and others.

INDEX

#

β

γ

A

B

C

D

E

F

G

H

I

K

L

M

N

O

S

T

W

Z

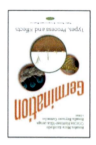